Television Broadcasting

Camera Chains

Television Broadcasting

Camera Chains

by

Harold E. Ennes

HOWARD W. SAMS & CO., INC.
THE BOBBS-MERRILL CO., INC.
INDIANAPOLIS · KANSAS CITY · NEW YORK

Preface

The manufacturers of television camera chains normally provide instruction manuals that range from preliminary (sketchy) data to elaborate coverage of circuit theory, operations, and maintenance of the specific equipment. Such manuals obviously cannot delve into certain training programs that are usually necessary before adequate comprehension of modern broadcast technology can be gained.

It is the purpose of this book to provide the fundamental and advanced training that is necessary if full benefit is to be obtained from the information in modern instruction books. To do this most effectively, where possible, complete, detailed schematics have been avoided, and, instead, use has been made of block diagrams with simplified diagrams of individual blocks under discussion. The overall system concept is stressed so that the reader can more readily grasp the meaning of a specific circuit adjustment in terms of its effect on system performance.

For the serious student, or for the practicing engineer, the information contained in the author's previous two books should be considered prerequisities to a study of this volume. These books are *Workshop in Solid State* and *Television Broadcasting: Equipment, Systems, and Operating Fundamentals*. Both are published by Howard W. Sams & Co., Inc. These books and the present volume serve a dual purpose, as basic textbooks for students or beginners and as factual guidebooks for practicing technicians.

The author extends his appreciation to the following organizations for providing information and photographs used in this book: Albion Optical Co.; Ampex Corporation; Amphenol Corporation; Belden Corporation; Cohu Electronics, Inc.; Hewlett-Packard Co.; International Video Corporation; Kliegl Bros. Lighting; Philips Broadcast Equipment Corp.; RCA; Shibaden Corp. of America; Tektronix, Inc.; TeleMation, Inc.; Telesync Corp.; Visual Electronics Corporation; WBBM-TV; and WTAE-TV.

HAROLD E. ENNES

Contents

CHAPTER 1

CHAPTER 2

CHAPTER 3

CHAPTER 4

CHAPTER 5

CHAPTER 6

CHAPTER 7

CHAPTER 8

CHAPTER 9

CHAPTER 10

CHAPTER 11

APPENDIX

Studio Lighting

The television camera depends primarily on the reflected light it "sees" through the optical system. Optimum performance is possible in the studio, where light can be controlled to shape the "taking" characteristics. In an "adequate" studio lighting system, we are concerned with the following basic factors:

1. Amount of light, and color temperature
2. Flexibility
3. Light control

It is assumed that the reader has basic knowledge of the types and techniques of lighting.[1]

1-1. AMOUNT OF LIGHT

It is possible to relate foot-candles (fc) to watts, lumens, etc., but such a study, while interesting from an academic viewpoint, is of little value to the practicing operator. Information regarding the amount of light at a given distance and direction from a given source is furnished by manufacturers of lighting equipment. Any conversion of watts to foot-candles to lumens is useless unless the particular angle of throw, type of lamp, and type of reflector are accounted for.

The number of lighting fixtures may vary from 20 to 25 in a small limited-program studio to more than 200 in a larger multiple-purpose studio. The amount of light required depends on the size and purpose of the studio, whether programs are in monochrome or color, and the degree of flexibility required. The key word for *good* studio lighting systems is *flexibility*.

---

[1]See, for example, Harold E. Ennes, *Television Broadcasting: Equipment, Systems, and Operating Fundamentals* (Indianapolis: Howard W. Sams & Co., Inc., 1971).

From previous experience, the foundation for an adequate amount of light can be laid as in Table 1-1. The higher values are representative of more complex productions, such as large variety shows with liberal use of effects and modeling lighting both foreground and background. Note in connection with this table that the quartz lamp has more light output per watt of input power over a broader area than the incandescent lamp has. Also, the size and shape of the quartz lamp permits design of reflectors of higher efficiency.

The net production area to which Table 1-1 applies is the actual area used within the cyclorama curtain. The net production area can be approximated by subtracting an area 4 feet wide all around the perimeter of the studio. For instance, a 40′ × 60′ studio (2400 square feet) has a net production area of 32′ × 52′, or 1664 square feet. However, smaller studios, such as the one in the following example (20′ × 30′), do not always employ cycloramas, and the wall-to-wall dimensions normally are used for these studios.

Table 1-1. Light Requirements for Net Production Area

	100% Incandescent	100% Quartz
Monochrome 100-125 fc Color 250-350 fc	25-35 W/ft² (Nominal: 30 W/ft²) 75-110 W/ft² (Nominal: 90 W/ft²)	10-20 W/ft² (Nominal: 15 W/ft²) 40-60 W/ft² (Nominal: 50 W/ft²)

A typical incandescent lighting package for a 20′ × 30′ studio with a 15-foot ceiling is illustrated in Fig. 1-1. On the basis of the nominal values shown for incandescent lights in Table 1-1, this 600-square-foot studio would require a total of 18,000 watts for monochrome, or 54,000 watts for color. Note from the equipment list in Fig. 1-1 that two wattages are listed for each fixture. In most instances, the lower-wattage lamps would be used for monochrome, and the higher-wattage lamps would be used for color.

Fig. 1-2A gives the photometric data for a 2000-watt incandescent unit with a 12-inch round lens. Note that for the flood operating position, about 65 foot-candles are available at 20 feet directly on the axis. Since the flood beam is broad (about 55°), essentially the same number of foot-candles exists up to about 7 feet off the axis at this 20-foot distance. In the spot operating position (beam about 10°), almost 760 foot-candles are available at 20 feet directly on the axis, but only about 80 foot-candles are available just 2 feet off the axis at the 20-foot distance.

For illustrative purposes, Fig. 1-2B gives the same general data for a 5000-watt unit with a 16-inch round-beam lens. The graph for the flood position is for a 30-foot throw, and the spot-position graph is for a 60-foot throw.

Fig. 1-2C gives the photometric data for a 1000-watt unit equipped with an 8-inch oval-beam lens. Note that because of the beam shape, horizontal and vertical coverage must be considered separately.

See Table 1-2 for multiplying factors for distances other than those shown in Fig. 1-2. To use this table to find photometric data for throws other than those shown on test charts, apply the following procedure:

1. Locate the throw distance of the test in the group headed "Throw Distances Shown in Tests."
2. Directly below this figure is a series of actual throw distances desired. Pick the figure in the vertical series which corresponds to the distance at which you want photometric data.

Table 1-2. Multiplying Factors for Lighting Data

Throw Distances Shown in Tests						Multiplying Factors	
10'	15'	20'	30'	50'	60'	Multiply Dimensions by	Multiply Illumination by
Actual Throw Distances Desired							
			10'		20'	0.33	9.0
		10'	15'	25'	30'	0.5	4.0
	10'					0.66	2.25
			20'		40'	0.67	2.25
7.5'			15'			0.75	1.8
				40'		0.8	1.5
			25'		50'	0.83	1.44
10'	15'	20'	30'	50'	60'	1.0	1.0
			35'		70'	1.16	0.73
				60'		1.2	0.69
		25'				1.25	0.64
	20'		40'		80'	1.3	0.56
15'		30'	45'	75'	90'	1.5	0.44
	25'		50'		100'	1.7	0.36
			35'			1.75	0.32
20'	30'	40'	60'	100'	120'	2.0	0.25
25'		50'				2.5	0.16
	40'					2.7	0.14
30'						3.0	0.11
	50'					3.3	0.09

Courtesy Kliegl Bros. Lighting

3. When this figure is chosen, refer to the "Multiplying Factors" column directly to the right for the proper factors to use in multiplying dimension data and illumination data.

4. Example: If a test was calculated at a throw of 30′ and you require data for this unit at a throw of 50′, find the column headed 30′ and go down this column until the figure 50′ is located. Then, by extending directly to the right, you will find the multiplying factors 1.7 for dimension and 0.36 for illumination.

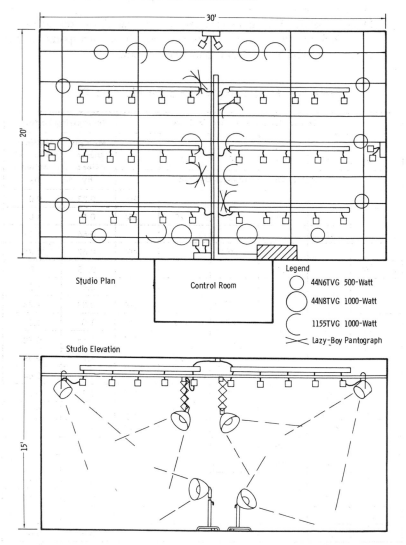

Legend
◯ 44N6TVG 500-Watt
◯ 44N8TVG 1000-Watt
⊂ 1155TVG 1000-Watt
✕ Lazy-Boy Pantograph

Studio Plan

Control Room

Studio Elevation

Fig. 1-1. Incandescent

Table 1-3. Performance of "Standard" Quartz Fresnel

Catalog Number	Lens	Wattage	Hours Life	Performance at 20 Feet				Unit Dimensions	
				Minimum Spot		Maximum Flood			
				Spread	FC	Spread	FC	Length	Diameter
3518	12"	1000	150	1' x 4'	500	15'	40	13¼"	13"
3518	8"	2000	150	3½' x 3½'	400	15'	55	13¼"	13"
3521	12"	2000	150	3'	500	17'	60	18"	18"

Courtesy Kliegl Bros. Lighting

The variation of foot-candles per watt from luminaires is shown by Tables 1-3 through 1-5. These tables illustrate not only the higher light output per watt for quartz lamps as compared to incandescent lamps, but also that the design considerably affects the actual foot-candles per watt of input power.

EQUIPMENT LIST

Fixtures

10—1155TVG 750/2000-Watt Standard Scoop 18"
1—44N3TVB 75/100-Watt Fresnel Camera Light 3" Lens
10—44N6TVG 500/750-Watt Fresnel Spotlight 6" Lens
4—44N8TVG 1000/1500-Watt Fresnel Spotlight 8" Lens
1—1365PTVG 500/750-Watt Pattern Projector **Klieglight**
1—1365TVG/Iris 500/750-Watt Iris **Klieglight**

Accessories

5—1080A 4-Way Barn Door for 44N62TVG
2—1081A 4-Way Barn Door for 44N8TVG
1—1097 Set for 16 Patterns for Projector
5—1078X 18" Diffuser Frame
2—1421 Caster Stand 25" Base
4—111TV Pantograph Hanger
4—10E955G Extension Cable 10'
2—25E955G Extension Cable 25'

Wiring

6—619G/10/5 Connector Strips 10' Long With 5 20 A Pigtail Outlets.
4—2433G/2 Wall Outlet Boxes

Control

3—2500-Watt Dimmers & Space For 3 Future Additions
9—2500-Watt Nondims
1—100 A 3-Pole Main Breaker
1—Cross-Connecting System for 34 20 A Circuits Using Either **Rotolector** Or Patching Devices

NOTE: Fixtures not shown are on portable floor stands and on camera.

Courtesy Kliegl Bros. Lighting

lighting plan for studio.

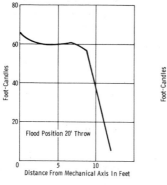
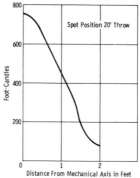

(A) Type 44N12.

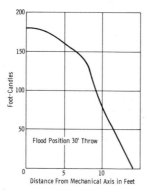
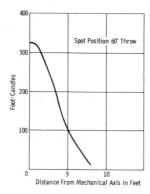

(B) Type 44N16.

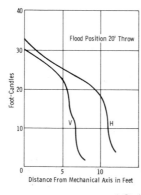
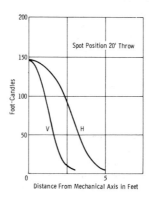

(C) Type 44N08.

Courtesy Kliegl Bros. Lighting

Fig. 1-2. Photometric data for representative luminaires.

When comparing the lamps shown by Tables 1-3 through 1-5 with the lamps of Fig. 1-2, it is necessary to note the wider angle of throw for the quartz luminaires. For example, the 2000-watt incandenscent lamp of Fig. 1-2A provides essentially 60 foot-candles (flood position) over a spread of about 6 feet at a distance of 20 feet. The "standard" quartz luminaire of Table 1-3 (2000 watts, 12-inch lens) has 60 foot-candles over a spread of 17 feet (in flood position) at a distance of 20 feet (an equivalent of almost three incandescent lamps for the same coverage). The equivalent luminaire in the high-efficiency design (Table 1-4) gives 225 foot-candles over a spread of 17 feet at a distance of 20 feet.

Table 1-4. Performance of High-Efficiency "XKE" Fresnel

| Catalog Number | Lens | Wattage | Hours Life | Performance at 20 Feet | | | | Unit Dimensions | |
| | | | | Minimum Spot | | Maximum Flood | | | |
				Spread	FC	Spread	FC	Length	Diameter
3507	6⅜"	400	2000	4'	100	20'	25	12"	8¼"
3507	6⅜"	650	100	4'	300	20'	50	12"	8¼"
3525	8"	1000	150	2'	300	17'	100	13¼"	13"
3525	8"	2000	150	2'	500	17'	175	13¼"	13"
3527	12"	2000	150	1'6"	600	17'	225	18"	18"

Courtesy Kliegl Bros. Lighting

Table 1-5 shows that two lamps are available for the same luminaire: One is a 30-hour lamp that has an additional 40 foot-candles, compared to the 150-hour lamp, under the condition shown. This is normal in all luminaires; shorter-life lamps such as a 110-volt bulb on a 120-volt circuit give higher light output at a reduced life expectancy.

Table 1-5. Performance of Follow Spots

| Catalog Number | Wattage | Hours Life | Performance at 75 Feet | | Unit Dimensions | |
			Spread	FC	Length	Diameter
1393	1000	30	6'	180	46"	16"
1393	1000	150	6'	140	46"	16"

Courtesy Kliegl Bros. Lighting

A 40' × 60' studio (net production area of 1664 square feet) lighted for color with incandescent lamps would require 1664 × 90, or 150 kW of power (Table 1-1). Fig. 1-3 illustrates a typical quartz studio-lighting package for a 40' × 60' studio. In this case, a total of 72.2 kW is required. Additional lighting for the cyclorama brings the total to about 100 kW. Note that these power inputs are close to those computed from the data of Table 1-1.

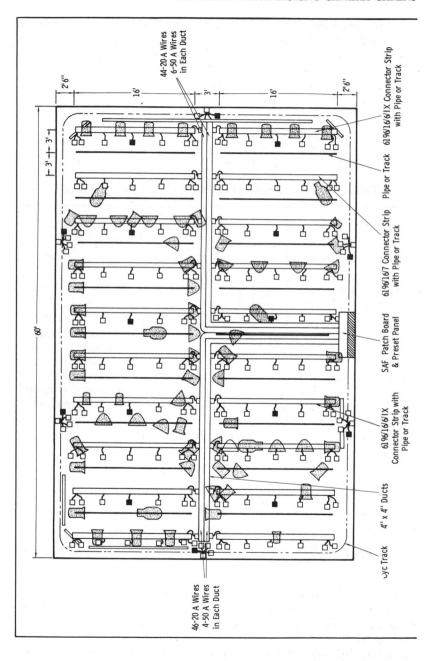

Fig. 1-3. Quartz-iodine

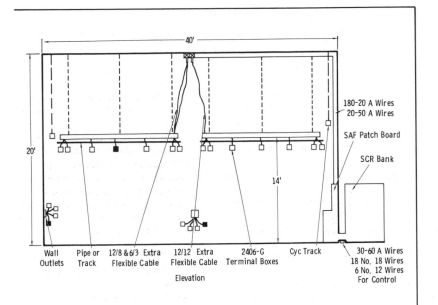

180-20 A Wires
20-50 A Wires

SAF Patch Board

SCR Bank

40'

20'

14'

| Wall Outlets | Pipe or Track | 12/8 & 6/3 Extra Flexible Cable | 12/12 Extra Flexible Cable | 2406-G Terminal Boxes | Cyc Track | 30-60 A Wires 18 No. 18 Wires 6 No. 12 Wires For Control |

Elevation

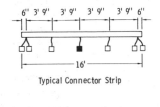

6'' 3' 9'' 3' 9'' 3' 9'' 3' 9'' 6''

16'

Typical Connector Strip

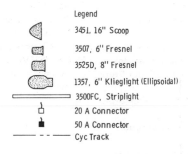

Legend

3451, 16'' Scoop

3507, 6'' Fresnel

3525D, 8'' Fresnel

1357, 6'' Klieglight (Ellipsoidal)

3500FC, Striplight

20 A Connector

50 A Connector

Cyc Track

Note: See equipment list on page 20.

Courtesy Kliegl Bros. Lighting

lighting plan for studio.

EQUIPMENT LIST

For 100% Quartz-Iodine Lighted 40' × 60' Color Studio (Package No. 27CQ)

Fixtures & Accessories

27—No. 3451 1000 W Quartz Scoop
2—No. 44N3TVB 150 W Cameralight
6—No. 3507 650 W Quartz Fresnel
36—No. 3525D 1000 W Quartz Fresnel
3—No. 1357P/6W 1000 W Quartz Pattern Klieg
2—No. 1357/6/1 1000 W Quartz Iris Klieg
4—No. 1106A 4-Way Barn Door for 650 W Fresnel
18—No. 1081A 4-Way Barn Door for 1000 W Fresnel
14—No. 585A 16-Inch Color/Diffuser Frame for Scoop
3—No. 1097 Sets of 16 Patterns for Kliegs
2—No. 1421 Castered Floor Stands
9—No. 111TV Pantograph Hangers
9—No. 10E955G 10-foot Extension Cables for Pantographs
2—No. 25E955G 25-foot Extension Cables for Floor Stands

Wiring Devices

10—No. 619G/16/7 Connector Strips (each 16 feet long and wired with 2 double and 3 single 3-foot pigtail outlets on five 20-ampere circuits)
10—No. 619G/16/6/1X Connector Strips (Each 16-feet long and wired with 2 double and 2 single 20-ampere 3-foot pigtail outlets and one 50-ampere outlet)
10—No. 2406G/10 Ceiling Terminal Boxes
10—No. 2406G/10X Ceiling Terminal Boxes
1—No. 12/8 100-foot Multiconductor Drop Cable
1—No. 6/3 100-foot Multiconductor Drop Cable
1—No. 12/12 100-foot Multiconductor Drop Cable
6—No. 24330/3/1X Wall Outlet Boxes (Each with three 20-ampere and one 50-ampere pigtail outlets)

Control Center

1—Composite one scene, two subscene lighting preset system containing: 108 20-ampere and 16 50-ampere counterweighted male plugs, 75 Automatic Cold-Patching 20-ampere female jacks with associated circuit breakers and 15 50-ampere jacks. 1 Boardlight, 8 7000 W SCR dimmers, 7 7000 W plug-in nondims which permit future insertion of dimmers, 1 300-ampere 3-pole main breaker, a preset section with 15 pots with selector switches and 2 submasters and lock and key switch

Lamp Package

27—No. Q1000T3/4 Quartz Lamps for Scoops
2—No. 150G16½/3DC Lamps for Cameralights
6—No. FAD-650 Quartz Lamps for Fresnels
36—No. DXW-1000 Quartz Lamps for Fresnels
5—No. 1000T6Q/RCL/1 1000 W Lamps for Kliegs

Cyc Package

(Sufficient to light an "L" shaped cyc covering one 40-foot wall, 1 curve, and ½ of the 60-foot wall)
8—No. 3500FC 7-foot Striplights With 4 Reflectors and Glass Filters
3—No. 3500AFC 3½-foot Striplights With 2 Reflectors (For Curve)
2—No. 453TVG/3 Adapters (To convert a 50-ampere outlet to three 20-ampere outlets)

Courtesy Kliegl Bros. Lighting

Fig. 1-3. Quartz-iodine lighting plan for studio. (Cont'd.)

There are three possible ways of lighting for color television:[2]

1. The 100-percent use of standard incandescent spotlights and flood-lights. This has been the general practice, but, although it produces satisfactory results, it has a high cost because of greatly increased power and air-conditioning requirements.
2. The 100-percent use of quartz-iodine spotlights and floodlights. This method requires a minimum increase in power and air-conditioning expenditures.
3. A mixture of incandescent and quartz-iodine lighting. Here the increase in power and air-conditioning requirements is somewhere between the foregoing extremes.

Before considering the relative merits of these three lighting methods, review two basic requirements of the color system. First, the lighting level: This is 250 foot-candles plus a "holey factor." In order to understand this factor, it is essential to recognize that, in color, shadows are more critical than in black and white. If the contrast range is not carefully controlled, shadows become an unacceptable dark color, and may also cause video noise. Therefore, it is desirable to fill in these holes (shadows) by using additional luminaires. These may be either Fresnel-lens spotlights or scoop floodlights as the situation may dictate. Since the Fresnel in the flood position and the scoop both deliver a broad and soft-edged light beam, their coverage not only lightens the shadow (fills the hole) but also overlaps the main object being lighted. With several such "holey" luminaires being employed, it is possible to have the total overlap add 75 to 100 foot-candles to the overall lighting level. This results, then, in a level from 300 to 350 foot-candles of total illumination on the subject.

The second requirement is the color temperature of the light. By going to 100-percent quartz lighting, one has the opportunity to obtain an entire family of lamps and luminaires in either 3000 or 3200 Kelvin temperature. By and large, the TV industry has shown a preference for 3200-K lamps for color work. Telecasters may now obtain scoops, Fresnels, pattern and iris projectors—all with the same color temperature. When this is done, cameramen may move from one set to another without worrying about the light mixtures, since they all will have the same color.

A direct comparison of light, life, and temperatures between standard incandescent and quartz luminaires may be helpful at this point; see Table 1-6. The first observation from this table may very well be that the quartz lamp, although it may produce two to four times the light of its standard

---

[2]Much of the information given here was first presented at the Symposium on Theatre-Television Lighting sponsored by the Illuminating Engineering Society (IES), at Chicago, Ill., May 1966. It appears here by permission of Kliegl Bros. Lighting.

Table 1-6. Incandescent and Quartz-Iodine Comparison

	Incandescent Scoop	Quartz-Iodine Scoop	Incandescent Fresnel	Quartz-Iodine Fresnel
Life (Hours)	1000	500	200	150
Light (Lumen Output)	Decreases With Age	Unchanged With Age	Decreases With Age	Unchanged With Age
Temperature (Color)	Decreases With Age (to Orange)	Unchanged With Age	Decreases With Age (to Orange)	Unchanged With Age

incandescent counterpart, has a shorter life. This may be true if the absolute term "life" is used, but it is not so if the limited term "useful life" is employed. Many TV studios, in order to avoid degradation of color and light output (see Fig. 1-4) as lamps age, make it standard practice to change lamps at 50 percent of the anticipated life. This practice gives quartz lamps a longer useful life, since such lamps maintain color as well as light output throughout their life. Furthermore, the production staff is relieved of the burden of compensation for color degradation when the lighting is 100 percent quartz.

In going to color with all incandescent lighting fixtures, approximately 150 percent is added to the power and air-conditioning load. Technically speaking, a 100-percent increase is sufficient, but the "holey factor" adds an additional 50 percent. The luminaire complement would roughly double, with some 5000-watt Fresnels added to the 1000 to 2000-watt Fresnels in use for monochrone (only). This method has been producing excellent color pictures for networks and many independent stations. However, unless the station already has a large luminaire complement and has installed the increased power and air conditioning in anticipation of the move to color, this is the most costly way of lighting for color.

By providing 100 percent quartz-iodine fixtures in place of the incandescent fixtures—and by adding 50 percent more luminaires for the

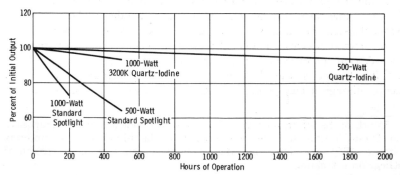

Fig. 1-4. Lumen-output fall-off curves for typical lamps.

"holey factor"—one may go to color with only a 50 percent increase in power and air conditioning. Not only does this method afford production of color TV pictures more inexpensively, it also makes it a far easier process for the production staff.

If existing equipment is reused rather than replaced, the method employed by many telecasters who convert to color is to use existing units for fill and background lighting and to add quartz units for key and back lighting. Providing for 75 percent quartz enables the station to meet color requirements. This adds 75 percent to the power and air-conditioning costs. Incidentally, this method produces excellent results, although it requires color balancing and mixing, and is subject to some of the color degradation exhibited by the old incandescent method.

1-2. AIR CONDITIONING

Air conditioning in the control room and studios is mandatory for most sections of the United States if optimum equipment performance is to be expected. Although the effect of temperature in individual components may be small as the temperature increases, the accumulative effects over the average operating day can become quite noticeable. Such troubles as poor focus, drifting linearity, waveform instability, and loss of resolution often can be attributed to excessive temperature changes occurring after adjustment and alignment of camera chains.

It is a fundamental truth that if equipment areas and studio areas are maintained within reasonably comfortable temperatures for personnel, rack equipment and components within camera and monitor housings will remain within optimum operating temperatures. If the room temperature is held no higher than 80°F and the humidity remains within the 40 to 45 percent range, both personnel and equipment should operate efficiently.

Obviously, the required air-conditioning capacity depends on the size of control rooms and studios, average seasonal temperature of the location, and the amount of equipment and lighting installed. A competent air-conditioning engineer as close to the community as possible should be consulted. Usually, it is necessary to provide sound-isolation baffles in ducts and outlets for the air stream to prevent sound leakage from control room to studio and between studios, and to prevent whistles or other noise caused by the forced air movement. This is a greater problem in television than in aural broadcasting because TV audio techniques require greater microphone-to-performer distances than the more intimate techniques used in radio broadcasting.

Table 1-7 gives the "rule-of-thumb" air-conditioning requirements for the conditions stated, and assuming a $40' \times 60'$ color studio. Note that a diversity factor of 60 percent normally is used for the design of the studio air-conditioning system. This would be 60 watts per square foot for incandescent lighting, 36 watts per square foot for quartz lighting, and about

Table 1-7. Comparison of Lighting Design Parameters

	Standard Incandescent	Quartz	Mixture
Lighting Levels (Foot-Candles)	250-350	250-350	250-350
Power (Kilowatts)	180	108	126
Watts per sq. ft. (NPA)*	100	60	75
Air Conditioning Heat Load** (kW)	108	62.4	75.6
(Tons)	31	17.1	21.6

*NPA—Net Production Area.
**Calculated on the basis of 60 percent of input power since the air conditioning is designed on a continuous 12-hour or longer basis, whereas the lighting is sporadic with peak lighting (heat) loads having only a 30-minute to 1-hour duration.

45 watts per square foot for a studio in which quartz and incandescent lamps are mixed.

1-3. POWER REQUIREMENTS FOR LIGHTING

In figuring the power required for studio lighting, a 100-percent utlization factor normally is used. The minimum diversity factor ever to consider is 80 percent, but the 100-percent factor is recommended. Thus, if the lighting system provides 100 watts/square foot for a net production area of 1000 square feet, the feeder requirement is 100 kW plus any additional power required for lighting control, rear-screen projectors, cameras and camera pedestals, etc. In general, however, the lighting-load feeders are separate from those for other studio equipment.

It is the usual practice to order a three-wire service with a grounded neutral. This is equivalent to two separate services, as indicated by Fig. 1-5. This arrangement would be adequate to handle all lights in the average installation. An extra service for office equipment, air conditioning, heating units, and the like is required.

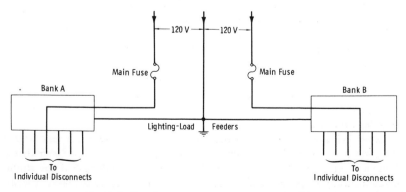

Fig. 1-5. Simple power arrangement for lighting.

The simple arrangement shown in Fig. 1-5 provides two individual breaker panels, A and B. Each light fixture usually is marked on the housing in large letters such as A-6, B-12, etc. This designates the panel and breaker involved for the individual luminaire. The individual breakers usually are of 20-ampere capacity, with some 50-ampere breakers necessary (for color) as specified in Fig. 1-3.

1-4. THE SUSPENSION SYSTEM AND POWER DISTRIBUTION

The most common system for providing current to suspended lights is the connector-strip method illustrated in Fig. 1-6. According to IES and SMPTE recommendations, ceiling (grid) outlets should be spaced evenly over the entire studio ceiling. The most convenient and least costly method involves the use of factory assembled and wired, UL-approved connector strips. These strips provide safe and convenient facilities for connecting spotlights, floodlights, etc., where units are hung from an overhead track or pipe batten. The particular strip illustrated in Fig. 1-6 has the following specifications: the front cover is removable for access to terminals; standard finish is flat black (circuit designations, stenciled on the front in 2-inch letters, sometimes are used); each section is furnished with proper hangers; strips are $2'' \times 4''$, 20-gauge steel; strips are furnished with 5 pigtails, each 3 feet long.

The sheet-steel wireway (with pigtail connectors) is mounted onto the pipe batten as in Fig. 1-6B. The pigtails are evenly spaced at $3'\text{-}9''$ intervals along the connector strip. Each light is clamped onto the pipe by means of a "C" clamp (Fig. 1-7), and the cord on the lighting unit is plugged into the pigtail connector. A terminal box provided at one end of the strip carries multiple-conductor flexible cable to a square duct which, in turn, contains conductors that carry the connection back to the lighting control center.

Another type of overhead light suspension is shown in Fig. 1-8. Both the fixtures and the individual cross tracks may be moved at will. Individual cross tracks travel on master support tracks, and the fixtures travel on the cross tracks. The ability to move individual fixtures and/or individual cross tracks gives full location flexibility to each lighting fixture. All movement of fixtures and track may be accomplished from the floor with the use of a pole. All fixtures have the same tilt and swing as when attached to the pipe batten with a standard pipe clamp.

In addition to overhead lighting, some provision normally is made for floor units. It is recommended that duplex or triple wall outlets be provided about every 30 running feet of wall. For monochrome, 20-ampere outlets are used with 50-ampere outlets provided in at least one position. For color, all outlets are usually of 50-ampere capacity. Such outlets provide power for floor-stand lights (such as low-level front lights), rear-screen projectors, follow spots, special-effect machines, etc.

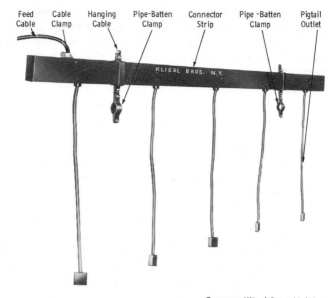

Feed Cable   Cable Clamp   Hanging Cable   Pipe-Batten Clamp   Connector Strip   Pipe-Batten Clamp   Pigtail Outlet

Courtesy Kliegl Bros. Lighting

(A) Typical unit.

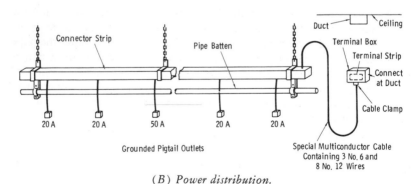

(B) Power distribution.

Fig. 1-6. Lighting connector strip.

Fig. 1-9 illustrates one type of wall box that is available with either 2, 3, or 4 pigtails 18 inches in length. Such pigtails permit the use of the standard heavy-duty lighting plugs required for television and theater luminaires.

1-5. LIGHTING CONTROL

A single circuit is brought to the lighting control center from each lighting unit. The "control center" for a small studio may be as simple as

Fig. 1-7. Double C clamp.

Courtesy Kliegl Bros. Lighting

a bank of circuit breakers, as described for Fig. 1-5. In a larger studio, breakers may be grouped and then mastered. Each load circuit is represented either by a cord and plug (similar to a telephone switchboard) or by a rotary switch. Individual loads then can be grouped into any one of 12 to 24 (or more) dimmer or nondim controls.

Fig. 1-10A illustrates the Kliegl *Saf-Patch* panel, in which "make" or "break" of a "hot" circuit is prevented automatically. The system consists of load plugs and female receptacles. Each receptacle is controlled by an individual magnetic-type circuit breaker. These breakers serve as on-off controls for their associated receptacles and may be used for individual changes without affecting other lights. Individual load plugs, one for each

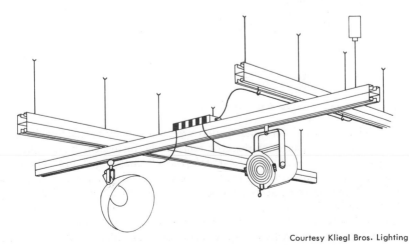

Courtesy Kliegl Bros. Lighting

Fig. 1-8. Double-track mobile system for overhead lights.

lighting circuit, are equipped with cable clamps that eliminate strain on internal connections. The special design of the load plugs makes it impossible for them to be inserted into a receptacle when the circuit breakers are in the on position (Fig. 1-10B). When the plugs are removed from the receptacles, they automatically trip the breakers to the off position *before* the plugs are completely withdrawn (Fig. 1-10C) eliminating any possibility of a "hot" break. As an additional safety feature, load plugs of different capacities cannot be interchanged and can be connected only to properly rated female receptacles. The plugs and patch cords are properly counterweighted for automatic restoring when not in use.

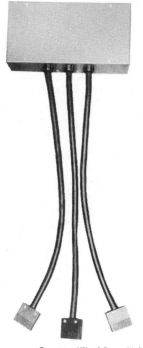

Fig. 1-9. Typical wall box.

Courtesy Kliegl Bros. Lighting

Individual standard *Saf-Patch* panels are furnished with six female receptacles, although panels of three or four can be furnished upon request. Groups of these panels form a *Saf-Patch Klieg-Board* (Fig. 1-11). The female receptacles can be of 20- or 42-ampere size and freely intermingled. Groups of female receptacles generally are wired together and fed by a dimming or nondim circuit.

Manually operated systems may be recommended for more simplified dimming applications where requirements are moderate. In this system, the dimming units and their controls are mechanically manipulated, and

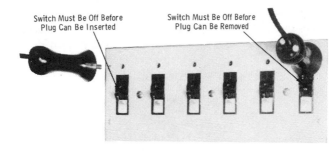

(A) Panel unit.

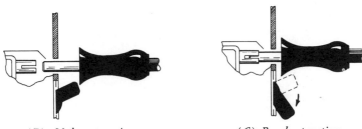

(B) Make operation.   *(C) Break operation.*

Courtesy Kliegl Bros. Lighting

Fig. 1-10. Kliegl *Saf-Patch* panel.

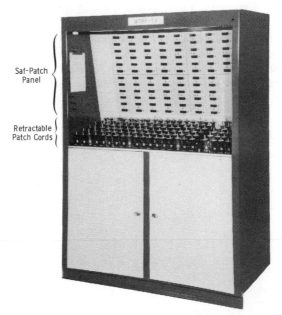

Courtesy Kliegl Bros. Lighting

Fig. 1-11. *Saf-Patch* lighting panel.

it should be remembered, therefore, that ease and flexibility of control will not duplicate those of a corresponding remote system. A mechanical system may include various forms of interlocking so as to provide for control

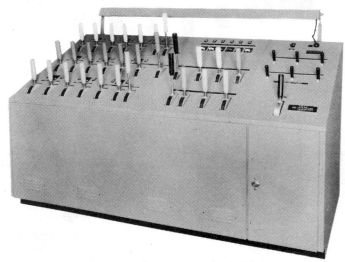

Courtesy Kliegl Bros. Lighting

Fig. 1-12. Autotransformer dimmer board.

of a few or all dimmers at any one time by means of submaster levers and a grand master lever. The autotransformer dimmer (Fig. 1-12) is the major component of a manually operated system. The diagram of Fig. 1-13A illustrates a typical manually operated autotransformer system. As you will note, a cross-connect circuit-selection panel for the interconnection

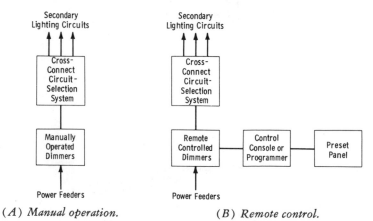

(A) Manual operation.   (B) Remote control.

Fig. 1-13. Two basic types of lighting control.

of lighting loads and dimmers is located between the dimmer bank and the lighting fixtures.

The use of remote operation is mandatory under certain circumstances: where presetting of scenes is desired; where two or more control stations are called for; where remote location of the control console is required. With a remote system, finger-tip controls are assembled in a control console. This control center then may be located away from the dimming bank in any desired location. At the same time, the power units can be located conveniently near the lighting fixtures so that feeder runs and other wiring may be kept to a minimum.

The diagram of Fig. 1-13B illustrates a typical remote-control system. As you will note, the control console and preset panel (if required) may be located at a good vantage point away from the dimmer bank. Located between the dimmer bank and the lighting fixtures is a cross-connect circuit-selection panel for the interconnecting of lighting loads and dimmers.

A cross-connect circuit-selection system that eliminates the use of patch cords and plugs is the Kliegl *Rotolector*. This device is a rotary circuit selector used to cross-connect lighting load circuits to dimming or nondim feeder lines. Each branch lighting circuit is wired directly to its own *Rotolector,* which, in turn, is connected to the various dimmer and nondim feeders. For example, a 12-point *Rotolector* can be fed by 10 dim and 2 nondim sources, or any other combination that totals 12. To patch the load to any one of the feeders, the operation is as follows:

1. The knob of the *Rotolector* is withdrawn. This automatically trips the associated circuit breaker before the load contact is broken.
2. The knob is rotated to the selected position, as indicated by the numbers on the dial, and then pushed in until contact is made.
3. When contact is made, the circuit breaker can be turned on to energize the circuit. "Hot" connections and arcing are eliminated automatically.
   Each 6½" × 6½ unit can be furnished with either 12 or 24 positions. Units are available to handle either 20- or 50-ampere lighting loads.

In manually operated autotransformer-dimmer *Klieg-Boards,* each autotransformer dimmer is controlled by its own operating handle, but all units may be coupled to submaster and grand master levers if desired. Control levers are color coded, labelled, and grouped for quick identification and use. Each dimmer is protected by a magnetic-type circuit breaker that can serve also as an on-off switch. Additional breakers are provided to protect and control secondary circuits. The board also can include circuit-breaker switches for control of nondim circuits, as well as remote-control switches and transfer switches. Manually operated autotransformer dimmers are used when there are strict budgetary restrictions and there is no need or desire for presetting of intensities or control from a remote location.

Fig. 1-14. SCR dimmer module.

Courtesy Kliegl Bros. Lighting

The majority of modern remote-control dimming installations utilize silicon controlled rectifier (SCR) dimmers. The wide-range capacity of these dimmers makes them especially suitable for auditorium lighting and other applications where large lighting loads are involved. They are small in size and require no outside cooling. Dimmer banks can be located remotely or as a part of a console and cross-connect system.

Fig. 1-14 shows an SCR dimmer module. This module is interchangeable with SCR nondim units, and can be safely inserted or removed from the board, while under load, without tripping any protective devices. The response to control is instantaneous, and no warm-up period is required. The units are designed for continuous operation at ambient temperatures up to 40°C. When loads in excess of the rated capacity of the dimmer are

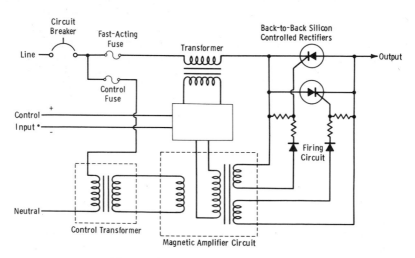

* The control input is completely isolated from the power input.

Courtesy Kliegl Bros. Lighting

Fig. 1-15. Circuit of SCR dimmer.

hot patched into the unit, it operates at reduced voltage until the overload is removed. Short-circuit protection is provided by means of silver-sand fuses that operate within 1/60 second to protect the rectifiers.

Fig. 1-15 is a schematic diagram of the SCR dimmer module of Fig. 1-14. The back-to-back SCR's are in series with the lighting load. The firing-circuit diodes for the SCR gates are activated from a low-voltage control input (never more than 28 volts) through a magnetic amplifier.

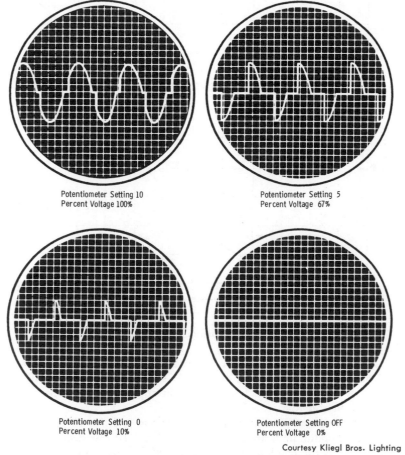

Potentiometer Setting 10
Percent Voltage 100%

Potentiometer Setting 5
Percent Voltage 67%

Potentiometer Setting 0
Percent Voltage 10%

Potentiometer Setting OFF
Percent Voltage 0%

Courtesy Kliegl Bros. Lighting

Fig. 1-16. Output waveforms of SCR dimmer.

Note that this allows complete isolation of the control input from the power input. Thus, the dimmer controls on the console are concerned only with the isolated low control voltage, about 28 volts dc at 12 milliamperes.

Fig. 1-16 shows the voltage waveforms at the output of the SCR dimmer module for four different settings of the control potentiometer. At a setting of 10, the output voltage is 100 percent, and the lights associated with that control are at maximum intensity. At a setting of 5, the gate voltage causes the SCR's to conduct later in the cycle, reducing the voltage output to 67 percent. At a setting of zero, the firing point in each cycle is still later, applying only 10 percent of full voltage. The control must be placed in the off position to remove the output voltage completely.

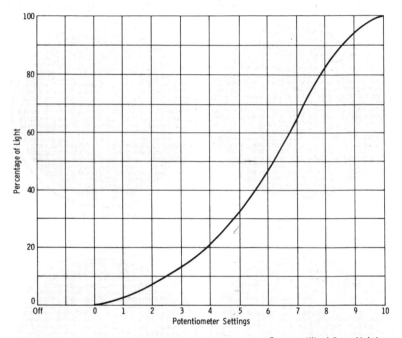

Courtesy Kliegl Bros. Lighting

Fig. 1-17. Light output resulting from SCR-dimmer control.

Fig. 1-17 gives the correlation between control setting and percentage of light output. Note that, for example, a setting of 5, which applies 67 percent of full voltage to a lamp, results in approximately 34 percent of maximum light output.

Models of these SCR dimmers are avaliable in 3, 6, 10, 12, and 15 kW capacities. Nondim units are available in 3, 6, and 10 kW sizes. A built-in device permits the nondim units to be set at any predetermined "on/off" position.

EXERCISES

Q1-1.   Give the nominal value of lighting, in watts/square foot, normally planned for (A) monochrome and (B) color when incandescent lighting is used.

Q1-2.   Give the nominal value of lighting, in watts/square foot, normally
         planned for (A) monochrome and (B) color when 100-percent
         quartz lighting is used.

Q1-3.   You have the lamp of Fig. 1-2B to be operated in the flood position.
         How many foot-candles would you expect at a throw distance of
         15 feet (A) on axis and (B) 10 feet off axis?

Q1-4.   You have the same lamp as above, operated in the flood position.
         How many foot-candles would you expect at a throw distance of 60
         feet (A) on axis and (B) 5 feet off axis?

Q1-5.   Approximately how many amperes does a 5000-watt lamp draw at
         a voltage of 115 volts?

CHAPTER 2

The System Concept

Fig. 2-1 illustrates the wide variance in fundamental television camera-chain systems. Because of the predominance of color telecasting, color systems will be emphasized in this training.

2-1. NTSC AND FCC COLOR STANDARDS IN PRACTICAL FORM

The FCC standards are based on the work of the National Television Systems Committee (NTSC) as originally filed with the FCC on July 21, 1953. We will often use the term "NTSC color" because the specific details of the system are far more inclusive than those spelled out in the FCC standards.

Since most colors can be duplicated by mixing correct amounts of three properly selected primary colors, it follows that a color TV system can be based on the transmission and reception of images in the three primary colors. The first step is accomplished in the television camera. The camera generates three different signals from the information contained in the image of the scene. There are three general types of color cameras, as follows:

1. Three cameras are operated from a single set of controls so that the view televised by each is identical. In front of each camera lens is placed a red, green, or blue filter. While the view imaged by all three cameras is identical, the light reaching the light-sensitive plate of each camera contains only the components passed by its respective filter. The three camera heads operate as one camera, and this camera produces a signal corresponding to the image in each of the primary colors. The brightness, or luminance, information is a function of the combination, and therefore the three images must be accurately registered to obtain a reasonably sharp picture, whether reproduced in monochrome or color.

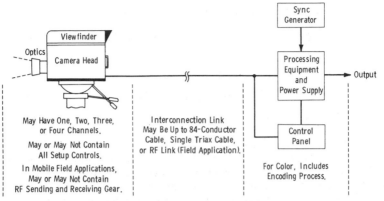

Fig. 2-1. Components of camera-chain systems.

2. Other cameras use a similar system, except that four pickup tubes are involved: three for the primary colors, and the fourth for luminance only. In this type of camera, only the luminance channel must have wide bandwidth; the three color channels can be relatively narrowband. With this type of camera, since the color channels carry very little brightness information, the effect of misregistration of images is slight on *monochrome* receivers.

3. A "convertible" system makes use of only two channels in the camera. One tube is used for a full-bandwidth luminance signal, and the second tube is used in the alternate red/blue channel. The fields are sequenced mechanically through a rotating red and blue filter wheel synchronized to the vertical rate of the main synchronizing generator. The green signal is obtained (before encoding) through subtractive matrixing of the red/blue and luminance signals. Special processing is required to convert the sequential color to the NTSC color signal.

In any of the basic types of camera mentioned above, we will be concerned with three signals, Y for luminance, and I and Q for chrominance. *Caution:* You will find a rather common application of the letter M (for "monochrome") to designate luminance information. In this book, the letter Y will be used for luminance, because in NTSC-FCC color specifications, the voltage of the composite color signal (which obviously contains both luminance information *and* chrominance information) is designated by the symbol E_M.[1]

---

[1]It is imperative that the reader have a color training background equivalent to that of Chapter 2 in *Television Broadcasting: Equipment, Systems, and Operating Fundamentals,* by Harold E. Ennes (Indianapolis: Howard W. Sams & Co., Inc., 1971).

The color picture signal has the following composition:

$$E_M = E_Y' + [E_Q' \sin(\omega t + 33°) + E_I' \cos(\omega t + 33°)] \quad \text{(Eq. 2-1.)}$$

where,

$$E_Q' = 0.41 (E_B' - E_Y') + 0.48 (E_R' - E_Y') \qquad \text{(Eq. 2-2.)}$$

$$E_I' = -0.27 (E_B' - E'_Y) + 0.74 (E_R' - E_Y') \qquad \text{(Eq. 2-3.)}$$

$$E_Y' = 0.30 E_R' + 0.59 E_G' + 0.11 E_B' \qquad \text{(Eq. 2-4.)}$$

The phase reference in Equation 2-1 is the phase of the color burst + 180°. The burst corresponds to amplitude modulation of a continuous sine wave.

Equations 2-1 through 2-4 have been developed in your previous training. Equation 2-5, which is in terms of the color-difference signals only (as demodulated by "narrow-band" color receivers), holds only for frequencies below 500 kHz (the region in which both Q and I are double sideband):

$$E_M = E_Y' + \left\{ \frac{1}{1.14} \left[\frac{1}{1.78} (E_B' - E_Y') \sin\omega t + (E_R' - E_Y') \cos\omega t \right] \right\}$$
$$\text{(Eq. 2-5.)}$$

This equation is as given in the NTSC-FCC standards; note that you can more clearly visualize this as:

$$E_M = E_Y + 0.493 (E_B - E_Y) \sin\omega t + 0.877 (E_R - E_Y) \cos\omega t$$

The I and Q channel bandwidths shown by Fig. 2-2 are specified by the FCC as follows:

I-Channel Bandwidth: Less than 2 dB down at 1.3 MHz
                          At least 20 dB down at 3.6 MHz

Q-Channel Bandwidth: Less than 2 db down at 400 kHz
                          Less than 6 dB down at 500 kHz
                          At least 6 db down at 600 kHz

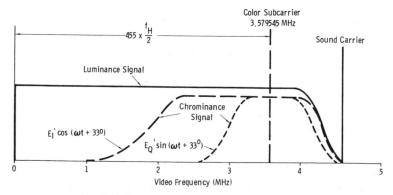

Fig. 2-2. Frequency distribution of color signal.

The prime signs in Equations 2-1 through 2-5 indicate gamma-corrected signals. In the discussion that follows, we will assume all system signals to be gamma corrected, and designate them as follows:

E_Y = corrected luminance-signal voltage

E_R = corrected red-signal voltage

E_G = corrected green-signal voltage

E_B = corrected blue-signal voltage

Caution: The method of gamma correction is not presently fixed by the FCC. In the three-camera color system, E_R, E_G, and E_B are all gamma corrected to make up the composite color signal. In the four-camera system, the luminance channel is gamma corrected, but the individual color channels may or may not be corrected in the same way. In this section, we assume that matrixing, encoding, and all following functions will be performed on properly corrected signals from the camera.

Fig. 2-3 is also a review, but we want to take a slightly different approach so that your thinking remains flexible. The luminance part of the complementary colors (yellow, magenta, and cyan) is formed as follows:

Yellow:

$$E_Y - E_B = 1.00 - 0.11 = 0.89$$

Magenta:

$$E_Y - E_G = 1.00 - 0.59 = 0.41$$

Cyan:

$$E_Y - E_R = 1.00 - 0.30 = 0.70$$

Be sure you can visualize this from your previous training. Thus, the luminance (only) part of the signal is as shown in Fig. 2-3A for 100-percent bars.

NOTE: Section 2-2 considers 75-percent color bars as normally used. However, the 100-percent bars are used for certain test procedures.

The values of I and Q for the primary colors and their complements are reviewed in Table 2-1. For example, the values for yellow are $I = 0.32$ and $Q = -0.31$.

$$
\begin{aligned}
E_{YELLOW} &= \sqrt{I^2 + Q^2} \\
&= \sqrt{(0.32)^2 + (-0.31)^2} \\
&= \sqrt{0.1024 + 0.0961} \\
&= \sqrt{0.1985} \\
&= 0.446, \text{ or simply } 0.45 \text{ as shown in Fig. 2-3B.}
\end{aligned}
$$

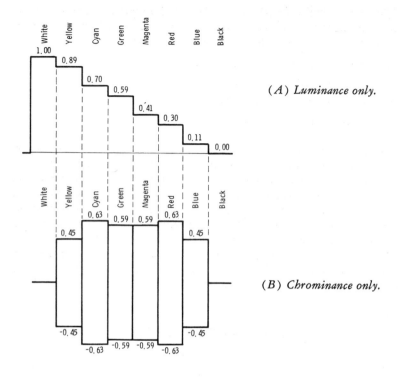

(A) Luminance only.

(B) Chrominance only.

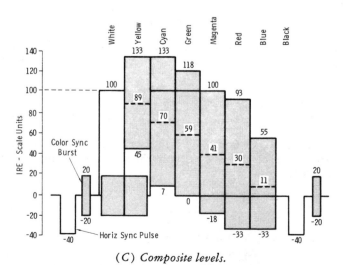

(C) Composite levels.

Fig. 2-3. Composition of 100-percent color bars.

You should be able to develop the remaining chroma amplitudes of Fig. 2-3B by the same reasoning. Remember that the subcarrier goes to zero for white and black (and all shades of gray).

Fig. 2-3C represents the composite signal (for maximum saturated chroma) on the IRE scale. (NOTE: The bar between green and red may be called "purple" or "magenta"; both are the same). Fig. 2-3C is for 100-percent color bars *without* inserted blanking pedestal.

The amplitude of the chroma signal interprets the degree of saturation of the particular hue. The fact that the yellow signal is 0.45 times the amplitude of the white luminance signal tells you this color is a fully saturated yellow; it contains a maximum unit of red and a maximum unit of green, with zero blue.

Table 2-1. Color System Relationships for Primaries and Complements

Transmitted Color	E_G	E_R	E_B	E_Y	G — Y	R — Y	B — Y	Q	I
Green	1	0	0	0.59	0.41	—0.59	—0.59	—0.525	—0.28
Yellow	1	1	0	0.89	0.11	0.11	—0.89	—0.31	+0.32
Red	0	1	0	0.3	—0.3	0.7	—0.3	+0.21	+0.60
Magenta	0	1	1	0.41	—0.41	0.59	0.59	+0.525	+0.28
Blue	0	0	1	0.11	—0.11	—0.11	0.89	+0.31	—0.32
Cyan	1	0	1	0.7	0.3	—0.7	0.3	—0.21	—0.60

Hue is determined by the phase angle of the signal with respect to a specified reference. (This is also a review, but, again, we want to take a slightly different tack so that you can "see" this system in its many facets. In this case, "familiarity breeds contentment.") See Fig. 2-4A. Suppose we have a carrier that starts with the phase of the R — Y vector, and we pass this carrier through a 90° delay line to obtain a carrier in quadrature (in phase with the B — Y vector). Now we can say that B — Y *lags* R — Y by 90°, or that R — Y *leads* B — Y by 90°. But recall that vectors "in action" rotate counterclockwise, and the starting point (0° or 360°) is generally taken at the B — Y axis on the right.

Now study Fig. 2-4B. It will be noted that when $E_R — E_Y$ is transmitted alone, it produces colors from bluish-red through white to bluish-green, which are located along its axis. When transmitted alone, signal $E_B — E_Y$ produces colors from purple through white to greenish-yellow. From the color triangle, it can be seen that varying amounts of both signals, when transmitted together, produce hue and varying saturations of any color located within the triangle. For example, Area 1 is enclosed by axes designated — $(E_R — E_Y)$ and — $(E_B — E_Y)$ and represents the colors that can be produced by the presence of these two signals.

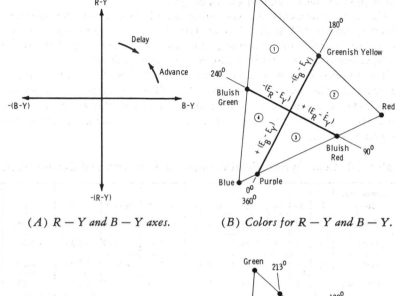

(A) $R - Y$ and $B - Y$ axes.   (B) Colors for $R - Y$ and $B - Y$.

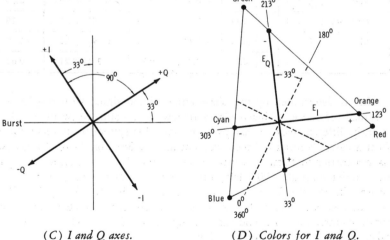

(C) I and Q axes.   (D) Colors for I and Q.

Fig. 2-4. Color phase angles.

You have learned previously that it is desirable to produce a "wideband" signal for the orange-cyan color regions. If color detail is transmitted by $E_R - E_Y$ alone ($E_B - E_Y$ being zero), the colors reproduced will be bluish-red and bluish-green instead of an orange-cyan mixture. So we set up a new pair of axes by starting with a vector advanced 33° from $R - Y$ as shown in Fig. 2-4C. This now becomes the "in-phase" vector (I). The quadrature component (Q) is advanced 33° from $B - Y$. The reference burst remains along the − (B − Y) axis. The new color gamut is now as shown on the color triangle of Fig. 2-4D. The respective bandwidth proportionments are as shown by Fig. 2-2. So the wideband I signal lies along the orange-cyan axis.

Now note that when R − Y was rotated to form the I axis, it was reduced in amplitude (Fig. 2-5A). For example, the red vector in terms of R − Y and B − Y is shown in Fig. 2-5B (see Table 2-1). Since R − Y was reduced by a factor of 0.877, the red vector as a result of I-Q modulation is as shown in Fig. 2-5C. This is to say that the original 0.7 of R − Y is now 0.7 (0.877), or 0.614, along the R − Y axis. The original − (B − Y) amplitude of −0.3 is now − 0.3 (0.493), or −0.14, along the − (B − Y) axis. This is the relationship you will see for the red vector on the station vectorscope as a result of I-Q modulation. We will cover this in detail later.

From trigonometric tables (or your slides rule), cos 33° = 0.839 and cos 57° = 0.544. If you use these values you arrive at the stated proportions of R − Y and B − Y for I and Q as given in Fig. 2-5A.

We have developed the chroma amplitudes for saturated primary and complementary colors in terms of I and Q. Now let us review the development of the chrominance-signal phase angle.

The graphic method for development of a specific phase angle (yellow) is shown in Fig. 2-5D. From previous training (tabulated by Table 2-1), the I and Q components of yellow are I = 0.32 and Q = − 0.31. The vector sum is 0.45 at an angle leading the +I axis by 44.1°. The angle, then, with the NTSC zero axis (B − Y) is 123° + 44.1°, or 167.1°.

To solve by trigonometric rather than graphic means:

$$\text{cotangent } \theta = \frac{\text{adjacent side}}{\text{opposite side}} = \frac{0.32}{-0.31} = -1.032$$

From trigonometric tables (or your slide rule), cos 33° = 0.839 and cos that, since the cotangent of the unknown angle is 1.032, you find the angle whose cotangent is this value. Without worrying about plus or minus values of the number, by visualizing the quadrant in which the I and Q sum must fall, you know whether the angle leads or lags the I or Q axis. Obviously, you can use any trigonometric function—sine, cosine, tangent, or cotangent—depending on which is most convenient in your computations.

A good way to remember the signal polarities associated with various colors is through the use of the color triangle. We have noted that an I, Q, or color-difference signal may have a negative polarity for some colors, and for other colors any one of these signals may have a positive polarity. By studying the color triangles of Fig. 2-6, you will find it easier to remember which colors produce an I, Q, or color-difference signal that is negative, and which colors provide a positive signal.

On the color triangle of Fig. 2-6A, the polarity of the I signal for each color is given. The colors that fall to the right of the Q axis are repre-

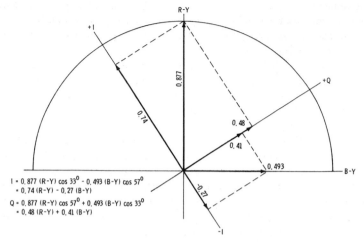

$I = 0.877 (R-Y) \cos 33^0 - 0.493 (B-Y) \cos 57^0$
$\;\;= 0.74 (R-Y) - 0.27 (B-Y)$
$Q = 0.877 (R-Y) \cos 57^0 + 0.493 (B-Y) \cos 33^0$
$\;\;= 0.48 (R-Y) + 0.41 (B-Y)$

(A) Rotation of R—Y to form I.

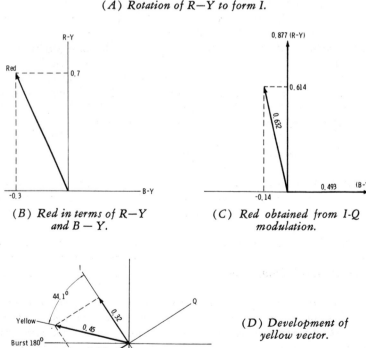

(B) Red in terms of R—Y and B—Y.

(C) Red obtained from I-Q modulation.

(D) Development of yellow vector.

Fig. 2-5. Color vectors.

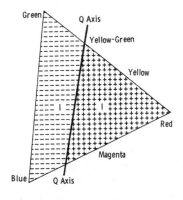

(A) I signal.

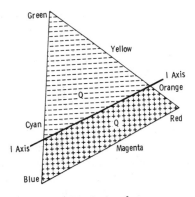

(B) Q signal.

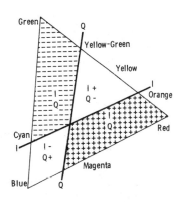

(C) I and Q signals.

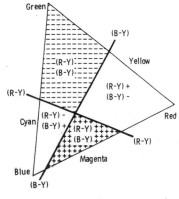

(D) R − Y and B − Y signals.

(E) Reference chart.

Color	R-Y	B-Y	I	Q
Green	−	−	−	−
Yellow	+	−	+	−
Red	+	−	+	+
Magenta	+	+	+	+
Blue	−	+	−	+
Cyan	−	+	−	−

Fig. 2-6. Polarities of color signals.

sented by a positive I signal, and the colors to the left of the Q axis are represented by a negative I signal. For instance, when blue, cyan, or green is transmitted, the polarity of the I signal is negative. When magenta, red, or yellow is transmitted, the I signal is positive in polarity. Fig. 2-6B shows the polarity of the Q signal for each color. As can be seen, the colors that lie above the I axis are represented by a negative Q signal, and those lying below the axis produce a positive Q signal. The polarity of the Q signal is negative when cyan, green, or yellow (above the axis) is transmitted; when blue, magenta, or red (below the axis) is transmitted, the Q signal is positive in polarity.

A composite drawing of the triangles in Figs. 2-6A and 2-6B is shown in Fig. 2-6C. Notice that the I and Q signals for colors which lie in the upper left-hand section of the triangle are negative and that the signals representing colors in the lower right-hand section are positive. Colors lying in the other two sections produce I and Q signals that are opposite in polarity. For instance, the Q signal is positive for blue, but the I signal is negative.

The key for determining the correct polarity for each of these signals is in knowing the location of the colors on the triangle and in remembering the negative and positive areas shown in Figs. 2-6A and 2-6B. With this knowledge, the polarity of each signal for any color can be determined easily.

Fig. 2-6D shows the polarities of the R − Y and B − Y signals for each color on the color triangle. This drawing can be used to determine the polarities of the R − Y and B − Y signals in the same manner that Fig. 2-6C can be used to determine the polarities of the I and Q signals. Fig. 2-6E lists the various signal polarities in tabular form.

The FCC states that:

"The [angles] of the subcarrier measured with respect to the burst phase, when reproducing saturated primaries and their complements at 75 percent of full amplitude,[2] shall be within ±10°, and [the] amplitudes shall be within ±20 percent of the values specified. . . . The ratios of the measured amplitudes of the subcarrier to the luminance signal for the same saturated primaries and their complements shall fall between the limits of 0.8 and 1.2 of the values specified for their ratios. Closer tolerances may prove to be practicable and desirable with advance in the art."

The relative tolerances specified are shown by Fig. 2-7. Note, however, that the FCC specifies "at 75 percent of full amplitude." A visual transmitter is never checked for FCC specifications at "full amplitude" for chroma signals, because this results in overshoots of about 33 percent, as you have found. But also note that the FCC specifies "saturated primaries and their complements." Observe in Fig. 2-7 that chroma amplitudes are

[2]Observe that color bars are saturated even when transmitted at 75 percent of full amplitude. This is further developed in following Section 2-2.

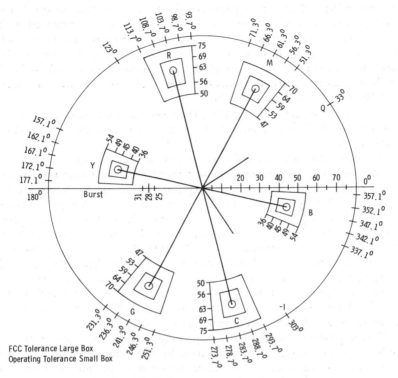

Fig. 2-7. Tolerances for color signals.

shown at their maximum saturated levels. We will cover the practical details of how to use this information as we progress.

The FCC fixes the frequency response of the color TV transmitter as follows: "For monochrome transmission only, the overall attenuation characteristics of the transmitter, measured in the antenna transmission line after the vestigial sideband filter (if used), shall not be greater than the following amounts below the ideal demodulated curve. . . .

> 2 dB at 0.5 MHz
>
> 2 dB at 1.25 MHz
>
> 3 dB at 2.0 MHz
>
> 6 dB at 3.0 MHz
>
> 12 dB at 3.5 MHz

"The curve shall be substantially smooth between these specified points, exclusive of the region from 0.75 to 1.25 MHz. . . ."

For color transmission, the above standard applies except as modified by the following: "A sine wave of 3.58 MHz introduced at those terminals of the transmitter which are normally fed the composite color picture signal shall produce a radiated signal having an amplitude (as measured with a diode on the rf transmission line supplying power to the antenna) which is down 6 ± 2 dB with respect to a signal produced by a sine wave of 200 kHz. In addition, between the modulating frequencies of 2.1 and 4.1 MHz, the amplitude of the radiated signal shall not vary by more than ± 2 dB from its value at 3.58 MHz. . . ."

Note here that the FCC refers to the response as measured by the "ideal detector" curve. This is treated further in the exercises at the conclusion of this chapter.

Envelope-delay tolerances set by the FCC are as follows:

"A sine wave, introduced at those terminals of the transmitter which are normally fed the composite color picture signal, shall produce a radiated signal having an envelope delay, relative to the average envelope delay between 0.05 and 0.20 MHz, of zero microseconds up to a frequency of 3.0 MHz; and then linearly decreasing to 4.18 MHz so as to be equal to -0.17 microsecond at 3.58 MHz. The tolerance on the envelope delay shall be ± 0.05 microsecond at 3.58 MHz. The tolerance shall increase linearly to ± 0.1 microsecond down to 2.1 MHz, and remain at ± 0.1 microsecond down to 0.2 MHz. (Tolerances for the interval of 0.0 to 0.2 MHz are not specified at the present time.) The tolerance shall also increase linearly to ± 0.1 microsecond at 4.18 MHz."

This requirement is illustrated graphically in Fig. 2-8. The operational techniques for meeting this specification for the transmitter must be left to more advanced study. However, the reader must be familiar with this aspect now, since it is necessary to consider the overall system (studio, transmitter, and receiver) in learning to recognize "color distortions."

Another FCC requirement is that E_Y, E_I, E_Q, and the components of these signals match each other in time to 0.05 microsecond. This requirement must be met by inserting proper delays in the Y and I channels

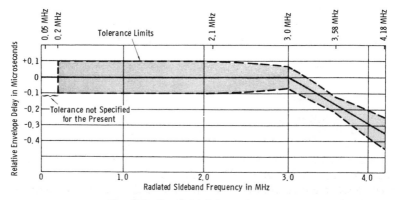

Fig. 2-8. Envelope-delay tolerances.

(both transmitter end and receiver end) to match the delay in the Q channel brought about by the narrow bandwidth in this channel.

Radiation of the transmitter more than 3 MHz outside the channel must be at least 60 dB below the visual transmitter power. The voltage of the upper sideband must not be greater than −20 dB for a modulating frequency of 4.75 MHz or greater. The voltage of the lower sideband must not be greater than −20 dB for a modulation frequency of 1.25 MHz or greater. For color, it must not be greater than −42 dB for a modulating frequency of 3.58 MHz (color subcarrier).

It does not necessarily follow that a transmitter that meets the sideband requirements when broadcasting monochrome will do so when radiating a color signal. This is because the chrominance subcarrier, when modulated on the picture carrier, may appear in the lower adjacent channel at a substantially higher level than the vestigial-sideband filter is designed to handle. An additional notch filter is required in that case; it provides attenuation at a frequency 2.33 MHz below the channel edge ($3.58 - 1.25 = 2.33$ MHz).

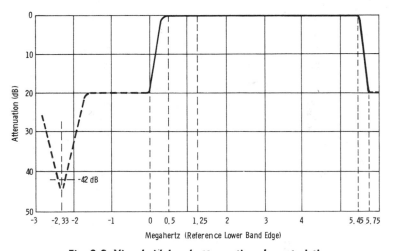

Fig. 2-9. Visual sideband attenuation characteristic.

The specification for visual-sideband attenuation is shown in Fig. 2-9. This characteristic must be measured with the color subcarrier present.

The FCC waveforms for horizontal, vertical, and color sync signals are illustrated in Fig. 2-10. It is common practice for the FCC to state all times in terms of H, where H is one television line, or 63.5 microseconds. In practice, times in microseconds usually are much more useful; these are given in Fig. 2-11 (horizontal pulses and color sync burst) and Table 2-2 (vertical pulses). With the aid of this information, you can use an oscilloscope to "standardize" your sync generator.

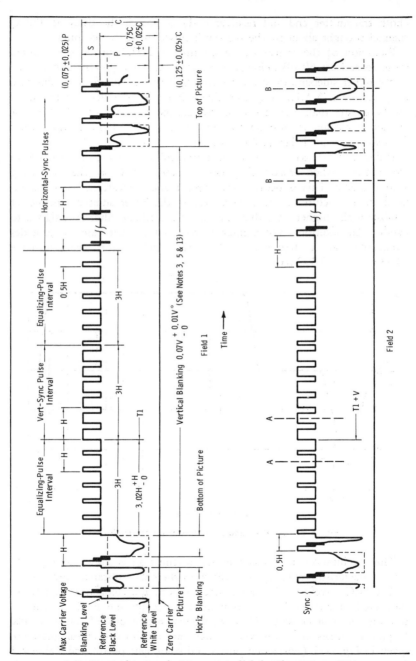

(A) Vertical interval. (See expanded details on page 52.)

Fig. 2-10. Standards for transmitted

NOTES

1. H = time from start of one line to start of next line.

2. V = time from start of one field to start of next field.

3. Leading edge and trailing edge of vertical blanking each should be complete in less than 0.1H.

4. Leading and trailing slopes of horizontal blanking must be steep enough to preserve minimum and maximum values of x + y and z under all conditions of picture content.

*5. Dimensions marked with asterisk indicate that tolerances given are permitted only for long time variations and not for successive cycles.

6. Equalizing-pulse area shall be between 0.45 and 0.5 of area of a horizontal-sync pulse.

7. Color burst follows each horizontal pulse, but is omitted following the equalizing pulses and during the broad vertical pulses.

8. Color bursts to be omitted during monochrome transmission.

9. The burst frequency shall be 3.579545 MHz. The tolerance on the frequency shall be ±10 Hz with a maximum rate of change of frequency not to exceed 1/10 Hz per second.

10. The horizontal-scanning frequency shall be 2/455 times the burst frequency.

11. The dimensions specified for the burst determine the times of starting and stopping the burst, but not its phase. The color burst consists of amplitude modulation of a continuous sine wave.

12. Dimension P represents the peak excursion of the luminance signal from blanking level, but does not include the chrominance signal. Dimension S is the sync amplitude above blanking level. Dimension C is the peak carrier amplitude.

13. For monochrome transmission only, the duration of the horizontal-sync pulse between 10 percent points is specified as 0.08H ± 0.01H, the period from the leading edge of sync to the 10 percent point on the trailing edge of horizontal blanking is specified as 0.14H minimum; and the duration of vertical blanking is specified as 0.05V, +0.03V and −0. All other dimensions remain the same.

synchronizing waveforms.

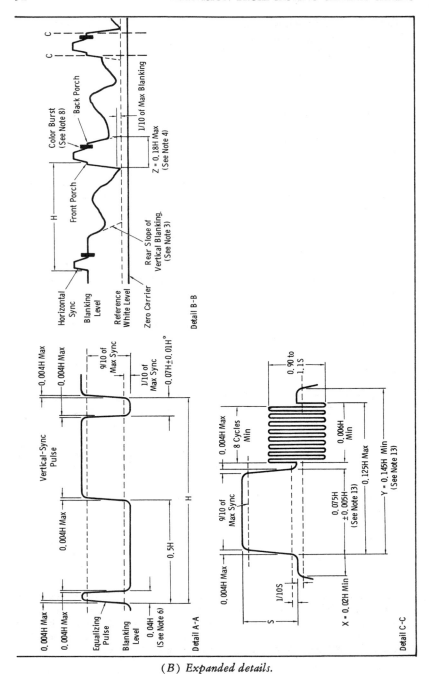

(B) Expanded details.

Fig. 2-10. Standards for transmitted synchronizing waveforms. (Cont'd.)

The FCC further states: "The color picture signal shall correspond to a luminance component transmitted as amplitude modulation of the picture carrier and a simultaneous pair of chrominance components transmitted as the amplitude-modulation sidebands of a pair of suppressed subcarriers in quadrature." Also, "The chrominance subcarrier frequency shall be 3.579545 MHz ±10 Hz with a maximum rate of change not to exceed 1/10 Hz per second."

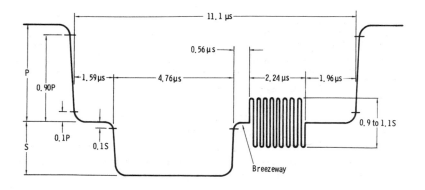

	Nominal Microseconds	Tolerance Microseconds
Blanking	11.1	+0.3 −0.6
Sync	4.76	±0.32
Front Porch	1.59	+0.13 −0.32
Back Porch	4.76	+0.96 −0.61
Sync to Burst	0.56	+0.08 −0.17
Burst	2.24	+0.27 0
Blanking to Burst[1]	6.91	+0.08 −0.17
Sync & Burst	7.56	+0.38 −0.49
Sync & Back Porch	9.54	±0.32

1. Blanking to burst tolerances apply only to signal before addition of sync.

Fig. 2-11. Time intervals for horizontal-sync pulse.

Table 2-2. Vertical-Pulse Widths

Pulse	Minimum (μs)	Nominal (μs)	Maximum (μs)
Equalizing	2.0	2.4	2.54
Vert Serration	3.81	4.5	5.08
Vert Blanking	1167	1250	1333

NOTES

1. Vertical-sync pulses are not specified. See Detail A-A, Fig. 2-10B. The width of the vertical-sync pulse is set by the tolerance on the width of the vertical serration. An interval of 63.5 microseconds (H) must exist from the leading edge of the last (leading) equalizing pulse to the trailing edge of the first serration.
2. The width of the equalizing pulse must be 0.45 to 0.5 of the horizontal sync width used.
3. Vertical blanking in terms of H is from 18.375 to 21 lines. Although 21 lines is shown as "maximum" in the above chart, this is the width of vertical blanking maintained by the networks. It allows vertical-interval test signals to be inserted (usually on lines 18 and 19 of vertical blanking) with a suitable "guard band" of blanking lines before the start of active line scan.
4. Horizontal and vertical blanking must be of proper ratio to establish the 4:3 aspect ratio.

The FCC describes the luminance modulation as follows: "A decrease in initial light intensity shall cause an increase in radiated power (negative transmission)." Further, "The blanking level shall be transmitted at 75 ± 2.5 percent of the peak carrier level. The reference white level of the luminance signal shall be 12.5 ± 2.5 percent of the peak carrier level. . . .

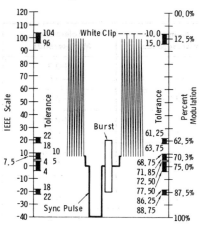

Note: Tolerance value specified at blanking level applies to sync amplitude only. The variation of blanking with respect to the tolerance of setup shown is assumed to be zero.

Fig. 2-12. Video-signal amplitudes.

The reference black level shall be separated from the blanking level by the setup interval, which shall be 7.5 ± 2.5 percent of the video range from blanking level to the reference white level."

The above specification is illustrated in Fig. 2-12. The studio scale is shown on the left, and the corresponding transmitter scale (percent modulation) is shown on the right.

2-2. EVOLUTION OF THE COLOR-BAR SIGNAL

Remember that the nominal input level to the encoding system from the camera is 0.714 volt. This is 100 IEEE units when the scope is calibrated for 1 volt over an IEEE graticule of 140 units. The output level of tube-type color-bar generators usually is changed to the 75-percent amplitude by switching the internal output impedance from 75 ohms to 75 percent of this value, or 56 ohms. Thus, a 220-ohm resistor switched across the output gives 56 ohms, with a resulting 25-percent reduction in output level. In solid-state generators, the output level sometimes is set by switching zener diodes with the proper reference voltage to set the clipper levels.

But there is another consideration. The camera outputs have the setup level established at a nominal 7.5 percent of white video. The color-bar output of Fig. 2-3 does not have blanking inserted. To convert to the standard transmission system, consider the signal in terms of IEEE units for a 1-volt (peak to peak) signal in 140 IEEE units. The setup is 7.5 percent from 0 IEEE units toward peak white. Sync is added and extends to −40 IEEE units. Burst is added on the blanking base (0 IEEE units) and extends between +20 IEEE units and −20 IEEE units (peak-to-peak). If maximum level (occurs in yellow and cyan) is to be no more than 100 IEEE units (reference white peak), we have the following considerations:

See Table 2-3. The chroma amplitude for 100-percent yellow bars is 0.447 (0.894 peak to peak), which, added to the yellow luminance of 0.89, gives a peak amplitude of 1.34, for an overshoot of 34 IEEE units above reference white. This chroma is now reduced to 75 percent, or $(0.75)(0.447) = 0.3352$. Since this level, added to the luminance level, must not exceed 1 (or 100 IEEE units), then $1.000 − 0.3352 = 0.6648$, and the new luminance level for yellow at 75 percent of full amplitude is 0.664 (Fig. 2-13). The white bar at 1.00 is then reduced to 0.746 by this process, or to approximately 75 percent of full amplitude. All other parts of the signal are reduced by the same proportion. (The "voltage" scale in Fig. 2-13 must be taken as proportional to the IEEE scale.)

As yet, we have not added the 7.5-percent setup level. When this is done, the video of the previous signal is reduced in peak-to-peak value by this same amount. (The blanking pedestal contains no video information. Peak picture black is now at 7.5 IEEE units instead of 0 IEEE units.)

So now for yellow (for example) to have the same "brightness" as before, its amplitude must be reduced toward white by the same amount it was reduced toward black by the 7.5-percent pedestal. You have found already that the difference between the 75-percent yellow luminance and peak white is (essentially) 0.336 $(1 − 0.664 = 0.336)$. To move this the appropriate amount toward white, the 0.336 is reduced by 7.5 percent of white: $(0.336)(92.5\%) = (0.336)(0.925) = 0.310$. This is the new difference (to allow for setup) between peak white and yellow luminance.

Table 2-3. Color-Bar Signal Chart

Signal	R	G	B	Y	Subcarrier Amplitude	Subcarrier Phase
R	1	0	0	0.30	0.635	103.4°
RG (Yellow)	1	1	0	0.89	0.447	167.1°
G	0	1	0	0.59	0.593	240.8°
GB (Cyan)	0	1	1	0.70	0.635	283.4°
B	0	0	1	0.11	0.447	347.1°
BR (Magenta)	1	0	1	0.41	0.593	60.83°
RGB (White)	1	1	1	1	0	—
B — Y ($R-Y=0$)	0.1571	0	1	0.1571	0.4135	— 0.367°
Q ($I=0$)	0.5371	0	1	0.2711	0.4265	33°
R — Y ($B-Y=0$)	1	0	0.3371	0.3371	0.5848	90.03°
I $Q=0$	1	0.4056	0	0.5393	0.4865	123°
(G — Y) $\angle$ 90° ($G-Y=0$)	1	0.7317	0	0.7317	0.4313	146.38°
Color Burst	—	—	—	—	0.20	180°
— (B — Y) ($R-Y=0$)	0.8429	1	0	0.8429	0.4135	179.6°
— Q ($I=0$)	0.4629	1	0	0.7289	0.4265	213°
— (R — Y) ($B-Y=0$)	0	1	0.6629	0.6629	0.5848	270.03°
— I $Q=0$	0	0.5944	1	0.4607	0.4865	303°
— (G — Y) $\angle$ 90° ($G-Y=0$)	0	0.2683	1	0.2683	0.4313	326.4°

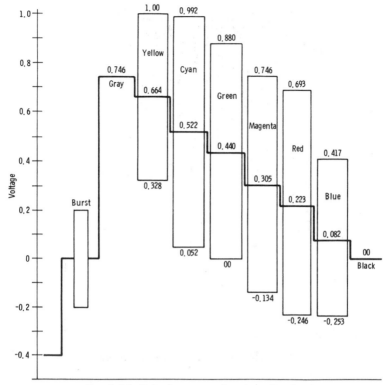

Fig. 2-13. Color bars at reduced amplitude.

Therefore the final new luminance value for yellow is $1 - 0.310 = 0.690$, or 69 IEEE units. See Fig 2-14. All other signals, except for sync and burst, are reduced (toward white) in like manner.

Fig 2-15 illustrates the newly developed 75 percent bar signal. The peak-to-peak chroma values are designated at the top of each bar in IEEE values. Bear in mind that although the bars are reduced in value to 75 percent of NTSC developed bars, they still represent fully saturated signals. For any color to be desaturated, some amount of all three primaries would have to be present.

2-3. DEFINING AND RECOGNIZING "DISTORTIONS" IN NTSC COLOR

In the consideration of color-transmission "distortions," it is necessary to consider the transmission and reception processes together; in fact, it is impossible to consider one without the other. For example, consider the matter of gamma correction. The pickup device must actually be made nonlinear in gray scale to match the "average" color picture tube. The sys-

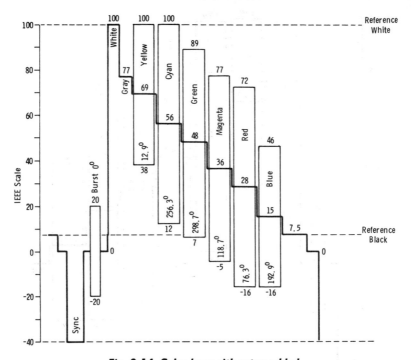

Fig. 2-14. Color bars with setup added.

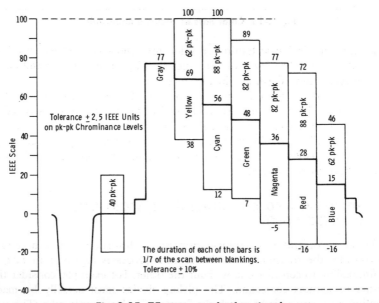

Fig. 2-15. 75-percent color-bar signal.

tem *following* the pickup device must be made as strictly linear as possible, so that all of the amplitude-versus-brightness correction can be made in the pickup device, and complicating factors are not introduced after this correction.

Gamma correction is a camera-head function, and as such will be covered in Chapter 6. The point to be emphasized here is that the system handling video information can be carrying a nonlinear function (from the pickup device) and that it is normal for this function to appear nonlinear. Nonlinearity of a certain kind is required (in the case of gamma correction) because of picture-tube characteristics, or (in other cases which we will point out) because of receiver or monitor design. In the cases just cited, you should understand that these nonlinearities are *not* "distortions"!

The Basic NTSC Color Problem

The number one problem is *phase sensitivity*. For accurate reproduction of a hue, the resultant color phasor must be held within a tolerance of $\pm 5°$. Hue error becomes noticeable at $\pm 10°$ from the proper position relative to burst. The transmission and reception process, to maintain such a tight tolerance, must operate with a signal time-base accuracy of a few nanoseconds.

Color information (hue and saturation) is affected by various types of amplitude and phase shifts in transmitting and receiving equipment. In addition, hue shifts can occur as a result of multipath transmission errors. Burst and subcarrier-sideband phase shift caused by variations in topography over different transmission paths can require the viewer to readjust his receiver with every station change. Although this can be annoying to the viewer, it is not a factor under the control of the station operator. However, it is necessary to adjust the station color gear to precise operating parameters (and this is possible!) and to *minimize,* as much as possible within the limitations of the equipment, any amplitude or phase distortion. When you receive complaints from viewers, you rest easy when you can prove the performance of your system to qualified technical agents.

In the case of multipath-reflection errors, if you are transmitting an accurately adjusted color picture, the viewer can readjust his hue control from the setting for a different station and still get a good color picture. Far worse is the situation in which there is a *differential phase* error. In this case, different colors can be shifted by different angles, depending on luminance. The viewer then can only attempt to strike a "happy medium" with his hue control.

Burst Phase Error

Remember this: The color monitor or receiver will "insist" that the burst phase not be in error. This results from the fact that the subcarrier-sideband information is synchronously demodulated with the burst as a reference. The angles clockwise from burst for various colors are as in Table 2-4.

Fig. 2-16A shows the proper vectors for burst, red, and magenta. Now we assume we have the burst-phase error (θ) shown in Fig. 2-16B. The color receiver requires the burst phase shown in Fig. 2-16A; to visualize this, put an imaginary pin at the center dot, and rotate the vectors as in Fig. 2-16C to place the burst at the correct phase. The other vectors become R + θ and M + θ, which represent a red shifted toward yellow and a magenta shifted toward red.

Table 2-4. Color Phase With Burst

Color	Phase Angle (Degrees)
Burst	0
Yellow	12.9
Red	76.6
Magenta	119.2
Blue	192.9
Cyan	256.6
Green	299.2

The effect of burst-phase error is to rotate all hues in the direction opposite to the burst-phase error. If the burst error in Fig. 2-16B had been counterclockwise, it would have been necessary to rotate the vectors clockwise to return the burst to the proper position. Red would then tend to go toward magenta, magenta would tend to go toward blue, etc., all around the color gamut.

You can visualize this most conveniently by using the color triangle as in Fig 2-16D. Note carefully that this triangle has been turned around from the position normally presented, to fit it into the NTSC phase diagram. The center of the circle is at illuminant C, where the color vectors collapse to zero value. Place the imaginary pin for the circle here; the color triangle must remain fixed in the position shown. If the burst slips *clockwise* by some angle, all reproduced colors shift *counterclockwise* by the same angle (and vice versa for counterclockwise slip of burst phase).

Note again the interdependence of the encoding and decoding processes. The hue control at the receiver is an operational adjustment. It has, in most modern receivers, a range of at least ±70°, whereas the tolerance is only ±10° for the *overall* transmission system. This tells you that any receiver in normal operating condition should be able to have its hue control adjusted to obtain proper colors from your station. Your responsibility at the sending end is to assure that the burst phase is as nearly correct as possible so that (theoretically) receivers need not be readjusted for proper color reproduction from different stations. Remember that if you *do*

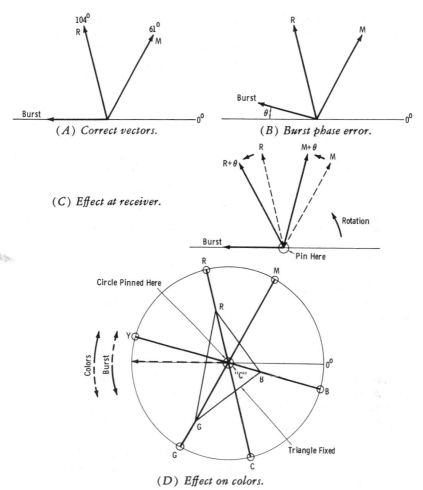

(A) Correct vectors.

(B) Burst phase error.

(C) Effect at receiver.

(D) Effect on colors.

Fig. 2-16. Effect of burst phase error.

transmit a burst phase error, the receiver exhibits a "locked phase error" so that colors reproduced are not exactly the same as in the original scene. Since the viewer does not have a direct comparison, he is not aware of this as long as he can obtain good flesh tones.

Quadrature Distortion

Quadrature distortion results from cross talk between the I and Q video information. It has more possible causes than most of the other types of color distortion. It does not involve an operational control at the receiver, but the receiver *can* cause this effect if the circuits, especially the quadrature-transformer adjustment, are faulty.

A symptom of quadrature distortion is color displacement; in severe cases, a girl's red lips can be in the middle of one cheek. In the more usual case, there is color fringing (not caused by camera misregistration or a misconverged picture tube) at the edges of color transitions.

The most obvious type of quadrature distortion occurs when I and Q are not phased exactly 90° in the encoder. Fig. 2-17 illustrates the case in which Q lags I by more than 90°. For simplicity, only the basic colors in the first and third quadrants are plotted. Now remember the polarities of I and Q in each quadrant. These are:

Quadrant 1 is bounded by +Q and +I (same polarity of I and Q).

Quadrant 2 is bounded by +I and −Q (opposite polarities of I and Q).

Quadrant 3 is bounded by −Q and −I (same polarity of I and Q).

Quadrant 4 is bounded by −I and +Q (opposite polarities of I and Q).

The relative values of I and Q for the colors plotted are as developed in Table 2-1. Note that colors with large amounts of Q are affected more than others. For example, red and cyan have relatively small amplitude and phase errors; green and magenta have larger amplitude and phase errors. By adjusting the hue control on the monitor or receiver, you cannot get a good red and cyan *simultaneously* with a good green (green will be yellowish) or magenta (magenta will be bluish).

Now if you will take the trouble to plot yellow and blue (second and fourth quadrants), you will find these increased in amplitude. For a Q lag greater than 90°, colors in the first and third quadrants are reduced in amplitude; those in the second and fourth quadrants are increased in amplitude. In each quadrant, the phase error is in the direction of the Q phase error.

This is emphasized further by Fig. 2-18, in which Q lags I by *less* than 90°. For simplicity, only magenta is shown in quadrant 1. Note that it is now increased in amplitude, whereas yellow and blue are reduced in amplitude. In either case, the phase error is in the direction of the Q error. Since the receiver separates signals with a 90° relationship, I will crosstalk into Q and vice versa.

Before going further, be sure to grasp the fundamentals of chrominance-signal transmission and reception for NTSC color. See Fig. 2-19 and the following analysis:

Fig. 2-19A represents the signal as transmitted. I and Q are double sideband in the region shown. The upper sideband of the wideband I chrominance is cut off at the transmitter to achieve a 20-dB roll-off at the sound carrier frequency. A portion of the lower sideband of the I chroma constitutes single-sideband information; no Q chroma exists there.

The outputs of the I and Q demodulators are equal in the double-sideband region (Fig. 2-19B).

Over the single-sideband region, the voltage output of the I demodulator is one-half that which occurs in the double-sideband region (Fig. 2-19C). Note also that a one-half-I voltage, shifted in phase by 90°, appears at the Q-demodulator output. This is E_I at its single-sideband frequencies of about 0.6 to 1.5 MHz.

The output of the I demodulator is boosted by 6 dB above 0.5 MHz to recover the gain lost in the single-sideband region (Fig. 2-19D). The Q demodulator is limited in bandwidth to 0.5 MHz.

The filtering and relative gain action of Fig. 2-19D results in voltages E_I and E_Q free of crosstalk (Fig. 2-19E). This assumes, of course, that I and Q are actually being transmitted in the proper quadrature relationship.

"Narrow-band" color receivers demodulate on the $R - Y$ and $B - Y$ axes. These receivers use the same bandwidth for all chrominance compo-

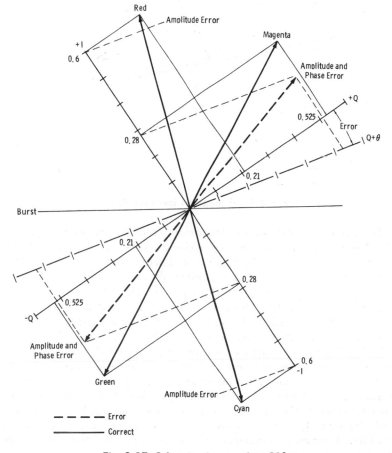

Fig. 2-17. Q lagging I more than 90°.

nents. With all chrominance channels of the same bandwidth, delay equalization is unnecessary, and no crosstalk occurs in a properly aligned receiver of this type. Again, this assumes that the I and Q chroma signals are being transmitted in the proper quadrature relationship. Also, there are other causes of lack of quadrature than misadjustment of the Q lag.

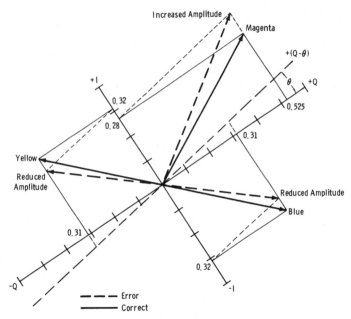

Fig. 2-18. Q lagging I less than 90°.

To investigate other causes of quadrature distortion, study Fig. 2-20. Fig. 2-20A shows the usual double-sideband representation of a modulated carrier. If, for example, the highest modulating frequency is 0.5 MHz, upper sidebands (f_U) extend 0.5 MHz above f_C, and lower sidebands (f_L) extend 0.5 MHz below f_C. This can be represented by equal-amplitude phasors (vectors) rotating in opposite directions as in Fig. 2-20B. The resultant amplitude is the vector sum of f_U and f_L added to the carrier vector so that the resultant always lies along carrier line YO.

In Fig. 2-20C, the amplitude-frequency response exhibits a rapid roll-off above the carrier frequency. Thus f_U is severely attenuated (Fig. 2-20D). The resultant vector no longer lies along line YO, but contains a quadrature component, as shown in Fig. 2-20E. If f_C is the color sub-carrier frequency, this type of sloping response will result in crosstalk in both the I and Q detected signals in the video-frequency range of 0 to 0.6 MHz. It makes no difference whether I-Q or color-difference demodulation is used; the result is the same as crosstalk among all the colors.

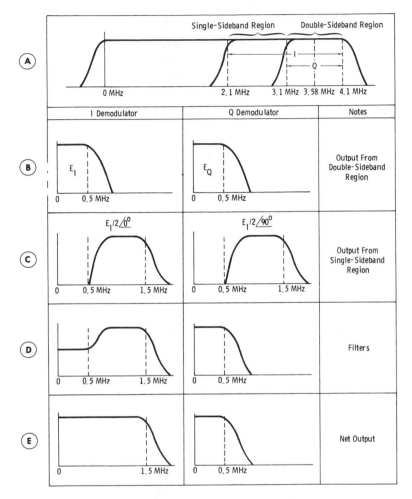

Fig. 2-19. Response of I and Q demodulators.

Fig. 2-20F shows this as applied to transmitter envelope response. A rapid rolloff too close to the color subcarrier frequency of 3.58 MHz will result both in desaturation of colors and in quadrature crosstalk. Fig. 2-20G shows rapid variations in response around 3.58 MHz. Although the FCC allows a ±2 dB variation, the Rules further state that this variation must be substantially smooth. Sudden dips or peaks must be avoided for good color transmission.

(Envelope-delay distortion at the transmitter is a major contributing factor in color misregistration. However, a study of this subject is more appropriate to a text on television system maintenance.)

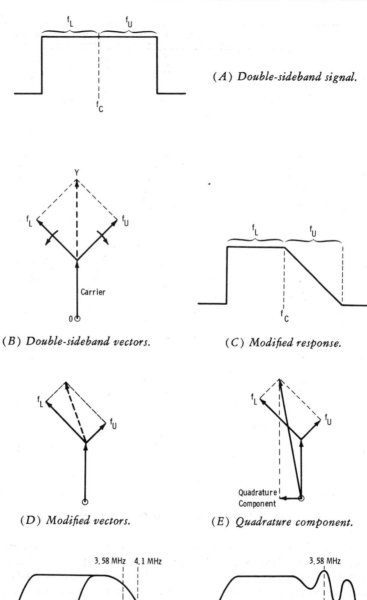

(A) Double-sideband signal.

(B) Double-sideband vectors.

(C) Modified response.

(D) Modified vectors.

(E) Quadrature component.

(F) Excessive rolloff.

(G) Uneven response curve.

Fig. 2-20. Causes of quadrature distortion.

Effect of Carrier Unbalance

We know that in a doubly balanced modulator, the carrier is suppressed so that only the sidebands remain. If this suppression is not perfect, the carrier appears in the output, and a condition known as *carrier unbalance* exists. Under this circumstance, the carrier adds itself vectorially to all vectors present in the encoder output. To visualize this, study Fig. 2-21A, which represents a carrier unbalance in the positive direction of the I modulator. Since the unbalance occurs in the I modulator, a new line, parallel to the I axis, is drawn from the proper color vector to the new vector representing the amount of carrier present. Since the unbalance is in the positive direction, the new vector is toward the +I axis. The resultant colors are shifted toward the orange axis of the +I vector, as well as being changed in amplitude.

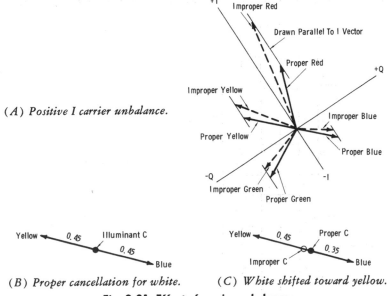

(A) Positive I carrier unbalance.

(B) Proper cancellation for white.   (C) White shifted toward yellow.

Fig. 2-21. Effect of carrier unbalance.

Now see Fig. 2-21B. Recall that primary and complementary colors have equal amplitudes but are opposite in phase. If both yellow and blue have the same amplitude, the result of their vector addition is illuminant C, or white. This is the proper complementary relationship. But note from the vectors of Fig. 2-21A that the blue amplitude has been reduced and the yellow amplitude has been increased. You see the result in Fig. 2-21C: white or gray areas become colored because of incomplete cancellation of the subcarrier. Remember that "white" or "gray" can occur only during an interval of zero subcarrier.

Carrier unbalance shifts all hues (as well as whites and grays) in the direction of unbalance. A positive I unbalance shifts toward orange; a negative I unbalance shifts toward cyan. A positive Q unbalance shifts toward yellow-green; a negative Q unbalance shifts toward purple.

Effect of Video Unbalance

Recall that double balancing of the modulators means that both the carrier and the modulating video are balanced out, leaving only the sidebands of the subcarrier frequency. If I and Q video suppression is not complete, the condition is known as *video unbalance*.

See Fig. 2-22A. The outputs of the I and Q channels for the indicated color bars are shown. Vector addition of the I and Q signals results in the amplitudes shown in Fig. 2-3. Fig. 2-22B shows the result of a ±Q video unbalance. Note that the axis for all colors with plus values of Q is shifted in the positive direction. However, the actual peak-to-peak values

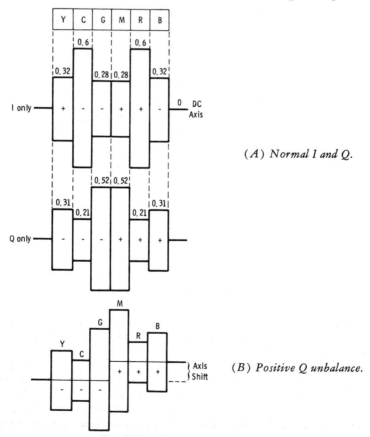

(A) Normal I and Q.

(B) Positive Q unbalance.

Fig. 2-22. Effect of video unbalance.

of these colors remain the same. The net result is that the unwanted video signal is added to the *luminance* signal after the chroma signal is combined with the luminance signal, and the gray scale of the picture is distorted. Note that the effect of a positive Q video unbalance is to brighten reds, blues, and purples, and to darken yellows, greens, and cyans.

For a negative video unbalance, the colors with negative amounts of Q would be shifted upward. In this case, reds, blues, and purples would be darkened, and yellows, greens, and cyans would be brightened.

Effect of Chroma Gains and Gain Ratio

The transmission paths from encoder input to receiver matrix must maintain a constant ratio for Y, I, and Q. A variation of gain in any one of the paths results in loss of color fidelity.

The noncomposite luminance level for a color-bar pattern (Fig. 2-3A) is 0.7 volt to peak white. When chrominance gain is correct (Fig. 2-3B) and of the proper I-to-Q gain ratio, and chrominance is added to the luminance signal of Fig. 2-3A, we have the following condition (see Fig. 2-3C):

1. Bars 1 and 2 overshoot by 33 percent.
2. Bars 5 and 6 undershoot by 33 percent.
3. Green (bar 3) just touches black level.

The above assumes that 100-percent bars are used, and that no blanking (pedestal) is inserted in the signal.

Now see Fig. 2-23. This is the same presentation as Fig. 2-3B except that the values of I and Q are shown for each color bar. Suppose that the ratio of I gain to Q gain is not correct. As you would expect, a deficiency of I gain would reduce the saturation of colors in the orange-cyan gamut, leaving greens and purples practically unaffected. Conversely, a deficiency of Q gain would reduce saturation of greens and purples without practical difference in the orange-cyan region. By noting the relative I and Q levels making up each color as in Fig. 2-23, you can understand how the pattern of Fig. 2-3C would show these deficiencies on a scope:

1. If I is high relative to Q, bar 2 will be higher than bar 1, and bar 5 will be lower (greater undershoot) than bar 6.
2. If Q gain is high relative to the I gain, bar 1 will be higher than bar 2, and bar 6 will have greater undershoot in the black region than will bar 5.

Effect of Differential Gain

Differential gain means that the gain of the 3.58-MHz chroma information is not constant with brightness level. This results in a change of saturation sensation with brightness.

Fig. 2-24A represents a stair-step signal with a superimposed 3.58-MHz sine wave. Fig. 2-24B shows the same signal observed through a high-pass filter to eliminate the low-frequency steps; a system with strictly linear amplitude response will result in sine waves of equal amplitudes for each step, as shown.

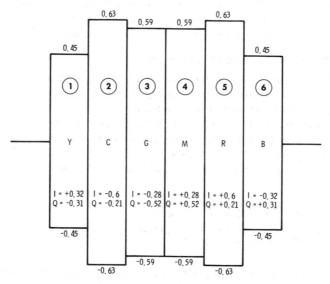

Fig. 2-23. I and Q chroma amplitude ratios.

Fig. 2-24C represents the type of nonlinearity in which black regions are compressed and white regions stretched. Normally, this condition will be apparent also on the steps, as shown, and would be evident when the signal is passed through a low-pass filter to observe the steps only. However, it is possible to have linear low-frequency amplitude response and nonlinear high-frequency amplitude response. This is why the use of low-and high-pass filters is convenient in observing test signals of this type.

Fig. 2-24D illustrates the opposite type of nonlinearity, and the same conditions apply.

The transfer curves of Figs. 2-24C and 2-24D are unusual. Generally, you will find that amplitude nonlinearity occurs either at one end or the other, or at both ends, with a relatively linear middle response. This tells you that those colors near the white or black extremes normally will be most susceptible to saturation changes, particularly with highly saturated colors.

Effect of Differential Phase

Fig. 2-25A shows the same signal as that pictured in Fig. 2-24A. Although the sine waves may look the same on each step, a phase displace-

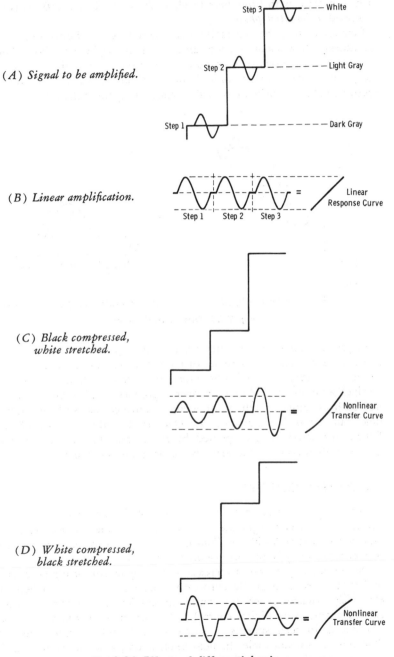

Fig. 2-24. Effects of differential gain.

ment can occur with brightness level, as shown by Fig. 2-25B. This error is termed *differential phase*.

Remember that the subcarrier phase carries *hue* information. A low-brightness yellow should be the same as a high-brightness yellow. This is not to say that the two yellows would appear the same on the receiver. But the point is, the observed color should be yellow and not (for example) green or red as the brightness of the yellow component changes.

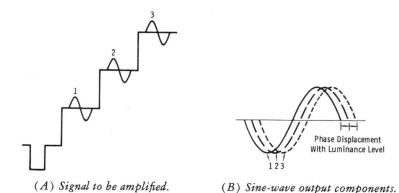

(*A*) *Signal to be amplified.*   (*B*) *Sine-wave output components.*

Fig. 2-25. Differential phase.

In practice, the effect of differential phase is judged best in the yellow and blue areas (the two extremes of the luminance scale). A system introducing as much as 10° of differential phase can result in a monitor or receiver adjustment that gives proper reproduction of a high-luminance hue such as saturated yellow, or a low-luminance hue such as saturated blue, but not both simultaneously. One or the other will be off-color. When this defect is accompanied by more than 10-percent differential gain (as often occurs), the error becomes quite noticeable.

2-4. DIGITAL CONCEPTS

The rate of change of technology in broadcasting requires technical personnel to spend a greater proportion of time in acquiring new knowledge to solve problems. Continuing education therefore is no longer incidental to the job, but an essential part of it.

One example is the rapid increase in the application of digital circuitry to broadcast equipment. Synchronizing generators, video switchers, and electronic character generators commonly employ such circuitry. Recently, camera chains using digital control, either wholly or partially, have been introduced. In such systems, a single triaxial cable replaces the bulky 85-conductor camera cable. In field applications, a simple rf link can be used to control a remotely located camera from a base station.

This section briefly reviews the common symbols and terminology of the basic logic functions.[3] Integrated-circuit (IC) "chips" are normally used, and "schematics" are usually just flow diagrams which the user must know how to interpret.

The digital concept recognizes only two numbers, or *conditions:* 1 (one) and 0 (zero). In *positive logic,* a 1 (high level) represents the true or more positive level, and a 0 (low level) represents the false or less positive level. This kind of logic is used most often. *Negative logic* (sometimes used) means that the voltage level assigned to logic 1 is *negative* with respect to the voltage level assigned to logic 0.

See Fig. 2-26. In row 1 is the symbol for a noninverting amplifier. Thus, if we have A on the input, we should have A on the output. Sometimes the output is identified by X, as shown, to distinguish between input and output signals. If the input is 1, the output is 1, and if the input is 0, the output is 0. *Truth tables* are input/output tables that show input conditions and the resultant output conditions.

Row	Symbol	Truth Table	Terminology
1	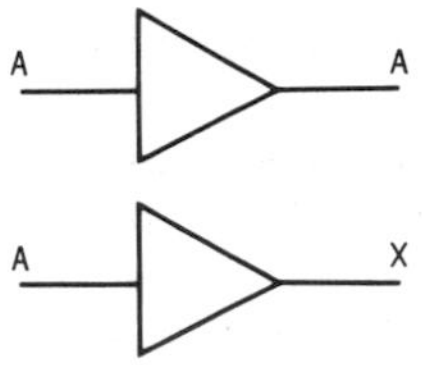	A / X 1 / 1 0 / 0	Noninverting Amplifier
2	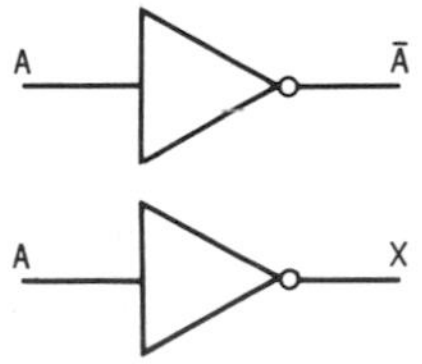	A / X 1 / 0 0 / 1	Inverting Amplifier Output (X) is Complement of Input (Phase Inversion). Significant input is high (1).
3	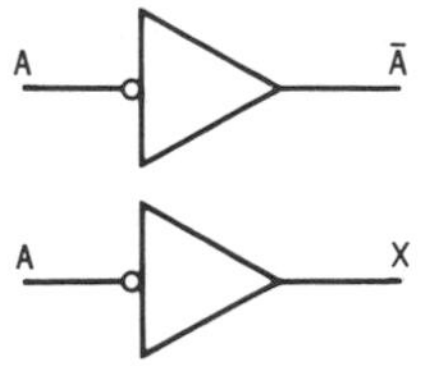	A / X 0 / 1 1 / 0	Same as (2), but significant input is low level (0).

A = Input or output signal.
Ā = Not A; complement of A; false if A is true.
O = Low (0) is the significant (reference) state.

Fig. 2-26. Common logic symbols (amplifiers).

---

[3]The reader should have basic background at least equivalent to that contained in Harold E. Ennes, *Workshop in Solid State* (Indianapolis: Howard W. Sams & Co., Inc., 1970).

In row 2 is the symbol for an inverting amplifier. A small circle at the input or output of a symbol indicates that 0, or the low state, is the significant state. If just the input or output has the small circle, the amplifier is inverting. Thus, for the symbol shown in row 2, we know that if the input is A, the output is inverted A, which is called *not A*. The bar over the letter (for example, as in $\overline{A}$) means "not A," inverted A, or the *complement* of A. All three terms are synonymous and simply mean that the input signal is inverted. Another way of saying this (see the truth table) is that if A is 1, then X is 0 (the significant state), and if A is 0, then X is 1.

Row 3 illustrates just the opposite of row 2. The action should be apparent from inspection.

Digital circuitry is made up of amplifiers, gates, and flip-flops. A 1 or 0 condition is analogous to an output controlled by a switch—it is either on (closed) or off (open). The digital stage is either fully conducting or fully cut off.

The AND gate is a basic logic circuit. It has two or more inputs and one output. The output will be 1 only when all inputs are 1 simultaneously. If one (or more) of the input signals is 0, the necessary condition for a 1 output is not fulfilled, and the output is 0.

The NAND gate is functionally equivalent to an AND gate followed by an inverter. Thus, the NAND gate produces a 0 at the output when all the input signals are 1 simultaneously.

The OR gate is another basic logic circuit. Like the AND gate, it has two or more inputs and a single output. The output is 1 when one *or* more of the inputs are at the 1 state. Thus, if any signal input is at the 1 state, the output is 1.

The NOR gate is functionally equivalent to an OR gate followed by an inverter. Thus, the NOR gate produces a 1 output only when all the input signals are 0 simultaneously. A 1 applied to any input results in a 0 output from the NOR gate.

The basic AND, NAND, OR, and NOR gates are reviewed in Fig. 2-27. Consider the AND gate in row 1. Since no circle is shown, we know that the significant output is 1 (high level) when the inputs are 1 simultaneously. This may be expressed as $X = AB$, which is read "X equals A and B." Note from the truth table that the output is 1 only when both A and B are 1.

Also note in row 1 that the OR gate, as symbolized, has the same truth table. We know from the symbology that the significant output is 0 (low level) when one or both of the inputs are 0. This may be expressed as $X = A + B$, which is read "X equals A or B." X is 1 only when both inputs are 1 simultaneously.

The reader should follow through the remainder of Fig. 2-27 and be sure he understands the truth table for each pair of AND (NAND) and OR (NOR) gates.

Row	AND (NAND)	Truth Table			OR (NOR)
1	A, B → X	A	B	X	A, B → X
	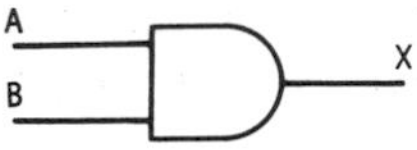	1	1	1	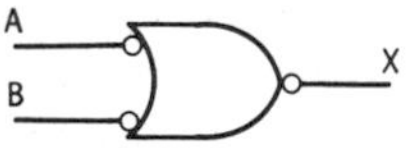
		1	0	0	
		0	1	0	
		0	0	0	
2	A, B → X	A	B	X	A, B → X
	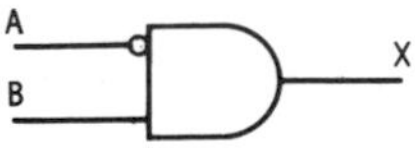	1	1	0	
		1	0	0	
		0	1	1	
		0	0	0	
3	A, B → X	A	B	X	A, B → X
	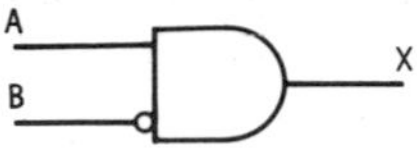	1	1	0	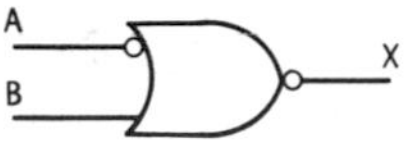
		1	0	1	
		0	1	0	
		0	0	0	
4	A, B → X	A	B	X	A, B → X
	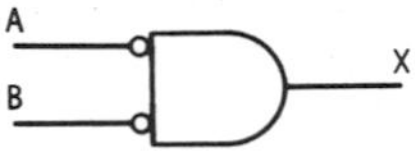	1	1	0	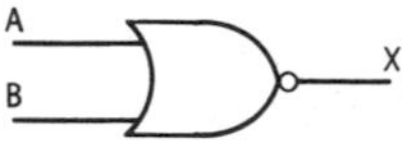
		1	0	0	
		0	1	0	
		0	0	1	
5	A, B → X	A	B	X	A, B → X
	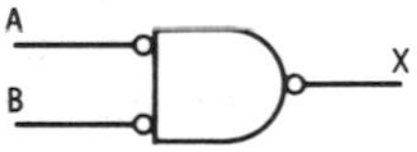	1	1	1	
		1	0	1	
		0	1	1	
		0	0	0	
6	A, B → X	A	B	X	A, B → X
	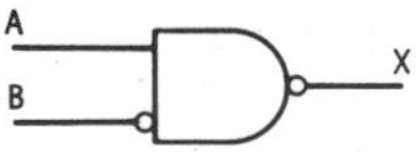	1	1	1	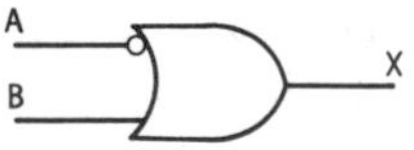
		1	0	0	
		0	1	1	
		0	0	1	
7	A, B → X	A	B	X	A, B → X
	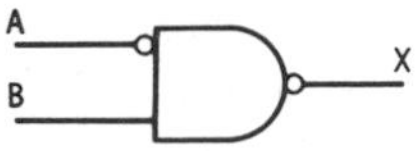	1	1	1	
		1	0	1	
		0	1	0	
		0	0	1	
8	A, B → X	A	B	X	A, B → X
		1	1	0	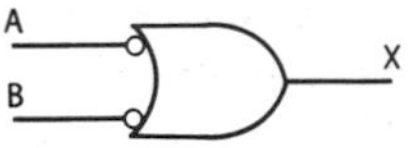
		1	0	1	
		0	1	1	
		0	0	1	

Fig. 2-27. Common logic symbols (gates).

Another basic block of digital circuitry is the *flip-flop*. This is a bistable multivibrator which remains in its most recent state until an input causes it to change states. The input trigger pulse usually is differentiated and then applied to a diode to polarize the pulse so that only the positive-going or negative-going edge causes the bistable circuit to respond. Sometimes a dc shift rather than a pulse is used.

For example, note the *toggle* flip-flop of Fig. 2-28A. Every time a negative-going transition occurs at input T, the bistable changes state. Recall that a multivibrator can have two output signals with a 180° phase relationship. Thus, if a trigger arrives when $Q = 0$ and $\overline{Q} = 1$, then Q changes to 1 and $\overline{Q}$ changes to 0.

With the toggle flip-flop, there is no predetermined state for the two outputs when the circuit is first turned on. Thus, the state of the outputs after a trigger pulse is applied cannot be predicted unless the present state is known. The circuit simply changes states each time a negative-going transition is applied. The *set-reset bistable,* or *RS flip-flop,* (Fig. 2-28B) overcomes this problem. The set-reset circuit has two inputs and the usual two complementary outputs. As indicated by the truth table, a 1 input to the set (S) terminal makes the Q output 1 and the $\overline{Q}$ output 0. A 1 input at the reset (R) terminal reverses the state: The Q output becomes 0 and the $\overline{Q}$ output becomes 1. Zero signals on both inputs do not change the state. If both inputs should receive simultaneous signals (1's), the next state cannot be predicted. Thus, simultaneous inputs are commonly termed *not allowed* or *forbidden* combinations. This simply says that the device cannnot be in both states simultaneously. The RS flip-flop cannot be used in logic situations which include the possibility of simultaneous set and reset inputs.

The action of the RS flip-flop is best understood by going momentarily to Fig. 2-29. Two NAND gates cross-connected as shown form a flip-flop. When power is applied, opposite states will exist; we will assume arbitrarily these are a 1 output for B and a 0 output for A. The 0 output of A at the input of gate B becomes a 1 at the output, and the 1 output of gate B at the input of gate A becomes a 0 at the output (phase inversion of NAND gate).

Now assume a 1 appears at the set (S) terminal. This 1 becomes a 0 at the output of gate B and drives the output of gate A to 1. Thus, a set input has set the significant output to the normal 1 and the previous 1 output to 0. A 1 input to R will now reset the device to the previous state. Follow this action again from the truth table for Fig. 2-28B. Note that the behavior of the circuit is predictable for three of the four possible input conditions.

Flip-flops may be *clocked* or *unclocked*. In the unclocked flip-flop just discussed, the outputs respond to the inputs as the inputs change. In the clocked flip-flop, a clock input must exist at the time the inputs change for the outputs to respond.

Given Present State		State After Trigger Pulse	
Q	Q̄	Q	Q̄
0	1	1	0
1	0	0	1

⌐ = Input responds to negative-going signal only.

(A) Toggle.

Input		Output	
R	S	Q	Q̄
0	0	No Change	
0	1	1	0
1	0	0	1
1	1	Not Allowed	

(B) Set-reset.

Input		Output	
R	S	Q	Q̄
0	0	Not Allowed	
0	1	0	1
1	0	1	0
1	1	No Change	

⌐_ = Input responds to positive-going signal only.

(C) Clocked set-reset.

Input		Output	
J	K	Q	Q̄
0	0	No Change	
0	1	0	1
1	0	1	0
1	1	Complement	

⌐ = Input responds to negative-going signal only.

(D) Clocked JK.

Fig. 2-28. Common logic symbols (flip-flops).

Fig. 2-28C illustrates a clocked flip-flop drawn for negative logic. The symbol at the clock-pulse (C_p) input indicates that the clock input responds only to a positive-going transition. The small circles at the R and S inputs indicate logic inversion at those points. Interpret this to indicate that a false level is inverted to become a true level within the block. Note that this circuit is the complement of that of Fig. 2-28B.

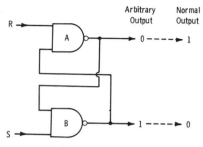

Fig. 2-29. Cross-connected NAND gates.

One of the most popular logic units is the JK flip-flop shown in Fig. 2-28D. There are no ambiguous states. When a 1 is applied to the J input, the Q output is 1 and the $\overline{Q}$ output is 0. (In a clocked flip-flop, the clock pulse must be present.) When a 1 is applied to the K input, the Q output flips to 0, and the $\overline{Q}$ output flips to 1. When 1's are applied to both the J and K inputs, the flip-flop switches to its complement state. Sometimes two or more J and K inputs exist. One J and one K input may be tied together for use as a clock input.

Special forms of gates are used in certain logic functions; special symbols are used to represent these functions. Fig. 2-30A indicates two NAND gates with outputs paralleled. The symbol indicating that this circuit actually performs as an OR circuit is shown at the junction of outputs f1 and f2. This is termed a *wired* OR, or sometimes a *phantom* OR. The truth statement for this circuit is: If f1 is true OR f2 is true, the output is true.

A circuit that produces a true output only when the input states are *not* identical is termed an EXCLUSIVE OR gate (Fig. 2-30B). Note from the

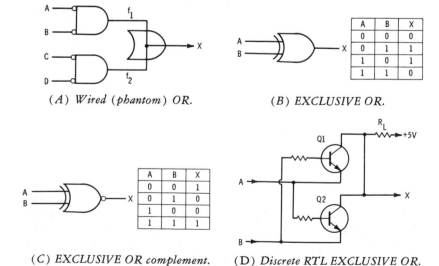

(A) Wired (phantom) OR.

A	B	X
0	0	0
0	1	1
1	0	1
1	1	0

(B) EXCLUSIVE OR.

A	B	X
0	0	1
0	1	0
1	0	0
1	1	1

(C) EXCLUSIVE OR complement.

(D) Discrete RTL EXCLUSIVE OR.

Fig. 2-30. Special gates for logic circuitry.

truth table that a 1 output is produced only if just one input is 0. If both inputs are of like polarity (either 0's or 1's), the output is 0.

The complement of this function is shown in Fig 2-30C. A 0 output is obtained only if the inputs are of opposite polarity. To make this clear, study the discrete resistor-transistor logic (RTL) circuit illustrated in Fig. 2-30D. If both inputs are 0's or 1's, both transistors have zero-biased base-emitter junctions, and neither can conduct. Under this condition, output X sees the full value of supply voltage (high level), because there is no current through R_L. Since the A input is tied to the emitter of Q1 and the base of Q2, while the B input is tied to the emitter of Q2 and base of Q1, *unlike* polarity (opposite logic levels) will turn one of the transistors on. The resultant voltage drop across R_L sends output X to a low level. Note again that this is the complement of the logic function in Fig. 2-30B, as indicated by the small circle at the output of the logic symbol shown in Fig. 2-30C.

A digital system contains numerous switching devices that have 0 or 1 outputs. Since operation is based on two states, the binary numbering system (based on the number 2) is a "natural" for this application.

Binary and decimal numbers are reviewed in Table 2-5. Only two symbols, 0 and 1, are used in the binary system. Note that when decimal 2 is reached, we move the binary 1 one place to the left to indicate we have counted to two one time. At the count of three, we use binary 11 to indicate one two plus one one, or 3. At the count of four, we are again out of symbols, so we write 100 which indicates one four plus no twos plus no ones. At the count of five, we write 101, which indicates one four plus no twos plus one one. We continue until at the count of seven we write

Position Value of Each Symbol	Decimal	Binary	Comments
1×2^3	8	1000	
1×2^2	4	100	
0×2^1	0	0	Any number times 0 is 0.
1×2^0	1	1	Any number to the 0 power is equal to 1.
Total	13	1101	Binary number is composed strictly of zeros and ones.

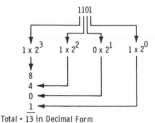

Total = 13 in Decimal Form

Fig. 2-31. Conversion of binary 1101 to decimal form.

Table 2-5. Binary and Decimal Numbers

Decimal	Binary
0	0
1	1
2	10
3	11
4	100
5	101
6	110
7	111
8	1000
9	1001
10	1010
11	1011
12	1100
13	1101
14	1110
15	1111
16	10000
17	10001
18	10010
19	10011
20	10100
21	10101
22	10110
23	10111
24	11000
25	11001
26	11010
27	11011
28	11100
29	11101
30	11110
31	11111
32	100000

111. Again we are out of symbols in all columns, so we write 1000, which indicates one eight plus no fours plus no twos plus no ones. The conversion of one example (binary 1101) to the equivalent decimal number is reviewed in Fig. 2-31.

The term *bit* means *bi*nary digi*t*. The term *character* refers to a group of bits. The term *word* refers to the total number of bits required for a particular system. A word may be defined either by the total number of bits or the total number of characters. For example, a certain system may use a 192-bit word. A character may consist of (for example) a group of four bits. Thus, this system uses a 48-character word.

Digital logic circuitry is easy to troubleshoot if the technician has a little experience and familiarity with a particular system. Fig. 2-32 is a simplified schematic diagram of a small portion of video logic circuitry in which raw sync is inserted at the input of IC4. Assume that sync pulses exist at the output of gate IC4 but not at the output of IC3. This does not necessarily mean that integrated circuit IC3 is faulty. Note that the symbols all indicate NAND gates. Pin 12 of IC3 must have +5 volts dc for the sync pulses to pass. It is obvious that for this to occur transistor Q1 must be cut off. Note that the hold-off bias for this stage is determined by a video clamp level, which may or may not be an internal adjustment. Checking the dc levels back from pin 12 of IC3 is necessary to determine the cause of the trouble. If pin 12 is receiving +5 volts when the sync output is lost, the IC3 chip probably is at fault. When a chip is definitely determined to be faulty, the entire chip is replaced. The important point to remember is that the input conditions to a chip must be correct for the proper output condition to exist.

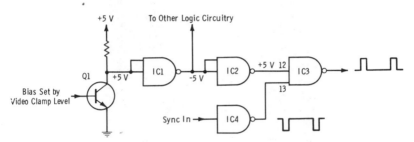

Fig. 2-32. Example of troubleshooting in logic circuit.

2-5. A DIGITALLY CONTROLLED COLOR CAMERA

A portable 3-*Plumbicon* color camera employing digital control is illustrated in Fig. 2-33. The camera was developed by CBS Laboratories, and, under agreement with CBS, it is being manufactured and marketed worldwide by Philips as the PCP-90 "Minicam." This camera, with back pack, produces an encoded signal intended for direct broadcasting. All signal processing is accomplished in the back pack so that separate red, blue, and green signals need not be sent to the base station. This reduces the possibility of noise pickup and cuts down on color errors caused by multipath effects.

The camera transmits signals to its base station (Fig. 2-33B) on the microwave frequency of 2 GHz (7 GHz and 13 GHz are optional) from an omnidirectional antenna on the back pack.

Remote-control signals to the camera are carried on a frequency of 30 MHz (for cable) or 950 MHz (for rf), with a 450-kHz subcarrier for the command signal and a low-frequency interphone carrier. The command system permits radio control of all functions from a base station located as far as 10 miles away, depending on the transmission path. The camera can be linked to its base station by a triaxial cable if terrain features interfere with wireless communication. However, cable losses limit the distance between camera and station to one mile, unless repeaters are used. For on-the-spot recording with a portable video recorder at the camera location, a local control box plugged into the back pack allows the operator to perform all functions of the digital command system.

A three-inch picture tube is used in the viewfinder along with a simple magnifying lens and a polarized filter to enhance contrast under outdoor

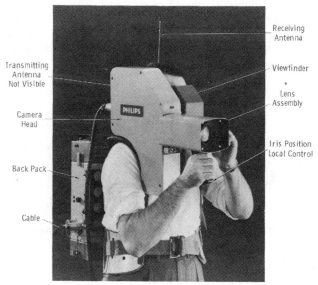

(*A*) *Camera and back pack.*

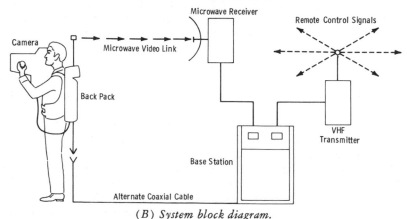

(*B*) *System block diagram.*

Courtesy Philips Broadcast Equipment Corp.

Fig. 2-33. PCP-90 digitally controlled portable color camera.

viewing conditions. A waveform is presented at the left of the picture display to enable the cameraman to adjust pedestal and iris locally when this is desirable. There is also a pulse to indicate field strength of the signal received at the base station.

The PCP-90 is available with either a Canon 6-to-1, $f/2.8$ zoom lens or an Angenieux 10-to-1, $f/1.8$ zoom lens. The camera head weighs 18½ pounds, and the back pack weighs 12 pounds (32 pounds with battery). The top section of the back pack (Fig. 2-33A) contains the uhf-vhf data receiver and microwave transmitter (in the wireless version of Fig. 2-33B). The main center section houses video processing (Chapter 6), including the NTSC encoder (Chapter 9), sync circuits, and command control circuits. At the bottom is the battery pack for wireless use, or the cable power converter.

For the digital command system, frequency modulation of a 950-MHz carrier is used. An audio tone at 5.4 kHz is phase-shifted 180° to identify a single bit of information. The resultant significant sidebands occur between 2.7 kHz and 8.1 kHz on the 450-kHz command subcarrier. This permits adding a 250-to-2500 Hz interphone voice channel to the main carrier without excessive cross talk. For cable operations, amplitude modulation of a 30-MHz carrier is used to transmit the video.

An example of a 24-bit command word is shown in Fig. 2-34A. Since up to six cameras can be controlled, the first three bits address the specified camera to be commanded in the immediately following word. This is followed by 15 bits of simultaneous commands, and then by six bits of one-at-a-time commands that allow any one of 64 individual functions to be controlled by the base-station operator. The complete command word for each camera is preceded by a 9-bit code of all 1's (Fig. 2-34B) to alert the camera decoder that an event is to take place. Also, the 24-bit command word is followed by a 24-bit parity, which is the complement (interchanged 1's and 0's) of the preceding command word. All of this is done for the purpose of reliable noise immunity of control. Thus, unless the command word is followed by the complement, nothing happens. This virtually eliminates false commands and erroneous functions of the command system.

The complete command word may be observed to consist of 57 bits (9-bit start plus 24-bit command plus 24-bit parity). The complete 57-bit command requires 10 milliseconds to transmit (Fig. 2-34C). At the end of this time, the base-station sequencer steps to the address of the next camera. Thus, the sequence time for six cameras is 60 milliseconds, allowing each camera control an interval of 10 milliseconds 16 times during each second.

A genlock module (horizontal, vertical, and color-subcarrier lock) is used to compare local sync with sync from each remote camera. Digital commands are transmitted to bring all cameras into exact synchronization with each other and with the local sync generator. This permits fades, lap dissolves, and special effects between any combination of remote and local signals.

It is quite natural that the principles of this design be extended into the area of studio operations. Such is the case with the Philips PC-100 illustrated in Fig. 2-35. The camera and studio control unit are shown; a portable control unit also is available.

The camera head is contained in a magnesium casting. Hinged covers are provided for access to interior components. The integral lens mounting includes internal mechanical drive shafts that automatically couple to internal shafts in the camera body. Available zoom lenses include fully servo-controlled lenses, as well as lenses with manually controlled focus and zoom and servo-controlled iris. Range extenders for these lenses are available. A remotely controlled, motor-driven filter wheel and two slide filters are provided between the lens and color beam splitter for color and neutral-density correction.

The color beam splitter is a prism block. Linear matrixing is included to allow the use of a more efficient beam-splitter prism. A one-inch separate-mesh *Plumbicon* tube is used; this tube provides a minimum of

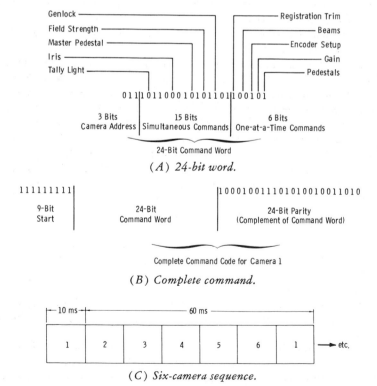

(A) *24-bit word.*

(B) *Complete command.*

(C) *Six-camera sequence.*

Fig. 2-34. Camera command code.

(A) Camera and studio control unit.

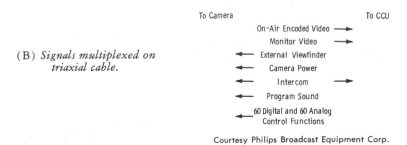

(B) Signals multiplexed on triaxial cable.

To Camera   To CCU

On-Air Encoded Video ⟶
Monitor Video ⟶
⟵ External Viewfinder
⟵ Camera Power
⟵ Intercom ⟶
⟵ Program Sound
⟵ 60 Digital and 60 Analog Control Functions

Courtesy Philips Broadcast Equipment Corp.

Fig. 2-35. PC-100 digitally controlled studio color camera.

35-percent modulation at 400 lines. Low target capacitance contributes to a 50-dB signal-to-noise ratio through the preamplifier. An anti-comet-tail gun increases the dynamic range of the tubes so that high lights of 20 to 30 foot-candles on the surface of the tubes can be handled without blooming. This gun effectively imparts a light-level saturation "knee" to the transfer curve.

The deflection-coil assembly is shielded by a spun *Mumetal* housing to eliminate the influences of external magnetic fields on the registration accuracy. This assembly is mounted in machined castings. The complete yoke assemblies are fixed into position in the RGB casting and factory aligned by fixing the yoke shell to the front housing. The yokes can be removed for servicing and replaced with only the normal line-up procedure. The "spider" casting provides long-term optical stability. Yoke rotation and focus are accomplished with separate knurled shafts.

Since the video-processing circuitry is in the camera head, the effects of temperature variations on a long camera cable do not influence the camera performance. With the exception of some presets, the camera does not contain any setup controls, so that the complete line-up of the camera chain can be carried out by the camera-control-unit (CCU) operator.

Information is transferred between the camera and the CCU by a triaxial cable; this cable weighs one-tenth as much as standard color cable. Transmission is accomplished by multiplexing three channels of information through the cable (Fig. 2-36). These channels include a video channel, for sending encoded video from the camera to the control location; a monitor channel, for sending monitor signals from various points in the video-processing chain to the control location; and a telecommand channel, for transmitting all control, registration, and setup signals from the CCU to the camera. In addition to these three channels, 100-volt dc power is supplied to the camera through the triaxial cable. The maximum cable length is one mile (more with the addition of repeaters).

The electronic view finder contains a 7-inch rectangular picture tube. It is tiltable, rotatable, and removable. Any combination of either RGB minus G or Y and external video signals can be selected for display. An electronic zoom indicator is superimposed on the top of the picture. The "on-air" tally light can be seen from all angles.

The camera control unit is separated into three major subassemblies; the monitor unit, the registration and operating panel, and the electronics unit. These subassemblies are linked by cables in the rear of the console and can be accommodated either in a standard 19-inch rack or in two 19-inch transit cases.

The electronics unit consists of a 7-inch-high rack that houses a two-level card bin. The card bin is mounted in a retractable, tiltable drawer for ease of servicing. Coaxial and multipin connectors are mounted on the rear panel. The circuitry includes the cable-drive demodulators, audio modulators, the analog-to-digital (A/D) converters, and the power

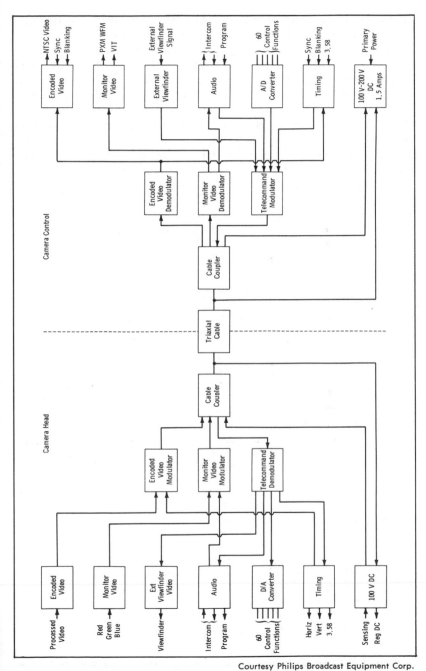

Courtesy Philips Broadcast Equipment Corp.

Fig. 2-36. Simplified block diagram of the PC-100 system.

supply. Final processing of the encoded video signal occurs in the electronics unit with the addition of studio sync and timing.

The registration and operating panels are mounted in a drawer assembly that can be placed in the cabinet at various positions. An overlay panel is provided to cover the registration controls after setup of the camera. To aid in camera setup, switches and associated lamps are interlocked to reduce the number of manual operations required. For example, a REGISTRATION push button is provided to switch the matrix and contours (Chapter 6) to off and the encoder input signals to the waveform monitor. At the same time, it also switches the waveform monitor to the RGB sequential mode and minus-green video to the picture monitor.

Other camera systems—the two-channel color camera and the more conventional cabled camera systems—are covered elsewhere,[4] and these descriptions will not be duplicated here. However, details of camera cables and other types of control and distribution cables are covered in the next chapter. Also included in Chapter 3 is coverage of the special types of camera-system regulated power supplies and power distribution.

EXERCISES

Q2-1.   The camera is delivering unity green, unity blue, and zero red signals. Calculate the resulting color amplitude and phase.

Q2-2.   The chroma signals on the black level immediately following the burst in Fig. 2-3C are special I and Q signals from a color-bar generator. They are simply special test pulses that are fed to the I and Q inputs of the encoder. They obviously are of zero luminance; the ac axis centers on black level. Why, then, are these signals actually visible on the color monitor?

Q2-3.   You may have learned that when (for example) the red chroma was less than 0.63 of unit luminance, this meant it was *not* fully saturated; some degree of "white" was present, meaning some degree of both of the other primaries was present. How can a fully saturated red be simulated at much less encoded amplitude?

Q2-4.   Where is the FCC requirement for the width of the horizontal front porch called out in Fig. 2-10?

Q2-5.   What is the proper amplitude of the color-sync burst?

Q2-6.   Does the color-sync burst occur following every horizontal-sync time in the complete composite color signal?

Q2-7.   How many cycles of color-burst signal should be present?

Q2-8.   What are the attenuation-vs-frequency requirements for a color transmitter?

Q2-9.   If you apply a linear stairstep signal to the first amplifier of a camera head, should the signal at the output of the color encoding system be linear?

---

[4]Harold E. Ennes, *Television Broadcasting: Equipment, Systems, and Operating Fundamentals* (Indianapolis: Howard W. Sams & Co., Inc., 1971).

Q2-10. If you apply a linear stairstep signal to the input of an encoder, should the signal at the output of the encoder be linear?

Q2-11. What is the basic problem in handling NTSC color?

Q2-12. What factor other than transmitting-system deficiencies can result in receiver color problems?

Q2-13. How would the following colors be reproduced in a properly adjusted monitor or receiver, if the burst in the encoder were at $+160°$ from zero reference on an NTSC polar diagram? (Assume all other adjustments to be correct.)
1. Yellow
2. Cyan
3. Green
4. Magenta
5. Red
6. Blue
7. White

Q2-14. What could cause white to be contaminated with a certain hue?

Q2-15. What would cause "washed-out" color in a properly adjusted receiver?

Q2-16. What is the major criterion in the adjustment of a receiver for "good color"?

Q2-17. What could cause a human face to be reproduced with unnaturally ruddy complexion and very dark red lips? Assume a properly adjusted receiver.

Q2-18. What would cause yellows to be "washed out," while blues are reproduced well?

CHAPTER 3

Camera Mounting, Interconnection
Facilities, and Power
Supplies

This chapter covers camera mounting facilities, cable interconnection methods, and power supplies and power distribution for the camera chain.

3-1. THE CAMERA PAN AND TILT CRADLE

The camera head normally mounts directly on the panning head to allow it to be moved vertically and horizontally by the camera operator. The panning head then is mounted on a pedestal (Fig. 3-1), a tripod that is either fixed or on wheels (Fig. 3-2), or a crane (Fig. 3-3). Any panning-head mounting on wheels is termed a *dolly*. Thus you may have a pedestal dolly, a tripod dolly, or a crane dolly.

The panning head includes controls that are adjustable to allow the camera operator to make pan and tilt operations smooth and stable with his individual touch. The panning head (or *cradle*) is the basic link in the chain of camera operations and must be mastered completely before the cameraman can do an adequate job.

The panning head normally engages the camera head by means of a slotted screw on the mounting plate. Once in place, the camera head is secured by tightening this screw. All lock and drag controls are then disengaged, and the panning handle is held loosely by the cameraman so that the center-of-gravity adjustment can be made. The adjustment usually is accomplished by means of a control on the front (or, in special cases on the rear) of the panning head. This adjustment allows the camera to be balanced properly (tilts neither forward nor backward) when the panning-head control handle is not held.

The adjustable friction (pan drag and tilt drag) controls must be set to suit the individual touch of the operator. Just sufficient drag must be used so that the movement of the camera does not "jiggle" or jerk the image. At rare intervals, the production may require a *whip shot* calling for a rapid pan. In this event, the drag must be light enough to allow rapid but smooth operation.

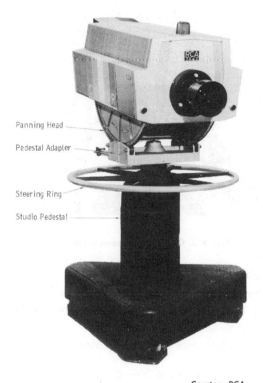

Panning Head

Pedestal Adapter

Steering Ring

Studio Pedestal

Courtesy RCA

Fig. 3-1. Camera mounted on studio pedestal.

A common error of new camera operators not familiar with the exact location and use of adjustments is to confuse the drag, brake, and lock controls. Sometimes the braking or even the locking of the panning head is done by means of the drag controls. This practice considerably shortens the life of the panning-cradle mechanisms.

The variety of panning heads and dollies in current use makes a detailed description of each type impractical. However, the nomenclature of adjustments should allow the cameraman to adapt the following procedures to his particular equipment. The information given is directly applicable to the panning head of Fig. 3-2.

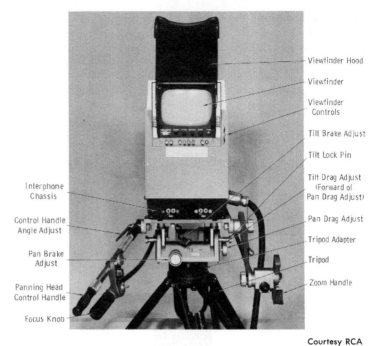

Viewfinder Hood
Viewfinder
Viewfinder Controls
Tilt Brake Adjust
Tilt Lock Pin
Tilt Drag Adjust (Forward of Pan Drag Adjust)
Pan Drag Adjust
Tripod Adapter
Tripod
Zoom Handle

Interphone Chassis
Control Handle Angle Adjust
Pan Brake Adjust
Panning Head Control Handle
Focus Knob

Courtesy RCA

Fig. 3-2. Controls on panning head.

Center of Gravity Adjustment

After the camera is installed, the center of gravity must be established for balanced operation. While firmly holding the handle, proceed as follows:

1. Release the tilt-lock pins and all tilt-motion drag on the cam-head mount.
2. Release the handle enough to note the direction in which the camera head tends to tilt.
3. Tilt the camera toward the required center of gravity adjustment and rotate the CG ADJUST control knob until the camera balances.
4. Apply the tilt brake with the TILT BRAKE lever, or secure with the tilt-lock pins.

Operation

The following procedures enable the operator to adjust the cam-head mount pan-and-tilt motion while the camera is operating.

Pan Motion—The PAN BRAKE knob is located at the lower rear center of the cam-head mount. To lock the mount in azimuth, rotate the PAN BRAKE knob clockwise. The PAN DRAG ADJUSTMENT knob is located on the right-hand side plate. To increase the drag in pan motion, rotate this knob clockwise.

Tilt Motion—The TILT BRAKE lever is located in the rear of the cam-head mount, below the top plate. To maintain the mount in the selected tilt position, press the lever to the left. To lock the mount vertically, rotate and release the spring-loaded tilt-lock knobs on both sides of the mount, and adjust the tilt until the pilot pins engage with the detents in the balance cams.

The TILT DRAG ADJUSTMENT knob is located just forward of the PAN DRAG ADJUSTMENT knob. To increase the drag in tilt motion, rotate the knob clockwise. *CAUTION:* Do not tighten the TILT DRAG ADJUSTMENT until the cam roller shaft is locked and not free to rotate. Otherwise, the cam will be damaged.

Lubrication—After every 100 hours of operation, three areas of the cam-head mount should be coated lightly with Dow Corning Silicone Lubricant, Compound No. 4, or equivalent. Apply the lubricant to the operating surfaces of the balance cams, the vertical stabilizer bars (just under the top plate of the panning head), and the CT ADJUST screw. The cam operating surfaces should be kept clean at all times.

NOTE: The cameraman should always lock the head when he leaves the camera for more than a few minutes.

3-2. PEDESTAL DOLLIES

The most common form of studio camera support is the pedestal dolly (Fig. 3-1). An example of such a dolly, the Houston Fearless Model PD-9 pedestal, is described in this section.

Description

This pedestal is designed to mount a standard friction head and a black-and-white or color television camera. It is designed for one-man operation. The pedestal acts as a firm, stable mount for the camera, and it also provides mobility for dolly shots and for raising and lowering the camera during operation. The rising column on which the camera is mounted is motor driven so that the camera is raised or lowered by operating a small switch attached to the head or camera control handle. The steering wheel, which is located directly below the camera at all heights, guides the three sets of dual wheels, one at each apex of the triangular base. The wheels are equipped with ball bearings.

Two types of steering are available, and by operating a foot pedal on the base of the pedestal, either may be chosen by the operator while the camera is on the air. *Synchronous steering,* in which all wheels are locked parallel and turn simultaneously, is best for tracking in a straight line. By operating the hand wheel and observing the positioning arrow on one of the spokes of the wheel, the operator can set the direction of the wheels before the pedestal is in motion. The pedestal will then travel in the direction indicated by the arrow. In *tricycle steering,* only the forward wheel is steered, and the back wheels are locked parallel. This enables the pedestal to follow a curved course and to turn sharply in any direction.

On the corners of the triangular base are nonslip step plates on which the cameraman can stand when the camera is in the raised position. The base of the pedestal is of arc-welded steel. The center column is seamless steel tubing. The trim and steering wheel are satin-chrome finish. All bearings in the equipment are packed and sealed at the factory, so no lubrication is required.

The specifications of the equipment are as follows:

Maximum Height (Not Including Friction Head)	55 inches
Minimum Height (Not Including Friction Head)	36 inches
Maximum Width	38¼ inches
Minimum Width	34½ inches
Electrical Requirements	115 volts, 60 hertz, 5.6 amperes
Weight	365 pounds

The motor-driven column consists of a seamless steel tube that is raised and lowered inside a larger, fixed seamless steel tube. It is held in position by two sets of three rollers, 120° apart, which guide and align the rising column. A rubber pad is provided at the center of the column so that the column will settle softly when it is moved to its lowest position. The column is driven up and down by a motor and gear box, the hoisting being done with a flexible steel cable. Limit switches prevent overtravel at either end of movement. The motor may be reversed instantly and is controlled by a switch at the end of a cable. The switch is fitted with a mounting bracket, and the mounting bracket may be attached to the control handle of either a standard tilt head or the RCA color camera (when panning head is not required). The switch usually is mounted so that it may be operated with the thumb of the left hand.

A large wheel is located just below the mount for the friction head. Turning this wheel steers the pedestal in the direction shown by an arrow on one of the spokes of the wheel. Operating the wheel drives a gear shaft that extends down through the center column of the pedestal. Since this shaft is telescoping, steering operation is available regardless of the height of the pedestal. The wheels are steered by a chain-and-sprocket arrangement that operates directly from the base of the telescoping shaft. The chain drive is directly to the front wheel of the triangular base. When the arrows on the wheel and column are in line and the STEER 3 pedal is depressed, all of the wheels are connected by the chain. Consequently, when the steering wheel is turned, all three wheels move in synchronism and remain parallel. When the arrows are aligned and the STEER 1 pedal is depressed, the two rear wheels are declutched from the chain drive and locked in their fixed positions. The steering then operates only on the front set of dual wheels and is of the tricycle type.

Cable guards are provided all around each pair of wheels to prevent the wheels from running into the cables. These guards have slotted holes and are adjustable to any height above the floor by pushing them up or down with the foot.

The head of the rising column is equipped to mount a standard friction head. To install the friction head, remove the lock nut from the head. Place the head in the circular hole in the top of the column, and insert the lock nut through the opening in the side of the column. Then attach and tighten the nut to hold the head in place.

A clip, located on the upper portion of the soundproof motor blimp, is provided as a retainer for the motor cable. To prevent the cable from dragging, it may be inserted into the clip after the motor-switch bracket is mounted on the control handle of the tilt head or camera.

Operation

The large wheel just below the tripod mount is turned to direct the wheels as desired by the operator. This may be done while the pedestal is in any position, as the operator normally operates the panning arm with one hand and moves and steers the pedestal with the other. The arrow on one of the spokes of the wheel indicates the direction in which the wheels are turned and the direction in which the pedestal may be expected to move when it is pushed by the operator. When synchronous steering is desired, the arrow on the wheel is aligned with the arrow on the column, and the STEER 3 pedal, on the back of the pedestal base, is depressed. Operating the hand wheel then turns all three sets of wheels simultaneously, and the sets of wheels remain parallel. This enables the pedestal to move in a straight line in any direction from its resting position. When the arrows are aligned and the STEER 1 pedal is depressed, the front wheel is steered as the front wheel of a tricycle, and the pedestal may be moved on a curved path determined by the setting of the steering wheel.

The cable guards are provided to push away cables as the pedestal is moved, to prevent the cables from jamming the pedestal or rocking the camera as it rolls over them. These cable guards are adjustable. On a level floor, they may be set so that they are barely above floor level. On uneven surfaces, they are set slightly higher. The cable guards are equipped with slotted holes so that they may be moved up and down with the foot to obtain the desired clearance.

Access to the column motor drive is obtained by releasing three buckles on the motor blimp. This housing should be removed on a regular basis, such as once a month, to permit inspection of the lifting cable for signs of fraying. Whenever fraying is observed, the defective cable should be replaced immediately.

3-3. THE CRANE DOLLY

The crane dolly is used for more elaborate productions and is rarely found outside the main network operating centers. The four-wheeled truck shown in Fig. 3-3 is pushed by one or more *trackers,* or *pushers,* who also may be responsible for controlling the height of the pivoted arm. The camera is mounted at the top of the arm, and a seat behind the camera position is provided for the cameraman.

The manual type of crane dolly is being replaced rapidly by motor-driven mechanisms, some of which can be operated solely by the cameraman. A recent model allows the cameraman to sit alongside the tracker; camera operation is by remote control. The viewfinder is replaced by a video monitor, and fingertip controls allow camera-aiming, lens-zoom, and focus adjustments to be made.

The elimination of the bulky camera cable by the techniques of digital, multiplexed systems (Chapter 2) could drastically affect camera dollying systems in the future. It is possible that lightweight color cameras, not inhibited by the more conventional 84-conductor cable, may be suspended on lever arms from the ceiling and remotely controlled to pan, tilt, raise, lower, and swing in any desired arc and at any height.

Courtesy Houston Fearless Corp.

Fig. 3-3. Crane dolly for television camera.

3-4. PROMPTING EQUIPMENT

This entire chapter is concerned mainly with components external to, but used in conjunction with, the actual camera or control units. One important system external to the camera is the means used to "prompt" the performer.

Formal speakers, newsmen, and commercial announcers "on camera" normally require a prompter. *Cue cards* (sometimes termed "idiot cards") are often used when the required information is limited. These are large cards, on which the information is printed in large letters, held by a person either near the camera or in some other suitable location.

Several types of prompters arc in use; one of these is illustrated in Fig. 3-4. A special typewriter with ⅜-inch type is used to prepare the copy on paper that has sprocket holes along each edge. The prompter unit normally mounts at the top front of the camera as shown in Fig. 3-4A. Eight lines of copy are visible at a time, and since the lines are 22 characters long, some twenty to twenty-five words are visible on one full frame. The sprocket holes along each edge of the paper engage sprockets on the prompter mechanism, and the sheet rolls upward at a speed controlled to suit the pace of the performer.

The prompter chassis shown in Fig. 3-4A is a one-piece aluminum structure, corner braced and welded. The copy-correction panel and spools are gold anodized to resist marks, scratches, and feed or take-up spool paper displacement burrs. The device is about 15 inches across the face, 11 inches high, and 5 inches thick.

The sync control (Fig. 3-4B) has a removable aluminum cover. Internal components are mounted on subchassis that are removable from the bed plate. The bed plate is secured to a one-piece foam-rubber shock mount attached to the bottom of the case. The inside of the case cover is acoustically lined with vinyl foam. An extendable hand control is shown lying on top of the case.

The external casing of the sync control, prompters, and hand control are electrically isolated from the ac supply. The cables connecting the sync control and the prompters are shielded, and the shield provides the continuous case ground between units. A 2-pin connector provides ac. The ground lead provided should be secured to a solid ground.

Line-for-line sync is provided by the control unit. In this application, "sync" means the rolls on two or more prompters, advancing in the same direction, move together, line for line with one another, regardless of speed or the amount of typed copy on each prompter. The dual function of drive and sync is performed by a synchro torque system consisting of a control transmitter coupled to 1, 2, 3, or more torque receivers as required. The transmitter is contained within the sync control and is coupled to the respective receivers in the prompters by five wires in the 7-core cable .The transmitter and the receiver are physically and electrically similar. Each consists

(A) Prompter on camera.

(B) Prompter control unit.

Courtesy Telesync Corp.

Fig. 3-4. Example of a prompter.

of a single-phase ac rotor surrounded by a 3-phase ac stator. Power to the rotor is fed through slip rings. If the rotors of two such machines are fed in similar phase with ac, transformer action causes a 3-phase voltage to be induced in the stators of the two machines. If these stators are connected to each other (delta configuration) in the correct phase, 3-phase currents are established in the three connecting wires. If one rotor is turned through 90°, the voltages induced in its stator are advanced or retarded by 90°, and this produces motor action in the other machine until it also has advanced 90°. If this is done smoothly, the two machines will move exactly together. If it is done abruptly, the second machine will, because of its inertia, tend to overshoot the 90° point and then return. In order to reduce this inherent action, a rotary damper is added to each receiver rotor shaft. It can be seen from this necessarily brief description that if one rotor is driven by the sync control transmitter, then the second rotor will revolve synchronously and will be able to drive a prompter. If one rotor is held so that it cannot revolve, the second rotor produces considerable resistance to any attempt to revolve it. It is this feature that provides the lock between one or more prompters and the sync control.

3-5. CAMERA-CHAIN POWER SUPPLIES

Camera-chain power supplies are normally rack-mounted units with means of distribution to rack equipment, the camera control console, and, through the camera cable, to the studio camera. Electronically regulated supplies are universal.

The principle of a vacuum-tube regulator is shown in Fig. 3-5. To review this function:

1. If the load draws more current or if the ac input to the rectifier section falls, the result normally would be lower terminal voltage.
2. Resistor R1, tube V2, and gas regulator tube V3 are in series across the rectifier filter output. Tube V3 holds the cathode of V2 at a con-

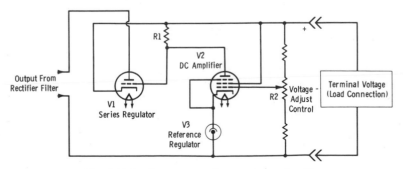

Fig. 3-5. Basic vacuum-tube regulator circuit.

stant positive potential with respect to ground. The setting of R2 determines the bias on dc amplifier V2.

3. A reduction in terminal voltage results in a more negative bias on V2, less current through V2, and hence less current through R1.

4. The decreased IR drop across R1 results in less negative bias on the series tube (V1). This, in turn, results in lowered series resistance; hence, it increases the terminal voltage to overcome the initial decrease in Step 1.

Most newer regulated power supplies employ transistors and (in some cases) zener diodes. The circuit shown in Fig. 3-6A is a common version of the basic emitter-follower (or common-collector) circuit. The power-supply load (symbolized by a variable resistance) is placed in series with

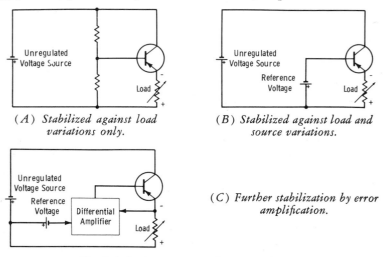

(A) Stabilized against load variations only.

(B) Stabilized against load and source variations.

(C) Further stabilization by error amplification.

Fig. 3-6. Basic transistor voltage regulators.

a transistor. The impedance of this transistor is controlled automatically in such a way that it tends to compensate for impedance changes (or current changes) in the load, thus maintaining an essentially constant voltage across the load. This action may be explained by noting that the voltage drop across the emitter-base junction of a transistor is usually negligible in comparison with the supply voltage (at least over reasonable operating ranges), so the emitter tends to remain near the potential established by the voltage divider in the base circuit. Since the base current is only a small fraction of the emitter (or load) current, the base voltage is not altered significantly by changes in load current, provided the resistors in the voltage divider are not too large.

An alternative approach to the explanation of the regulating action of Fig. 3-6A is to point out that the output impedance of an emitter follower is inherently low, and it approaches the impedance of the emitter-to-base junction alone as the base impedance decreases to zero. The output impedance never decreases to zero, however, so the regulation never becomes perfect with this simple circuit.

Note in passing that a pnp regulating transistor is more conveniently placed in series with the negative side of the load, rather than the positive side as would be the case with most vacuum-tube regulators. The transistor itself must, of course, be capable of handling the maximum load current. In practical transistorized power supplies, it frequently is necessary to mount the large series regulators on radiators or other types of heat sinks to keep the temperatures of the transistor junctions within safe limits.

While the simple circuit in Fig. 3-6A is reasonably effective in stabilizing the output voltage against load variations, it does not remove variations caused by voltage changes in the unregulated source. This is because the voltage on the base of the transistor is changed in proportion to the unregulated voltage. The circuit in Fig. 3-6B overcomes this problem through the use of a separate, stabilized reference voltage source at the base. Although a battery symbol is shown, the reference-voltage source in a practical circuit could be a *reference diode* (zener diode), which is a semiconductor diode with enough reverse bias to operate in the *breakdown region*. A diode operated in this manner behaves very much like the familiar glow tube, or gaseous voltage regulator; that is, the voltage drop across the device is essentially independent of the current over a rather wide range. Reference diodes are preferable to gaseous voltage-regulator tubes for most transistorized power supplies, because they operate at lower voltages (usually from 5 or 6 volts up to 60 volts or more) and because they are generally superior in stability and inherent regulation.

The degree of regulation attainable with the circuit in Fig. 3-6B is determined by the emitter-to-base impedance of the transistor itself, which might be of the order of a few ohms. Even better stabilization (or lower output impedance) can be provided by the use of additional gain in the control circuit to supplement the gain of the regulating transistor itself. Such an approach is illustrated in simplified form in Fig. 3-6C. The voltage across the load may be compared with a stabilized reference voltage in a differential amplifier, which can be designed with enough gain to make the voltage variation at the load as small as required.

In a reverse-biased junction diode (zener diode), at a certain value of reverse-bias voltage the current increases rapidly while the voltage across the diode remains essentially constant. This *breakdown voltage*, which may be any value between 2 and 60 or more volts, depends on the construction of the diode. This characteristic is similar to that of the gas-tube regulator, which begins conduction at a certain voltage and continues to conduct varying amounts of current while maintaining constant voltage between the elements. The zener diode is used in transistorized regulated power supplies to hold an element of the transistor at a given reference voltage.

A basic diagram of a constant-voltage regulated power supply is shown in Fig. 3-7. A regulated reference voltage is obtained from a full-wave rectifier; a bridge rectifier supplies the series-regulator transistor, which receives at the base a feedback voltage from a comparison circuit. This voltage is an error signal of such magnitude and polarity as to change the conduction of the series regulator, hence changing the current through the load resistor until the output voltage (E_O) equals the voltage across the voltage control (R_{EO}). Note that since the series regulator is in the positive side of the circuit, an npn transistor is used.

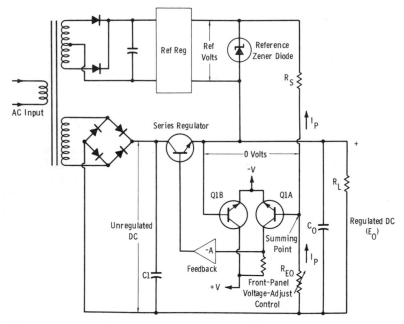

Fig. 3-7. Basic regulated power supply for constant voltage.

The difference between the two voltage inputs to the differential amplifier (Q1A and Q1B) is held at zero by feedback action. Thus, the voltage across summing resistor R_S is held equal to the reference voltage. The current through R_S and the output-voltage control (R_{EO}) is termed the *programming current*, I_p. The input impedance of the differential amplifier is high, and essentially all of the current (I_p) through R_S also passes through R_{EO}. Because the voltage across R_S is constant, I_p is constant. Since R_{EO} is variable, the output voltage is directly proportional to the resistance of this control. Thus, the output voltage is the same as the voltage drop across R_{EO} and will become zero if this control is reduced to zero ohms. This variable control is sometimes in series with a fixed resistor to hold the output voltage to a given minimum value.

Most reference supplies for units employing npn power transistors, as in Fig. 3-7, are referred to the positive output (or positive sensing, to be described) as a circuit common. The reference auxiliaries for supplies using pnp power transistors are referred to the negative output terminal.

The differential amplifier contains matched transistors (Q1A and Q1B) placed in thermal proximity or contained in a single "chip." This markedly improves the drift performance of the supply.

We can follow the feedback regulating action by assuming that the regulated dc output has momentarily increased. We will regard the positive output terminal as "common"; then the output-voltage increase causes the summing point to become instantaneously more negative. The resultant decrease in current through Q1A causes its collector to become more positive; hence, a more negative voltage is applied through the inverting feedback amplifier (−A) to the base of the series regulator. The resultant decreased conduction of the series regulator reduces the output voltage by an amount proportional to the momentary increase, and the error voltage between the bases of Q1A and Q1B is reduced to zero.

Camera-chain regulated power supplies are never as simple as the basic diagram of Fig. 3-7. For even moderate power outputs, the dissipation requirement for the series-regulator circuit is such that multiple transistors sometimes are used in parallel to provide adequate power-handling capacity. Most recent power supplies employ some means of *preregulation* in the rectifier path. The purpose of a preregulator is to allow the rectifier output to change in coordination with the output voltage so that minimum voltage drop occurs across the series regulator, and power dissipation is reduced to a small value in all series regulator elements. Silicon controlled rectifiers (SCR's) usually are used in the preregulator so that firing time can be controlled for required conduction angles.

Another feature found in modern supplies is the use of two extra wires between the supply and the load (Fig. 3-8). This results in optimum regulation at the load terminals rather than at the power-supply output terminals, compensating for the IR drop across the resistance of the wire. The current through the sensing lead is so small that, in spite of the resistance

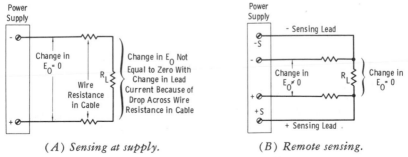

(A) Sensing at supply.   (B) Remote sensing.

Fig. 3-8. Principle of remote sensing.

of these leads, the voltage drop is negligible. This automatic arrangement eliminates the need for an adjustable tapped-transformer switch in the camera head to compensate for a change in cable length.

A block diagram of a power-supply module incorporating preregulation and remote sensing is shown in Fig. 3-9. Basic analysis by blocks is as follows:

Bridge Rectifier: The bridge rectifier provides full-wave rectification of the ac from the power transformer.

Switching Preregulator: The switching preregulator switches on and off twice during each half cycle to charge the filter capacitors. It maintains across the series regulator a small voltage drop that varies little with ac line-voltage changes. Thus, it is possible to reduce the power dissipated in the series regulator to the minimum required for adequate ac ripple filtering.

Preregulator Control: The preregulator control determines the conduction angle of the switching preregulator by comparing the full-wave rectified output from the bridge rectifier with the output from the series regulator.

Slow Turn-On: The slow turn-on circuit causes the output voltage to increase from zero to the rated output in about one second. This gradual build-up of dc reduces possible damaging current and voltage surges within equipment that obtains power from this unit, as well as high current surges within the supply itself.

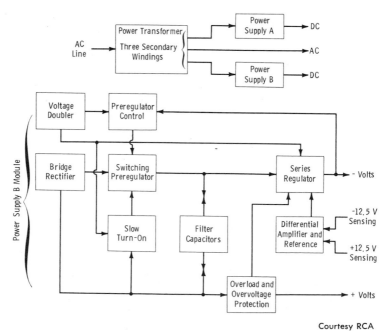

Courtesy RCA

Fig. 3-9. Block diagram of power-supply module.

Filter Capacitors: The filter capacitors are charged twice during each half cycle through the switching preregulator. During the time interval between charges, the energy that has been stored is discharged through the series regulator to the load.

Voltage Doubler: The voltage doubler provides the current source for the control-signal input to the series regulator. The use of the voltage doubler greatly reduces ripple on the dc output and minimizes the change in output voltage with changes in the ac line input. The voltage doubler also provides power to operate the switching preregulator.

Series Regulator: The series regulator filters out almost all of the ripple that appears across the filter capacitor. It compensates for changes in line voltage and load current; it also compensates for moderate voltage drops across the resistances of connector contact and intrarack wiring.

Differential Amplifier: The differential amplifier and reference provide the control signal for the series regulator. A differential amplifier rather than a single-ended amplifier is used because of its inherent temperature stability. Since the supply voltage must be extremely accurate, a precision reference is used instead of a voltage-adjustment control.

Overload and Overvoltage Protection: The overload-protection circuit prevents component damage, especially to transistors, because of an overload or short-circuited output. This circuit does not have to be reset after an overload; the output returns to its correct voltage as soon as an overload or short circuit is removed. The overvoltage-protection feature prevents the series regulator and load from being damaged (as a result of excessive power dissipation) in case the switching preregulator should develop a short circuit.

Fig. 3-10 illustrates the basic idea of overload protection. The value of R_B is such that, during normal regulator operation, Q3 is saturated. This places negligible resistance in series with the negative return. Potentiometer R_E is adjusted so that it produces sufficient voltage drop to cause X1 to conduct if the load should develop a short or a specific value of overload

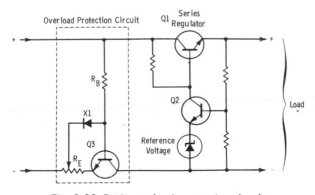

Fig. 3-10. Basic overload-protection circuit.

(low resistance, excessive current). Conduction of X1 reduces the bias on Q3 so that it appears as increasing series resistance in the regulator circuit. Relatively high voltages are required for the pickup tube or tubes in the camera head. The well-regulated low voltages required for transistor circuitry (such as plus 12 and minus 12 volts) are used to supply a square-wave oscillator. Usually, this oscillator is line-locked to one-half the horizontal scanning frequency so that any radio-frequency interference (RFI) components that should escape the heavy filtering do not appear as beat patterns in the picture.

A block diagram of the basic high-voltage supply is shown in Fig. 3-11A. This supply normally is located in the camera head. The binary counter receives either camera horizontal-drive pulses or separated horizontal-sync pulses from the composite camera blanking signal. Each input pulse is differentiated to obtain a trigger for the binary stage. Thus, the square-wave oscillator is synchronized to one-half the horizontal-scanning rate. This prevents any ripple or transients from appearing in the active scan interval. The square-wave oscillator is driven from cutoff to saturation. Since the voltage supply to this stage is extremely well regulated, excellent regulation of the generated high voltage is obtained.

The source of the high voltages is a step-up torroid transformer at the oscillator output. All "low sides" of the voltages are tied together for a common reference to insure tracking (Fig. 3-11B). The +1600 volts is developed by doubler X1-X2. The +800 volts is provided by half-wave

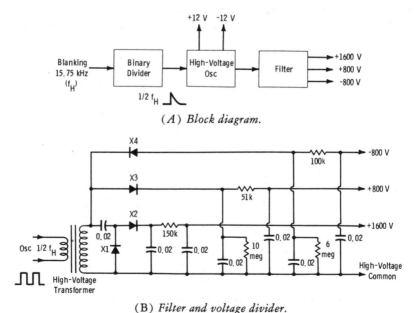

(A) Block diagram.

(B) Filter and voltage divider.

Fig. 3-11. Typical camera-head high-voltage supply.

rectifier X3, and the -800 volts is provided by X4. The current requirement is extremely small, allowing simple RC filtering. The slight alteration
in output voltage because of the load current is a negligible factor, since all
voltages to the pickup tube(s) track from the common reference.

3-6. CAMERA-CHAIN POWER DISTRIBUTION

Camera-chain power supplies normally are rack-mounted units from
which cabling extends to the camera control console and then to the camera
head. Fig. 3-12 illustrates a specific example involving the RCA TK-11
(monochrome) camera, a WP-15 power supply, and the focus-current regulator. J2 and J3 are paralleled output receptacles on the WP-15 regulator
chassis. J2 supplies the required voltage to the focus-current regulator and
incorporates an interlock circuit. If the focus-current regulator becomes
inoperative, B+ is removed from the power-supply output by means of the

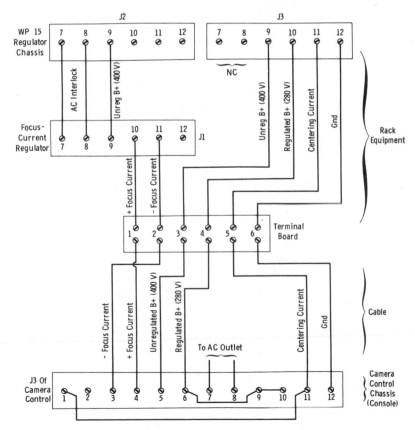

Fig. 3-12. Power distribution to camera control unit.

interlock relay. A customer-installed terminal board collects the outputs of the focus-current supply and the power supply, and directs these to the camera control console through the power cable.

Fig. 3-13 gives the complete picture of the power distribution in the TK-11 camera chain. The incoming power at J3 on the camera control unit is routed to the master monitor through an individual power cable, and to the camera through the camera cable. Some of the indicated voltages in the power cable are from internal supplies. For example, −500 volts is taken from the high-voltage supply in the camera head, and is fed through conductor 21 of the camera cable back to the control unit to be used for image focus control (photocathode potential). It is fed back to the camera head through conductor 15 of the camera cable.

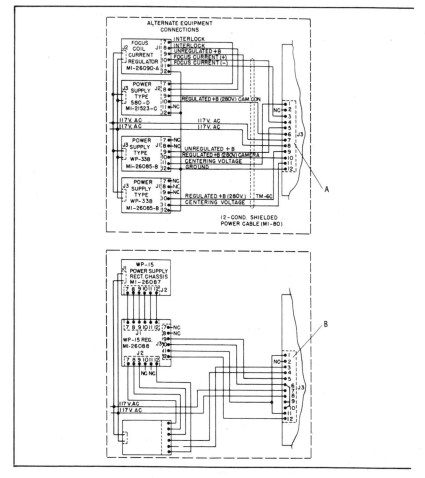

Fig. 3-13. Interconnection diagram,

Typical voltage distribution for a modern color camera chain is shown in simplified form in Fig. 3-14. The main power supplies are rack mounted, and all operating and fixed voltage distribution is by way of the control cable from the rack to the control panel and the camera cable from the rack to the camera head.

Fig. 3-15 represents a portion of a typical control panel for a four-channel color camera; voltage distribution is shown for the operating controls designated. The master white-level control is common to all channels, and the master chroma-level control is common to the three color channels. Individual color-channel controls are level controls to permit proper white balance on a neutral gray-scale (chip) chart. Distribution is then made to the respective circuitry in the camera head.

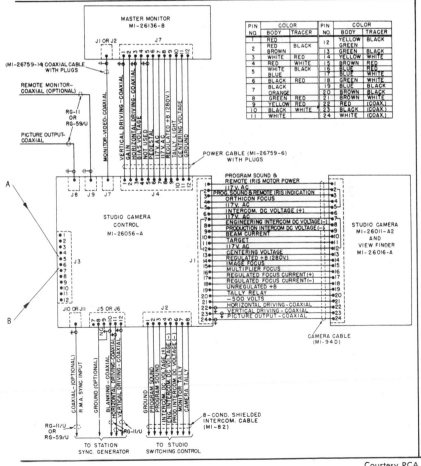

Courtesy RCA

TK-11A camera system.

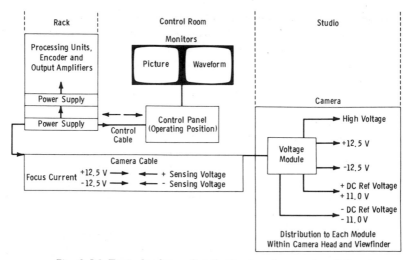

Fig. 3-14. Typical voltage distribution in color camera chain.

As shown in Fig. 3-14, the camera head normally contains a voltage module that receives the regulated voltage and serves as a main distribution point for all other modules in the camera. Fig. 3-16 illustrates basic reference-voltage generators in the voltage module for supplying other modules in the camera. Note that in Fig. 3-16A the regulated +12.5 volts is divided to + 11.8 volts at the base of transistor Q1. Since the transistor is silicon

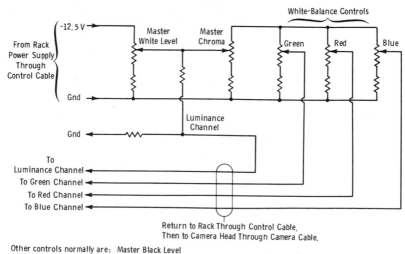

Fig. 3-15. Portion of control panel for four-tube color camera.

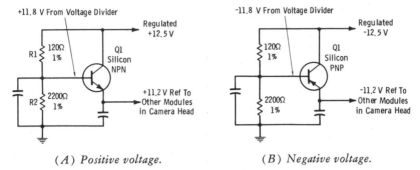

(A) Positive voltage.   (B) Negative voltage.

Fig. 3-16. Basic reference-voltage generators.

npn, the emitter is at 11.8 −0.6, or +11.2, volts for reference-voltage distribution. The circuit for the negative reference-voltage generator (Fig. 3-16B) is identical, except that a silicon pnp transistor must be used.

Circuitry in all other individual modules is decoupled from both the main power supply and the camera-head voltage module by decoupler circuits. Basic negative-voltage decoupler circuits are shown in Fig. 3-17. (Positive decouplers are identical except that npn transistors are used). The circuit of Fig. 3-17A employs a regulating zener diode for the base-voltage reference. The circuit of Fig. 3-17B uses the common reference-voltage supply from the camera-head voltage module.

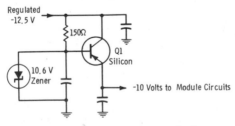

(A) Reference from zener diode.

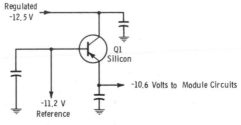

(B) Reference from generator.

Fig. 3-17. Basic module decoupler circuits.

3-7. POWER-SUPPLY MAINTENANCE

Power supplies, in addition to the overvoltage and overcurrent protection circuitry described previously, often include thermal relays and fuses.

Thermal-Relay Shutdown: The thermal relay opens the input circuit only when the power-supply output current exceeds the current rating specified for the operating *ambient temperature.* When the temperature decreases to normal, the thermostat will reset automatically. If this occurs often, forced-air cooling may be required.

Shutdown from Blown Fuse: Fatigue failure can occur as a result of mechanical vibrations combined with thermally induced stresses that weaken the fuse metal. Many times, fuse failures can be caused by a temporary condition, and replacing the blown fuse will make the fuse-protected circuitry operative again. Never replace a fuse with the unit turned on (power applied). The resulting temporary loose connection before solid contact is made may open the fuse again through no fault of the equipment. Always inspect fuse holders for tightness and cleanliness. Never substitute a fuse with a rating higher or lower than the original rating.

When a power supply shuts down from causes other than a thermal relay or blown fuse, it is necessary to determine whether the fault is internal to the supply or in the load. Suitable dummy loads should be made up and kept available so that this problem can be solved readily. Fig. 3-18 illustrates the use of a dummy load for such checks. The plus and minus sensing terminals should be connected to their respective outputs for internal sensing. Automotive-type lamps, such as the type 1073 (1.8 amperes at 12.8 volts), serve as excellent substitute loads. As many as needed should be paralleled to approach the maximum or nominal load of the supply. For example, if the nominal load is 6 amperes, three such lamps in parallel are quite suitable.

A typical power supply of this kind might have a nominal load of 6 amperes, an overload-protection limit of 10 amperes, and an overvoltage-protection limit of 16.5 volts. With the dummy load substituted for the normal load, an internal fault will cause the lamps to burn brighter than normal for an instant, and then the supply will shut down. The higher-

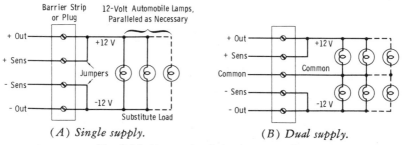

(A) *Single supply.*   (B) *Dual supply.*

Fig. 3-18. Dummy loads for power supply.

than-normal voltage is caused by lack of regulation in the supply. If the lamps light at the normal output voltage and remain lighted, the trouble is obviously in the load. In the case of the dual power supply of Fig. 3-18B, usually only one of the supplies or loads will be faulty. Thus, if the trouble is internal, one bank of lights will glow brightly and then go out. If all lamps remain lighted, the trouble is isolated to the load.

In the case of a 280-volt supply, a power resistor of the proper value to bring the power-supply output current near maximum normally is used as a dummy load. For example, if the specifications call for a maximum load current of 1.5 amperes, two 500-ohm, 200-watt resistors connected in parallel should be used. Such dummy loads should always be available and ready to connect to the barrier strip or receptacle at the power-supply output.

Older tube-type regulated supplies without modern protective circuitry do not shut down automatically in case of trouble. The output voltage simply wavers in value around nominal voltage. Such a supply sometimes can be out of regulation with almost no noticeable shift on an external voltmeter. This condition is best checked with an oscilloscope set on ac input at high sensitivity. The change in dc on the coupling capacitor in the scope will show up as a "bouncing" line on the free-running scope trace, usually with a large ripple component.

In the case of intermittent or unusual camera problems which seem to "chase themselves back" to some kind of power-supply problem, use of the scope is mandatory. However, certain precautions *must* be observed to make such tests valid. The importance of proper connection of load and monitoring leads to the power-supply output terminals cannot be overemphasized, since the most common errors associated with the measurement of power-supply performance result from improper connection to the output terminals. Failure to connect the monitoring instrument to the proper points will result in measurement of the characteristics not of the power supply, but of the power supply plus the resistance of the leads between its output terminals and the point of connection. Even connecting the load by means of clip leads to the power-supply terminals and then connecting the monitoring instrument by means of clip leads fastened to the load clip leads can result in a serious measurement error. Remember that the power supply being measured probably has an output impedance of less than 1 milliohm, and the contact resistance between clip leads and power-supply terminals will, in most cases, be considerably greater than the specified output impedance of the power supply.

All measuring instruments (oscilloscope, ac voltmeter, differential or digital voltmeter) must be connected *directly* by separate pairs of leads to the monitoring points. This is necessary in order to avoid the mutual coupling effects that may occur between measuring instruments unless all are returned to the low-impedance terminals of the power supply. Twisted pairs (in some cases shielded cable will be necessary) should be used to avoid pickup on the measuring leads.

Care must be taken that the measurements are not unduly influenced by the presence of pickup on the measuring leads or by power-line frequency components introduced by ground-loop paths. Two quick checks should be made to see if the measurement setup is free of extraneous signals:

1. Turn off the power supply and observe whether any signal is observable on the face of the CRT.

2. Instead of connecting the oscilloscope leads separately to the positive and negative sensing terminals of the supply, connect both leads to either the positive or the negative sensing terminal, whichever is grounded to the chassis.

Signals observable on the face of the CRT as a result of either of these tests are indicative of shortcomings in the measurement setup.

In measuring the input voltage, it is important that the ac voltmeter be connected as closely as possible to the input ac terminals of the power supply so that its indication will be a valid measurement of the power-supply input, without any error introduced by the IR drop present in the leads connecting the power-supply input to the ac line-voltage source.

Use an autotransformer of adequate current rating. If this precaution is not followed, the input ac waveform presented to the power supply may be severely distorted, and the rectifying and regulating circuits within the power supply may be caused to operate improperly.

A regulated power supply beginning to "slip" in performance usually has an increased amount of ripple over that specified by the manufacturer as the maximum. Fig. 3-19A shows an incorrect method of measuring peak-to-peak ripple. Note that a continuous ground loop exists from the third wire of the input power cord of the power supply to the third wire of the input power cord of the oscilloscope. This path is through the grounded power-supply case, the wire between the negative output terminal of the power supply and the scope, and the grounded scope case. Any ground current circulating in this loop as a result of the difference in potential (E_G) between the two ground points causes an IR drop that is in series with the scope input. This IR drop, normally having a 60-Hz (line-frequency) fundamental, plus any pickup on the unshielded leads interconnecting the power supply and scope, appears on the face of the CRT. The magnitude of this resulting noise signal can easily be much greater than the true ripple developed between the output terminals of the power supply, and can completely invalidate the measurement.

The same ground-current and pickup problems can exist if an rms voltmeter is substituted in place of the oscilloscope. However, the oscilloscope display, unlike the meter reading, tells the observer immediately whether the fundamental period of the signal displayed is 8.3 milliseconds (1/120 Hz) or 16.7 milliseconds (1/60 Hz). Since the fundamental ripple frequency present at the output of a supply is 120 Hz (as a result of full-wave rectification), an oscilloscope display showing a 120-Hz fundamental component is indicative of a "clean" measurement setup, whereas the presence of a 60-Hz fundamental usually means that an improved setup will result in a more accurate (and lower) measured value of ripple.

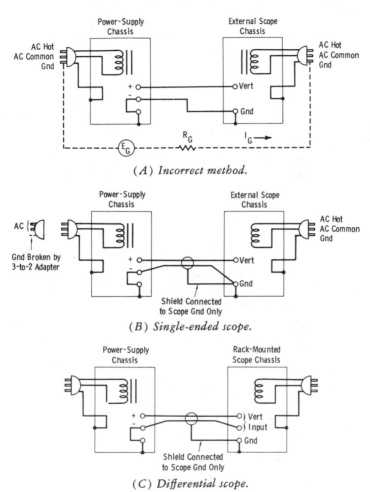

(A) *Incorrect method.*

(B) *Single-ended scope.*

(C) *Differential scope.*

Fig. 3-19. Ripple measurements with oscilloscope.

Fig. 3-19B shows a correct method for using a single-ended scope to measure the output ripple of a constant-voltage power supply. The ground-loop path is broken with a 3-to-2 adapter in series with the ac line plug of the power supply. Notice, however, that the power-supply case still is connected to ground through the power-supply output terminals, the leads connecting these terminals to the scope terminals, the scope case, and the third wire of the scope power cord.

Either a twisted pair or (preferably) a shielded two-wire cable should be used to connect the output terminals of the power supply to the vertical-input terminals of the scope. When a twisted pair is used, care must be taken that one of the two wires is connected both to the grounded terminal of the power supply and the grounded input terminal of the oscilloscope. When shielded two-wire cable is used, it is essential for the shield to be connected to ground at one end only so that no ground current can exist in this shield and induce a noise signal in the shielded leads.

To verify that the oscilloscope is not displaying ripple that is induced in the leads or picked up from the grounds, the plus scope lead should be touched to the minus scope lead at the power-supply terminals. The ripple value obtained when the leads are in this position should be subtracted from the actual ripple measurement.

In most cases, the single-ended scope method of Fig. 3-19B will be adequate to eliminate extraneous components of ripple and noise so that a satisfactory measurement may be obtained. However, in more stubborn cases, or in measurement situations in which it is essential that both the power-supply case and the oscilloscope case be connected to ground (e.g. if both are rack-mounted), it may be necessary to use a differential scope with floating input, as shown in Fig. 3-19C. If desired, two single-conductor shielded cables may be substituted for the shielded two-wire cable with equal success. Because of its common-mode rejection, a differential oscilloscope displays only the difference in signal between its two vertical-input terminals, thus ignoring the effects of any common-mode signal introduced because of the ac difference in potential between the power-supply case and scope case. Before a differential-input scope is used in this manner, however, it is imperative that the common-mode-rejection capability of the scope be verified by shorting together its two input leads at the power supply and observing the trace on the CRT. If this trace is a straight line, the scope is properly ignoring any common-mode signal present. If the trace is not a straight line, the scope is not rejecting the ground signal and must be realigned in accordance with the manufacturer's instructions until proper common-mode rejection is attained.

The complete hookup for checking a regulated power supply for comparison to manufacturer's specifications is shown in Fig. 3-20. Most modern 12.5-volt regulated supplies maintain the rated output voltage within plus or minus 1 percent up to maximum rated load with an ac line input of from 90 to 130 volts. The most severe test is at maximum output current with minimum line-voltage input. The ripple voltage is normally around 5 millivolts peak-to-peak minimum for a 12.5-volt supply. Note that with the variable load, the overcurrent-protection circuitry can be checked conveniently in this same setup.

The peak-to-peak ripple voltage and waveshape may be measured readily with the oscilloscope, but some manufacturers give the ripple specification in terms of rms voltage. Fig. 3-21 shows the relationship between the peak-

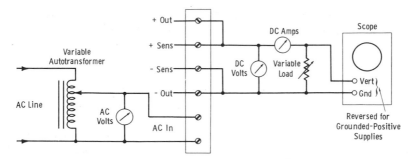

Fig. 3-20. Setup for checking power-supply performance.

to-peak and rms values of three common waveforms. The output ripple of a dc power supply usually does not approximate the sine wave of Fig. 3-21A; in many cases the output ripple has a waveshape that closely approximates the sawtooth of Fig. 3-21B. In this case, the rms ripple is 1/3.464 of the peak-to-peak value displayed on the oscilloscope. The square wave (Fig. 3-21C) is included because this waveshape has the highest possible ratio of rms to peak-to-peak values. Thus, the rms ripple and noise present at the output terminals of a power supply cannot be greater than one-half the peak-to-peak value measured on the oscilloscope. In most cases, the ripple waveshape is such that the rms value is between one-third and one-fourth of the peak-to-peak value.

When a high-frequency spike measurement is being made, an instrument of sufficient bandwidth must be used; an oscilloscope with a bandwidth of 20 MHz or more is adequate. Measuring noise with an instrument that has insufficient bandwidth may conceal high-frequency spikes that are detrimental to the load. The test setups illustrated in Figs. 3-19A and 3-19B generally are not acceptable for measuring spikes; a differential oscilloscope is necessary. Furthermore, the measurement concept of Fig. 3-19C must be modified if accurate spike measurement is to be achieved.

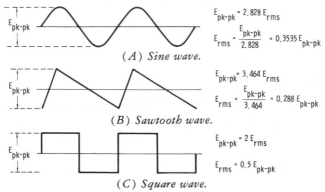

$$E_{pk-pk} = 2.828\, E_{rms}$$

$$E_{rms} = \frac{E_{pk-pk}}{2.828} = 0.3535\, E_{pk-pk}$$

(A) *Sine wave.*

$$E_{pk-pk} = 3.464\, E_{rms}$$

$$E_{rms} = \frac{E_{pk-pk}}{3.464} = 0.288\, E_{pk-pk}$$

(B) *Sawtooth wave.*

$$E_{pk-pk} = 2\, E_{rms}$$

$$E_{rms} = 0.5\, E_{pk-pk}$$

(C) *Square wave.*

Fig. 3-21. Conversion of peak-to-peak to rms.

The Hewlett-Packard Company suggests the following procedure for checking their regulated supplies for noise spikes:

1. As shown in Fig. 3-22, two coaxial cables must be substituted for the shielded two-wire cable.

2. Impedance-matching resistors must be included to eliminate standing waves and cable ringing, and capacitors must be used to block the dc current path.

3. The lengths of the test leads outside the coaxial cables are critical and must be kept as short as possible; the blocking capacitor and the impedance-matching resistor should be connected directly from the inner conductor of the cable to the power-supply terminals.

4. Notice that the shields at the power-supply ends of the two coaxial cables are not connected to the power-supply ground, since such a connection would give rise to a ground-current path through the cable shield, resulting in an erroneous measurement.

5. The measured noise-spike values must be doubled, since the impedance-matching resistors constitute a 2-to-1 attenuator.

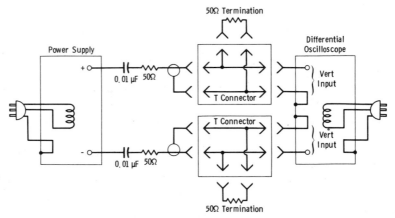

Fig. 3-22. Arrangement for measuring noise spikes.

The circuit of Fig. 3-22 also can be used for the normal measurement of low-frequency ripple and noise. Simply remove the four terminating resistors and the blocking capacitors, and substitute a higher-gain plug-in preamplifier in place of the wide-band plug-in module required for spike measurements. Notice that with these changes, Fig. 3-22 becomes a two-cable version of Fig. 3-19C.

It may happen that a camera chain loses transistors in certain modules on a rather consistent basis. Sometimes this trouble is attributable to the power supply. Modern regulated supplies have a slow turn-on and also a certain time constant for turn-off to prevent excessive transients from damaging delicate transistors. Transistors actually can be damaged upon turn-off of the power supply; this damage obviously does not show up until the equipment is turned on again.

The checking for on and off transients must be done with an oscilloscope, but many pitfalls exist in such measurement. All power-supply output terminals have a small amount of inductance, and the load can be either capacitive or inductive. The best that can be done without laboratory-type equipment and setups is to compare the on and off transients with those of another camera chain that has not exhibited a problem from this cause. When such a check does reveal a power supply with excessive on or off transients, all filter time constants and all "overshoot" or transient protection circuitry should be checked thoroughly.

The preventive maintenance of regulated power supplies is extremely important to overall stability. The following four tests enable the maintenance engineer to keep a running check on the condition of his regulated supplies, particularly the older tube-type variety:

Test 1. Determine the voltage output range at fixed load. Use a fixed load that will draw at least two-thirds of the maximum rated load current. Rotate the voltage-adjust control to its extremes, and record the minimum and maximum voltages. For example, the normal available range of a 280-volt regulated supply might be from 270 to 300 volts, at a given load current. Failure to reach the normal maximum voltage is usually the result of a weak dc-amplifier tube or voltage-adjust tube (or transistor).

Test 2. In vacuum-tube regulators, check the currents in the series regulator tubes for balance. Most regulated supplies incorporate a meter-selector switch on the panel for measuring individual regulator-tube currents. Table 3-1 shows the application of such readings. Notice that the total load in this example is 1014 milliamperes; therefore, the ideal average for each of the six tube sections is 1014/6, or 169 mA. Since maximum tube life and stability can be expected when these currents are balanced within ± 10 percent (20 percent total variation), a record of individual currents is kept, and it is compared to the minimum and maximum

Table 3-1. Tabulated Data for Test 2

Tube	Current (mA)
V1A	168
V1B	160
V2A	180
V2B	182
V3A	164
V3B	160

Operating Data
Total = 1014 mA
Average/Section = 169 mA
Lowest Desirable = 152.1 mA
Highest Desirable = 185.9 mA

values that should occur for the given load. This indicates the need for a tube change before trouble occurs, barring any sudden failure.

Test 3. Measure the input-voltage regulation (voltage output with fixed load and varying input ac line voltage). The setup for this and the following test is shown in Fig. 3-23. Adjust the power supply to be tested to 0.5 volt above the reference supply, and connect a voltmeter between the two outputs to measure this voltage difference. By means of the variable autotransformer, make measurements at the reference line voltage (usually 117 volts) and over a specified range, such as 100 to 130 volts ac input. Table 3-2 shows the data recorded at station WTAE-TV for an RCA WP-15B supply. Note the excellent input voltage regulation of this supply with a fixed load.

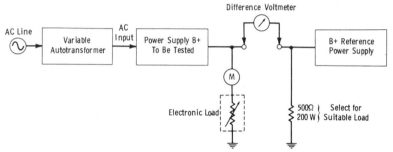

Fig. 3-23. Test setup for checking power-supply regulation.

Test 4. Measure the output-voltage regulation under varying loads (fixed ac line voltage with varying load current). In this test, the variable autotransformer is adjusted for an ac input of 117 volts, and the electronic load on the supply under test is varied over a specified range. Table 3-3 shows the corresponding data for the WP-15B of Test 3. Note that the results indicate low internal power-supply resistance and good regulation under varying loads.

Table 3-2. Tabulated Data for Test 3

Line Volts	Output Volts	Actual Volts Variation
117 (Ref)	280.5	0 (Ref)
100	280.94	+ 0.44
105	280.75	+ 0.25
110	280.62	+ 0.12
115	280.5	0
120	280.42	− 0.08
125	280.37	− 0.13
130	280.31	− 0.19

Table 3-3. Tabulated Data for Test 4

Load Current (mA)	Output Volts	Actual Volts Variation
1000 (Ref)	280.54	0 (Ref)
400	280.52	— 0.02
600	280.5	— 0.04
800	280.52	— 0.02
1200	280.54	0
1400	280.56	+ 0.02
1500	280.51	— 0.03

NOTE: For stability in video levels, the associated power supply should have very low internal resistance, theoretically zero. (This is never attained in practice.) The internal dc output resistance may be found as follows:

$$R_O = \frac{\Delta V_O}{\Delta I_L}$$

where,

R_O is the dc output resistance,
ΔV_O is the change in output voltage,
ΔI_L is the change in output current.

For example, if the output voltage changes 0.1 volt with a load-current change of 1000 mA (1 ampere):

$$R_O = \frac{0.1}{1} = 0.1 \text{ ohm}$$

The condition of power-supply filters and general regulation efficiency should be checked several times yearly by observing the ripple content of the output voltage on an oscilloscope. An increase in ripple content indicates the need for filter replacement or better regulation efficiency before deterioration reaches troublesome proportions. Typical commercial power supplies have a maximum of 2 millivolts peak-to-peak ripple content on a 280-volt regulated output. The ripple on an unregulated voltage may be as high as 2 volts (peak-to-peak) on a 400-volt dc output.

The operating condition of a zener diode (or any other type of diode) is most reliably checked with an oscilloscope and the simple associated circuitry shown in Fig. 3-24. The upper trace in Fig. 3-25 illustrates a typical curve obtained by this method. If desired, the forward trace may be eliminated by means of an added silicon diode, as shown by the dash lines in Fig. 3-24. In this case, a curve such as the lower trace illustrated in Fig. 3-25 results.

Fig. 3-26 is an interpretation of the trace in Fig. 3-25. As the variable ac voltage (Fig. 3-24) is increased from zero, the voltage is traced horizon-

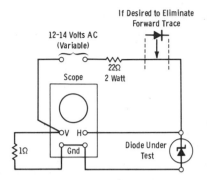

Fig. 3-24. Test setup for checking semiconductor diodes.

tally (A-B) along the scope graticule, which may be calibrated in volts/ centimeter. When the zener breakdown voltage is reached, the horizontal trace should remain the same length as the current curve increases. If desired, the vertical scale (B-C) may be calibrated in milliamperes/centimeter. Care should be taken not to exceed the maximum current specification (wattage = voltage applied across the diode times the diode current) of the zener diode being tested.

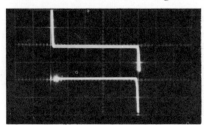

Fig. 3-25. CRO traces showing diode operation in test circuit.

The same test circuit should be employed to check a regular diode at the operating potential encountered in the circuit in which it is used. Some diodes (such as the common 1N34) have a natural *hysteresis loop*, as shown by the upper trace in Fig. 3-27. This loop should remain stable without jitter or erratic "looping" as the voltage is varied around the normal operating level. Other diodes (such as the Type 1N279) do not reveal

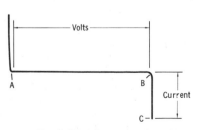

Fig. 3-26. Interpretation of diode trace.

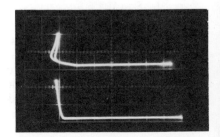

Fig. 3-27. Traces with and without hysteresis.

a loop (lower trace in Fig. 3-27). There should be no instability of trace as the voltage is varied around the normal operating level.

3-8. INTERCONNECTING FACILITIES

There are two basic types of camera interconnections: (A) Camera cable between camera and rack or control units, with multiconductor cable between rack equipment and remote-control panels, and (B) Rf links, employing either air link or triaxial-cable link (Chapter 2). In the second method, multiconductor cable may be employed between base-station rack equipment and remote-control panels.

Representative types of camera cable are shown in Fig. 3-28. Monochrome camera cable usually consists of three coaxial conductors and 21 (Fig. 3-28A) or 25 (Fig. 3-28B) single conductors. In the typical application of a 24-conductor cable in Fig. 3-13, note that some conductors are paralleled for proper current capacity, and the three coaxial elements normally provide drive-pulse feed to the camera and carry the video output current from the camera to the control unit. The surge impedance of the coaxial portions of camera cable is 50 ohms.

For assembly purposes, conductors in a camera cable are divided into groups. For the 24- or 28-conductor cable, the single wires are divided into three groups. Together with the three coaxial conductors, these groups are assembled around a waterproof jute core. The entire assembly is then taped and covered with a woven shield over which are placed a cotton braid and a neoprene outer jacket. Color cameras may require cables with 82 (or more) conductors, as shown in Fig. 3-28C.

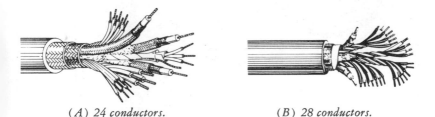

(A) 24 conductors.   (B) 28 conductors.

(C) 82 conductors.

Fig. 3-28. Examples of camera cables.

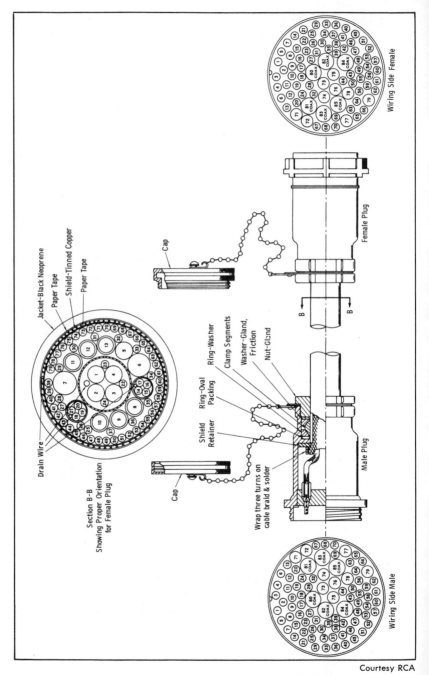

Fig. 3-29. Cable and connectors for RCA TK-42 color camera.

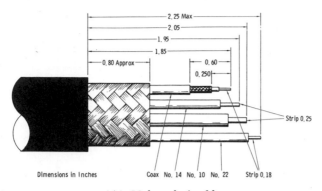

(A) *Male end of cable.*

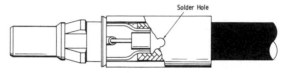

(B) *Male end of coax.*

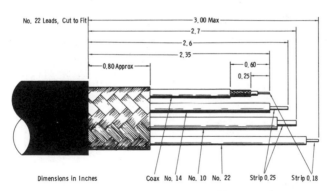

(C) *Female end of cable.*

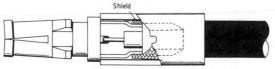

(D) *Female end of coax.*

Fig. 3-30. Preparation of cable ends.

Table 3-4. Conductors in TK-42 Camera Cable

Pin	Wire	AWG	Color	Pin	Wire	AWG	Color
1	74	22	Wht-Blu-Orn	44	24	22	Wht-Blk-Vio
2	28	22	Wht-Brn-Red	45	53	22	Wht-Yel-Blk
3	73	22	Wht-Blu-Red	46	54	22	Wht-Yel-Brn
4	72	22	Wht-Blu-Brn	47	35	22	Blu
5	Shield Drain	22	Tinned Copper	48	60	22	Wht-Yel-Vio
				49	14	22	Wht-Blu
6	71	22	Wht-Blu-Blk	50	Quad Drain	22	Tinned Copper
7	42	22	Wht-Red-Vio				
8	40	22	Wht-Red-Grn	51	56	22	Wht-Yel-Orn
9	39	22	Wht-Red-Yel	52	57	22	Wht-Yel-Yel
10	77	22	Wht-Blu-Blu	53	59	22	Wht-Yel-Blu
11	75	22	Wht-Blu-Yel	54	15	22	Wht-Vio
12	76	22	Wht-Blu-Grn	55	17	22	Wht-Blk-Blk
13	70	22	Wht-Grn-Gry	56	64	22	Wht-Grn-Red
14	43	22	Wht-Red-Gry	57	58	22	Wht-Yel-Grn
15	41	22	Wht-Red-Blu	58	16	22	Wht-Gry
16	27	22	Wht-Brn-Brn	59	63	22	Wht-Grn-Brn
17	38	22	Wht-Red-Orn	60	61	22	Wht-Yel-Gry
18	79	22	Wht-Blu-Gry	61	34	22	Grn
19	78	22	Wht-Blu-Vio	62	62	22	Wht-Grn-Blk
20	30	22	Brn	63	32	22	Orn
21	44	22	Wht-Orn-Blk	64	33	22	Yel
22	Quad Drain	22	Tinned Copper	65	66	22	Wht-Grn-Yel
				66	65	22	Wht-Grn-Orn
23	21	22	Wht-Blk-Yel	67	69	22	Wht-Grn-Vio
24	Core Drain	18	Tinned Copper	68	68	22	Wht-Grn-Blu
				69	31	22	Red
25	26	22	Wht-Brn-Blk	70	67	22	Wht-Grn-Grn
26	18	22	Wht-Blk-Brn	71	7	10	Grn
27	20	22	Wht-Blk-Orn	72	NC		
28	29	22	Wht-Brn-Orn	73	2	14	Wht
29	45	22	Wht-Orn-Brn	74	1	14	Blk
30	25	22	Wht-Blk-Gry	75	3	14	Orn
31	19	22	Wht-Blk-Red	76	4	14	Blu
32	23	22	Wht-Blk-Blu	77	5	14	Brn
33	47	22	Wht-Orn-Orn	78	NC		
34	46	22	Wht-Orn-Red	79	6	10	Red
35	37	22	Wht-Red-Red	80	10	Coax	Wht-Red
36	49	22	Wht-Orn-Grn	81	11	Coax	Wht-Orn
37	48	22	Wht-Orn-Yel	82	9	Coax	Wht-Brn
38	52	22	Wht-Orn-Gry	83	12	Coax	Wht-Yel
39	36	22	Wht-Red-Brn	84	8	Coax	Wht-Blk
40	50	22	Wht-Orn-Blu	85	13	Coax	Wht-Grn
41	51	22	Wht-Orn-Vio				
42	55	22	Wht-Yel-Red				
43	22	22	Wht-Blk-Grn				

Courtesy RCA

Table 3-5. Conductor Functions in TK-42 Camera Cable

Pin	Function	Pin	Function
1	Spare	44	A.C. Sen.
2	Tally	45	Spare
3	Spare	46	Brg. Cap
4	Spare	47	Polarity Sw
5	Shield Drain	48	Green White Level
6	Spare	49	V.F. Relay (Color Blanker)
7	Gamma	50	Quad Drain
8	Mono Blk. Level Relay	51	Blue White Level
9	Lens Cap	52	Blue Black Level
10	Spare	53	Red White Level
11	Test Switch	54	Spare
12	White Level Cont	55	Interphone-Cue Lo.
13	Spare	56	Blue Hor Cent.
14	Aperture	57	Red Black Bal.
15	Hor. Advance	58	Interphone-Cue Hi
16	Red Hor. Cent.	59	−12.5V Sense
17	Blue Vert. Cent	60	Time Const. Relay
18	+12.5V Sense	61	Green Hor Cent.
19	Black Level Cont	62	Red Vert. Cent.
20	Spare	63	Spare
21	Power Switch	64	Spare
22	Quad Drain	65	Relay Gnd. Return
23	Interphone Prod Coil	66	Green Vert. Cent.
24	Core Drain	67	Sensitivity Cont.
25	V.F. Relay (M-Blanker)	68	D.C. Gnd Sense
26	Spare	69	Spare
27	Interphone-Common	70	Mono White Level
28	Sensitivity Sw	71	+ 12.5V
29	Spare	72	Spare
30	+250 V	73	A.C. Reg.
31	Interphone-Eng. Coil	74	A.C. Reg.
32	50/60 Hz Sel.	75	A.C. Unreg.
33	Spare	76	A.C. Unreg.
34	Spare	77	Gnd.
35	V.F. Relay K2	78	Spare
36	Green Selector	79	− 12.5V
37	V.F. Relay K5	80	Red Video
38	Blue Selector	81	Mono Video
39	Red Selector	82	Blue Video
40	Spare	83	V.F. Video
41	R.B.G.M. Sel. Com	84	Green Video
42	M Selector	85	Camera Sync
43	A.C. Sen.		

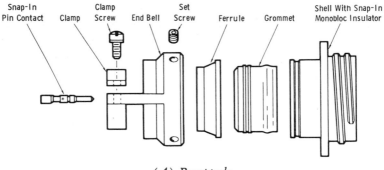

(A) Receptacle.

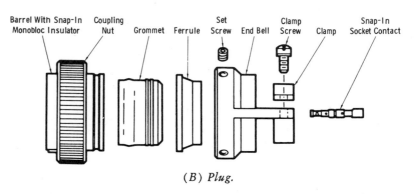

(B) Plug.

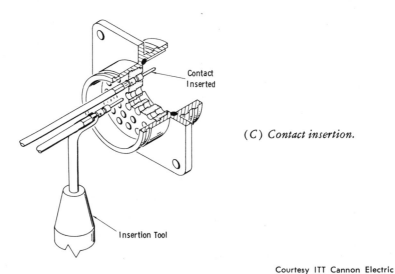

(C) Contact insertion.

Fig. 3-31. Connectors for camera cable.

To acquaint the reader with a typical color-camera cable installation, Figs. 3-29 and 3-30 and Tables 3-4 and 3-5 are included. At the top of Fig. 3-29, Section B-B shows a typical grouping of conductors by shielding isolation and drain-wire combinations. The numbers in Section B-B are "wire numbers" only, not connector pin numbers. Note that at the center are four No. 14 AWG, tinned-copper, vinyl-plastic insulated, color-coded conductors, along with three No. 22 AWG conductors (Table 3-4). An aluminum *Mylar* shield is wrapped around this group and a standard, tinned-copper, No. 18 AWG core drain wire. Two quad groupings with associated drain wires are contained in the cable, and the overall shield drain wire is connected to pin 5 of the connector and in turn to frame ground.

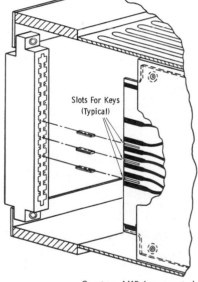

Fig. 3-32. *AMP-LEAF* connector for printed circuit boards.

Slots For Keys (Typical)

Courtesy AMP Incorporated

Note that pins 72 and 78 of the connector have no wires attached, but the pins still must be installed. The largest wires used are No. 10 AWG (pins 71 and 79); these wires are used to carry the +12.5- and −12.5-volt power-supply current to the camera (Table 3-5). Note from the correlation of Tables 3-4 and 3-5 that the core shielded group is employed for ac, and the shielded quad groups are used primarily for interphone (Chapter 8).

Since the rack-mounted end of the camera cable receives power, the receptacle at the rack is female, and the male end of the camera cable goes to this receptacle. The camera head normally has a male receptacle to receive the female end of the camera cable. Fig. 3-30 shows the preliminary conductor preparation for each end.

Camera cables and multiconductor control cables normally are terminated in a plug or receptacle of the general type shown in Fig. 3-31. There are many individual types of connectors, and the camera technician should obtain the correct assembly procedures for every type of connector he may become concerned with in assembly or disassembly of cables. ITT Cannon Electric, AMP Incorporated, and Amphenol all supply instruction sheets for their respective connectors; these sheets specify proper crimping tools to use and give all other necessary information. Sometimes such sheets are included in the instruction books for camera chains.

Figs. 3-31A and 3-31B give the general nomenclature of plug and receptacle parts. After the pin or socket has been crimped properly to the conductor, it usually is inserted with a special tool, as illustrated in Fig. 3-31C. It is most important that exact instructions be followed for the particular assembly or disassembly involved. Special pin and socket extraction tools usually are available for disassembly.

(*A*) *Studio termination.*

(*B*) *Patch rack for cameras.*

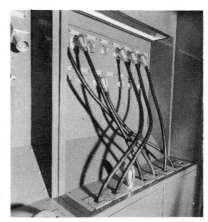

(*C*) *Patch rack for control units.*

Courtesy WBBM-TV

Fig. 3-33. Camera-cable connections at WBBM-TV.

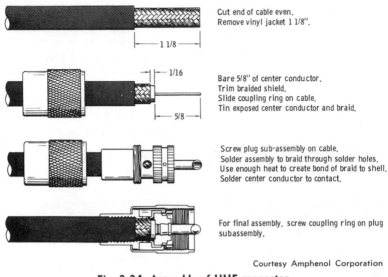

Cut end of cable even.
Remove vinyl jacket 1 1/8".

1 1/8

Bare 5/8" of center conductor.
Trim braided shield.
Slide coupling ring on cable.
Tin exposed center conductor and braid.

1/16

5/8

Screw plug sub-assembly on cable.
Solder assembly to braid through solder holes.
Use enough heat to create bond of braid to shell.
Solder center conductor to contact.

For final assembly, screw coupling ring on plug
subassembly.

Courtesy Amphenol Corporation

Fig. 3-34. Assembly of UHF connector.

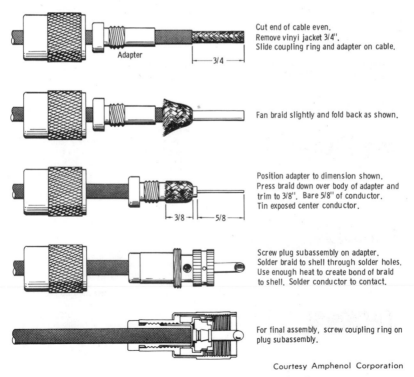

Cut end of cable even.
Remove vinyl jacket 3/4".
Slide coupling ring and adapter on cable.

Adapter

3/4

Fan braid slightly and fold back as shown.

Position adapter to dimension shown.
Press braid down over body of adapter and
trim to 3/8". Bare 5/8" of conductor.
Tin exposed center conductor.

3/8   5/8

Screw plug subassembly on adapter.
Solder braid to shell through solder holes.
Use enough heat to create bond of braid
to shell. Solder conductor to contact.

For final assembly, screw coupling ring on
plug subassembly.

Courtesy Amphenol Corporation

Fig. 3-35. Assembly of UHF connector with adapter.

Individual modules of modern color camera chains normally employ a quick connect-disconnect arrangement (Fig. 3-32) for both rack-mounted and camera-head boards. The particular arrangement shown is the *AMP-LEAF* connector. The boards are slotted to mate with keying plugs that are positioned in the connector housing. Thus, accidental insertion of a contact board other than in the specified position is prevented.

Since camera-control units and rack-mounted gear are normally in the control room, which is some distance from the tightly enclosed studios, the camera cable seldom connects directly from camera head to rack or

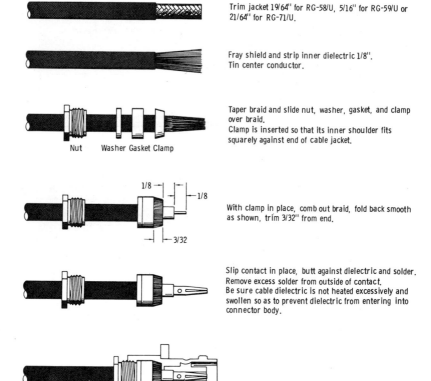

Trim jacket 19/64" for RG-58/U, 5/16" for RG-59/U or 21/64" for RG-71/U.

Fray shield and strip inner dielectric 1/8". Tin center conductor.

Taper braid and slide nut, washer, gasket, and clamp over braid.
Clamp is inserted so that its inner shoulder fits squarely against end of cable jacket.

Nut   Washer Gasket Clamp

1/8 — 1/8
— 3/32

With clamp in place, comb out braid, fold back smooth as shown, trim 3/32" from end.

Slip contact in place, butt against dielectric and solder. Remove excess solder from outside of contact.
Be sure cable dielectric is not heated excessively and swollen so as to prevent dielectric from entering into connector body.

Female Contact

Push assembly into body as far as it will go.
Slide nut into body and screw in place with wrench until tight.
For this operation, hold cable and shell rigid and rotate nut.

Male Contact

Courtesy Amphenol Corporation

Fig. 3-36. Assembly of BNC connector.

control unit. The camera-head cable normally is connected to a receptacle on a wall in the studio, as shown in Fig. 3-33A. (Note how the outlets in Fig. 3-33A are tilted downward so that the cable reaches the floor with a minimum of strain.) A patch rack sometimes is used, as shown installed at WBBM-TV in Fig. 3-33B. By the arrangement illustrated, seven cameras can be made available to any one of fourteen outlets in four different studios. Fig. 3-33C shows the patch rack for control units assigned to studio distribution of the cameras. Note the portion of the camera-cable patching rack on the left.

The preparation of miniature coaxial cable for plug-in module connectors or camera-cable connectors is a special technique for the particular type of connector, as mentioned. However, the distribution cable (nominal surge impedance of 75 ohms) is usually RG/59U or the larger RG/11U. Fig. 3-34 shows the technique for assembling RG/11U cable to the UHF-type (83-1SP) connector. Fig. 3-35 includes the adapter used with the UHF plug for the smaller RG/59U cable. Fig. 3-36 illustrates the BNC connector assembly as normally used for RG/59U cable.

EXERCISES

Q3-1. Upon what does the camera head normally mount?

Q3-2. Name the basic controls found on a camera panning head, and describe the basic function of each.

Q3-3. What are "synced prompters"?

Q3-4. What is the purpose of the preregulator section of a regulated power supply?

Q3-5. What is remote sensing, and why is it used in modern regulated power-supply systems?

Q3-6. Why are oscillators in camera high-voltage supplies usually synchronized to one-half the horizontal-scanning frequency?

CHAPTER 4

The Camera Pickup Tube, Yoke Assembly, and Optics

In this chapter, consideration will be given to the more advanced characteristics of the pickup tube, yoke, and optical assemblies associated with television cameras. To understand this material, the reader should have a fundamental background in these subjects.[1]

We will consider the following characteristics of the image orthicon, vidicon, and lead-oxide tubes:

1. Sensitivity
2. Spectral response
3. Dark current
4. Gamma (transfer characteristics)
5. Resolution
6. Signal-to-noise ratio
7. Lag
8. Dependence on target voltage
9. Temperature dependence
10. High-light handling
11. Operational techniques

4-1. THE IMAGE ORTHICON

Fig. 4-1 illustrates the three basic types of image-orthicon (I.O.) tubes. Fig. 4-1A shows the 3-inch non-field-mesh tube. In the 3-inch field-mesh type (Fig. 4-1B), two electrodes, the field mesh and suppressor grid, are

---

[1]Harold E. Ennes, *Television Broadcasting: Equipment, Systems, and Operating Fundamentals* (Indianapolis: Howard W. Sams & Co., Inc., 1971), Chapters 1 and 4.

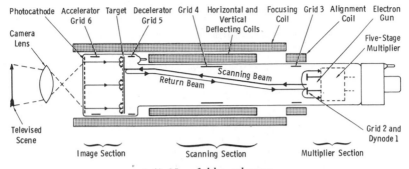

(A) Non-field-mesh type.

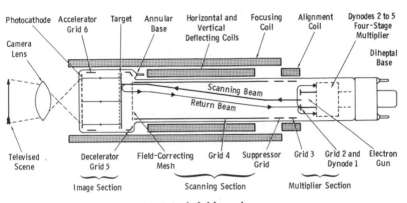

(B) 3-inch field-mesh type.

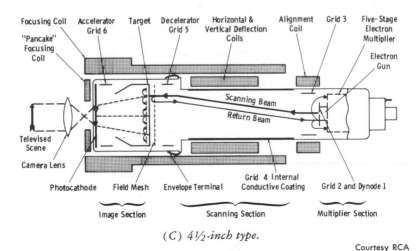

(C) 4½-inch type.

Courtesy RCA

Fig. 4-1. Basic construction of image orthicon.

added. In this tube, the additional electrodes normally are operated at a fixed potential. The 4½-inch tube (Fig. 4-1C) always incorporates a field mesh that is provided with a variable bias potential. Note also the addition of the "pancake" (magnifier) coil at the front of the tube. This magnifier coil is connected in series aiding with the main focusing coil, and provides about 120 gausses in the plane of the photocathode. The useful area on the photocathode is identical to that of the 3-inch I.O. (1.6-inch, or 40.7-millimeter, diagonal). The pancake coil magnifies the image 1.5 times (to a 2.4-inch, or 61-millimeter, diagonal) to fill the target.

The image orthicon consists of three primary sections: the image section, the scanning section, and the electron-multiplier section. A basic review of these functions is provided in the following subsections.

Image Section

The image section contains a semitransparent photocathode on the inside of the faceplate, a grid to provide an electrostatic accelerating field, and a target that consists of a thin glass disc with a fine-mesh screen very closely spaced to it on the photocathode side. Focusing is accomplished by means of a magnetic field produced by an external coil, and by varying the photocathode voltage.

Light from the scene being televised is picked up by an optical lens system and focused on the photocathode, which emits electrons from each illuminated area in proportion to the intensity of the light striking the area. The streams of electrons are focused on the target by the magnetic and accelerating fields.

On striking the target, the electrons cause secondary electrons to be emitted from the glass. The secondary electrons thus emitted are collected by the adjacent mesh screen, which is held at a definite potential of about 2 volts with respect to target cutoff voltage. Therefore, the potential of the glass disc is limited for all values of light, and stable operation is achieved. Emission of the secondary electrons leaves on the photocathode side of the glass a pattern of positive charges that corresponds to the pattern of light from the scene being televised. Because of the thinness of the glass, the charges set up a similar potential pattern on the opposite, or scanned, side of the glass.

Scanning Section

The side of the glass away from the photocathode is scanned by a low-velocity electron beam produced by the electron gun in the scanning section. This gun contains a thermionic cathode, a control grid (grid 1), and an accelerating grid (grid 2). The beam is focused at the target by the magnetic field of an external focusing coil and by the electrostatic field produced by grid 4.

Grid 5 serves to adjust the shape of the decelerating field between grid 4 and the target in order to obtain uniform landing of electrons over the entire target area. The electrons stop their forward motion at the surface of the glass and are turned back and focused into a five-stage signal multiplier, except when they approach the positively charged portions of the pattern on the glass. When this condition occurs, electrons are deposited from the scanning beam in quantities sufficient to neutralize the charge pattern on the glass. Such deposition leaves the glass with a negative charge on the scanned side and a positive charge on the photocathode side. These charges neutralize each other by conduction through the glass in less than the time of one frame (Fig. 4-2).

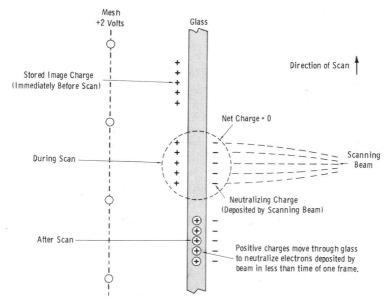

Fig. 4-2. Target charge and discharge cycle of image orthicon.

Alignment of the beam from the gun is accomplished by a transverse magnetic field produced by an external coil located at the gun end of the focusing coil. Deflection of the beam is accomplished by transverse magnetic fields produced by external deflecting coils.

The electrons turned back at the target form the return beam, which has been amplitude modulated by absorption of electrons at the target in accord with the charge pattern. (The more positive areas of the charge pattern correspond to the high lights of the televised scene.)

One of the basic problems with the image orthicon is shown in Fig. 4-3. When the photocathode is excited by a very bright object in the scene, it emits a very large number of photoelectrons. These electrons are attracted toward the target because the target screen is at about +2 volts (above cutoff) and the photocathode is at a much higher negative potential. When the photoelectrons strike glass target B, secondary electrons are emitted and attracted to target mesh A, where they normally are passed to ground through the target-potential supply. In extreme high-light areas, the target glass rapidly assumes a potential higher than that of the mesh, and collection of additional electrons on the mesh is prevented. These free electrons return to the target in areas adjacent to the bright spot, there nullifying the positive charges in areas C. Thus, the scanning beam is repelled in these areas, and a large return beam results, Since this represents black, a *black halo* surrounds the excessive high light in the image. We will see later in this discussion how the combined effects of target area, target spacing (capacitance), and the field mesh can minimize this effect.

Multiplier Section

The return beam is directed to the first dynode of a five-stage electrostatically focused multiplier. This multiplier utilizes the phenomenon of secondary emission to amplify signals composed of electron beams. The electrons in the beam impinging on the first dynode surface produce many other electrons, the number depending on the energy of the impinging electrons. These secondary electrons are then directed to the second dynode and release more new electrons. Grid 3 facilitates a more complete collection by dynode 2 of the secondary electrons from dynode 1. The multiplying process is repeated in each successive stage, with an ever-increasing stream of electrons until those emitted from dynode 5 are collected by the anode and become the current in the output circuit. The multiplier section amplifies the modulated beam between 500 and 1000 times.

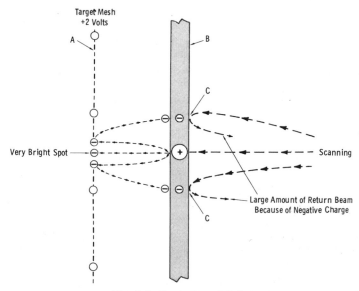

Fig. 4-3. Formation of halo.

The signal-to-noise ratio of the output signal from the image orthicon is high. The gain of the multiplier raises the output signal sufficiently above the noise level of the video-amplifier stages that these stages contribute a negligible amount of noise to the final video signal. The signal-to-noise ratio of the video signal, therefore, is determined only by the random variations of the modulated electron beam.

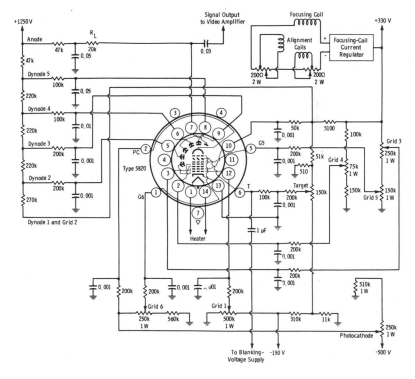

Courtesy RCA

Fig. 4-4. Typical voltage-divider and control circuits for 3-inch image orthicon.

It can be seen that when the beam moves from a less-positive portion of the target to a more-positive portion, the signal-output voltage across the load resistor (R_L in Fig. 4-4) changes in the positive direction. Hence, for high lights in the scene, the grid of the first video-amplifier stage swings in the positive direction.

Fig. 4-4 is a typical schematic diagram of the control circuitry associated with a 3-inch non-field-mesh tube such as the Type 5820. The dynode supply is almost universal, as can be noted from the operating potentials in Table 4-1. This table illustrates the similarities and differences between four representative types of I.O.'s.

Table 4-1. Typical Image-Orthicon Operation and Characteristics

Tube Type	5820 3-Inch	7293 3-Inch Field-Mesh	4492 4½-Inch	4536 4½-Inch
Photocathode Volts (Image Focus)	−300 to −500	−325 to −475	−600	−600
Grid 6 (Accelerator) Volts (70 to 80% of PC Volts)	−240 to −400	−210 to −360	−370 to −470	−370 to −470
Target Volts (Above Cutoff)	+2	+2	+2.3	+2.3 to +3
Grid 5 (Decelerator) Volts	0 to 125	0 to 40	40	40
Grid 4 Volts (Beam Focus)	140 to 180	140 to 180	70 to 90	70 to 90
Grid 3 Volts	225 to 330	260 to 300	250 to 275	250 to 275
Grid 2 (Dynode 1) Volts	300	300	280	280
Grid 1 Volts (For Picture Cutoff)	−45 to −115	−45 to −115	−45 to −115	−45 to −115
Dynode 2 Volts	600	600	600	600
Dynode 3 Volts	800	800	800	800
Dynode 4 Volts	1000	1000	1000	1000
Dynode 5 Volts	1200	1200	1200	1200
Anode Volts	1250	1250	1250	1250
Anode Signal Current (pk-pk)	3 to 24 µA	5 to 30 µA	20 µA (approx)	15 to 100 µA
Minimum Blanking Voltage (pk-pk)	5	5	5	8
Focus-Coil Current*	75 Gausses	75 Gausses	60 Gausses	60 Gausses

*Specified Field Strength at Center of Focus Coil (Approx 1 mA/Gauss)

Basic Problems

Table 4-2 shows still more similarities and differences between representative types of image orthicons. A "close" target-to-mesh spacing is about 0.001 inch, a "medium" spacing is about 0.002 inch, and a "wide" spacing is about 0.003 inch. The closer the spacing and the greater the target area, the more the capacitance is.

In designing an image orthicon for television, there are certain "trade-offs" in performance. The main drawbacks in earlier I.O.'s may be listed as follows:

1. Noise
2. Black halo around high lights
3. Untrue edge transitions on vertical image lines
4. Limited and poor gray-scale reproduction
5. Target sticking (retention of image over many frames)

There are several approaches to help minimize such drawbacks, and hence there are differences among tubes in target-to-mesh spacing, target material, and target size.

For example, the signal-to-noise ratio of an I.O. varies in direct relationship to the stored charge at the target. That is, as the target charge increases (greater capacitance), the signal-to-noise ratio increases. There are three possible ways to increase target charge:

1. Reduce the spacing between the target and target mesh. This reduction must not be carried to the point at which microphonics (picture noise caused by vibration of elements under camera motion) are induced, or the onset of burn-in (picture retention) occurs.
2. Increase the target-mesh potential. This must not be carried to the point at which spurious effects are induced by deflecting the scanning beam away from its proper point of target incidence, causing loss of resolution.
3. Increase the overall area of the target and target mesh, as in the 4½-inch image orthicon as compared with the 3-inch tube. Note from Table 4-2 that this approach results in the only effective increase in signal-to-noise ratio.

Increasing the target capacitance (by closer spacing between target and target mesh) results in a somewhat longer linear transfer curve. It also increases the amplitude of the high-light signal current, helping to reduce the black-halo effect. Higher capacitance also decreases the sensitivity of the tube. Note from Table 4-2 that, as the target capacitance increases, more light is required on the photocathode in order to reach the knee of the transfer curve.

Poor edge transitions are caused by poor beam landing. Since the beam approaches the target at near zero velocity, it may be attracted from its normal point of incidence to a more positive area, resulting in the convergence of highly positive (white) areas into darker areas. By incorporat-

Table 4-2. Comparison of Image Orthicons

Type	Field Mesh	Spectral Response	Target Material*	Target-to-Mesh Spacing (Inches)	Target-to-Mesh Spacing (Millimeters)	Target-to-Mesh Capacitance (pF)	Operating Point**	Illumination on Tube Face at Operating Point (Lumens/Ft²)	Amplitude Response at 400 TV Lines (Percent)	Limiting Resolution TV Lines (2% Response)	S/N 2 V	S/N 2.3 V	S/N 3.0 V
8685 (2-Inch)	Yes	Bialkali	E	0.001	0.0254	75	At Knee	0.075	25	600	—	—	30
5820 (3-Inch)	No	S-10	G	0.0022	0.056	100	At Knee	0.01	55	600	34	—	—
7293 (3-Inch)	Yes	S-10	E	0.0022	0.056	100	1 Stop Over Knee	0.02	60	600	34.5	—	—
							1 Stop Over Knee	0.02	60	675	34	—	—
8673 (3-Inch)	Yes	Bialkali	E	0.001	0.0254	200	At Knee	0.015	60	625	35	—	—
							1 Stop Over Knee	0.03	65	625	35.5	—	—
4492 (4½-Inch)	Yes	S-10	G	0.002	0.051	300	At Knee	0.025	75	800	—	38	—
4536 (4½-Inch)	Yes	S-10	E	0.001	0.0254	600	At Knee	0.04	60	800	—	—	40.5

*G = standard glass (ionic conduction); E = electronically conducting glass.
**This operating point is not necessarily that used for all operations; it sets the reference point for the following parameters of sensitivity and resolution.
***Signal-to-noise ratio with amplifier bandwidth of 10 MHz, no aperture correction (peak-to-peak signal high-light current to rms noise current).

ing a field-correcting mesh (Figs. 4-1B and 4-1C), which straightens and stiffens the electrostatic field on the scanned side of the target, much better beam landing results. This eliminates the "overemphasized" outlines seen from non-field-mesh tubes.

There are two types of target glass (Table 4-2), standard glass and electronically conducting glass. Conduction in standard glass is ionic, and this type of target is subject to burn-in and sticking. Electronically conducting glass targets have higher stability, greater resistance to burn-in, and less granular effect. With this target material, no orbiter is required.

Bialkali photocathodes (Table 4-2) are characterized by long-term stability in sensitivity and resolution.

Resolution

It can be shown[2] that each megahertz of bandwidth corresponds to 80 TV lines (approximately) of resolution. Thus, an amplifier with 10-MHz bandwidth would allow a picture resolution of 800 TV lines. But the

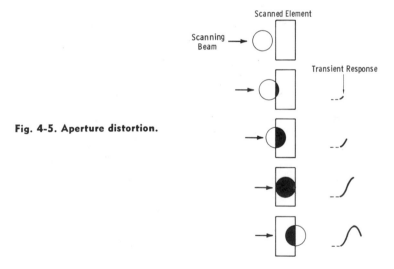

Fig. 4-5. Aperture distortion.

pickup tube is limited in its amplitude response at high frequencies (thin vertical lines in the image) by the *aperture effect* (Fig. 4-5). The scanning beam is round and is of finite size. Note from Fig. 4-5 how the tube output voltage is proportional to the cross-sectional area of the scanning beam; a somewhat sinusoidal transition is produced when the ideal transition would be a square wave.

---

[2]Harold E. Ennes, *Television Broadcasting: Equipment, Systems, and Operating Fundamentals* (Indianapolis: Howard W. Sams & Co., Inc., 1971), Section 3-3.

Curve B of Fig. 4-6 is typical of the Type 5820 I.O. The increased amplitude response at low line numbers (below approximately 250 lines) results from signal redistribution effects on the target when the lens is operated one or two stops above the knee of the transfer curve. The field-mesh 3-inch tube and the 4½-inch tube exhibit this characteristic to a much lesser extent because of the improved beam landing and lowered spurious response. However, many practicing engineers are at a loss to explain why the image appears "sharper" from a non-field-mesh tube (such as the 5820) than it does from a field-mesh type (such as the 7293), even though they can "see" the same horizontal *resolution* on the test pattern from either tube.

With a properly set up Type 5820 I.O., the "sharpness" and "snappiness" resulting from the overemphasized outlines of the picture elements cannot be denied. However, it is this very spurious response characteristic that is hard to control on "glints" from sequins on a dress at the studio, or when panning across unexpected light sources on remotes. We are all familiar with the black halos that result, sometimes giving the appearance of sending the entire scene into an unlighted coal mine. The bad-landing effects of the non-field-mesh image orthicon prohibit its use in color cameras.

The curves of Fig. 4-6 show the typical uncompensated response of the three pickup tubes in most common use today. These curves represent the condition in which the camera is focused on a square-wave test pattern and the signal is measured with an amplifier of 10-MHz bandwidth.

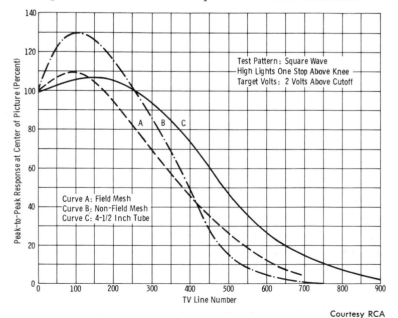

Fig. 4-6. Uncompensated response of image orthicons.

The "aperture effect" is similar to passing the signal through a low-pass filter with no phase distortion. Thus, aperture correction is ideally made with a device that produces a rise in high-frequency response that corresponds to the slope of the aperture loss, without introducing phase shifts. This is normally achieved by using an "open-ended delay line." The input capacitance of the preamplifier stage does produce some phase shift. Thus, we will always find some means of correcting the phase of the video signal. These circuits are variously termed *high-peakers* or *phase correction*, and there is no sharp line of demarcation between the two terminologies.

NOTE: High-peaker and phase-correction circuitry is covered in Chapter 5, since this is generally incorporated in video preamplifiers. Aperture correction is discussed in detail in Chapter 6.

Note from Fig. 4-6 the improvement in amplitude response of the 4½-inch tube compared to the 3-inch tube at higher line numbers such as 400 TV lines. The image size at the target of the 4½-inch tube is 1.5 times

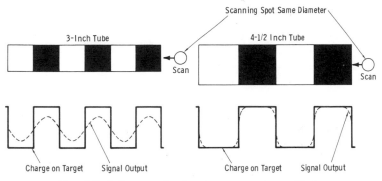

Fig. 4-7. Influence of tube size on aperture effect.

that of the 3-inch tube (Fig. 4-7). Since the scanning spot can be made just as small in the larger tube as it is in the smaller tube, the net effect is equivalent to obtaining a smaller scanning spot per picture element. Thus, the larger tube has better resolution and requires less aperture correction. Since aperture correction in turn introduces noise, particularly in the dark areas of a scene, better signal-to-noise ratio, as measured through the 10-MHz amplifier, results.

Tube Setup Controls

Fig. 4-8 illustrates setup controls and typical electrode voltages for the 4½-inch I.O. Actual terminology used by various manufacturers for corresponding setup controls differs. For example, photocathode focus may be termed "image focus." On the other hand, the image accelerator (G6) control is termed "image focus" on some cameras; on others, it may be called "acc" or "G6 volts." Beam focus sometimes is termed "wall focus" or "orth

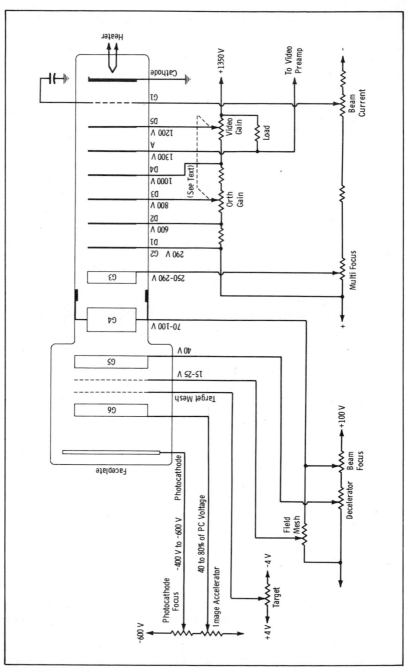

Fig. 4-8. Controls and typical voltages for 4½-inch image orthicon.

focus." Every operator must familiarize himself with the terminology for his particular camera, since this may vary not only between manufacturers, but also between systems of the same manufacturer.

The target control normally incorporates a "target set" switch that biases the target to the specified operating voltage below cutoff. With the minimum amount of light to be used, and maximum $f/$ stop on the lens, the target-voltage control is then set so that an image is just visible on the view finder. Restoring the target-set switch to normal then automatically sets the target at the calibrated voltage above cutoff. During operation, the sensitivity is controlled with a knob that controls either the lens iris or a neutral-density wheel in front of the optical system to vary the amount of exposure.

The beam control sets the G1 bias, and therefore determines the amount of beam current. This is adjusted to just discharge the maximum high lights, and must be readjusted continually during initial setup of the tube for all other operating parameters. After this, it is left alone. Too little beam results in white clipping or a reversed-polarity picture. Excessive beam results in loss of resolution from beam spreading, and excessive noise in the picture.

PC (photocathode) focus and beam focus are adjusted for maximum resolution. The voltage on G6 also affects resolution and normally is adjusted, consistent with the highest PC voltage that can be obtained, to eliminate "S" distortion. This distortion is evident as wavy horizontal image lines as the camera is panned.

The decelerator (G5) voltage is adjusted for minimum corner shading and maximum corner resolution.

The multiplier-focus (sometimes labelled MULTI FOCUS) control for G3 is adjusted for the most uniformly shaded picture that occurs near maximum signal output.

Note that ganged controls for dynodes D3 and D5 are shown in Fig. 4-8. Sometimes only one control (D3) is used, and this normally is labelled the ORTH GAIN control. It is set to give a specified signal high-light current into the input of the video preamplifier.

When the voltage on the field-mesh electrode is adjustable (4½-inch tubes), it usually is adjusted to eliminate mesh beat interference patterns in dark areas of the picture.

Setup techniques for pickup tubes are covered extensively elsewhere.[3] The following general outline is a composite of manufacturers' tube instructions, and is presented as a review.

The setup procedure for non-field-mesh tubes is as follows: After the tube has been inserted in its sockets and the voltages have been applied,

---

[3]For example, see: Harold E. Ennes, *Televison Broadcasting: Equipment, Systems, and Operating Fundamentals* (Indianapolis: Howard W. Sams & Co., Inc., 1971).

allow it to warm up for ¼ to ½ hour with the camera lens capped. Uncap the lens momentarily while adjusting the grid-1 voltage to give a small amount of beam current. This procedure will prevent the mesh from being electrostatically pulled into contact with the glass disc. Make certain that the deflection circuits are functioning properly to cause the electron beam to scan the target. Adjust the deflection circuits so that the beam overscans the target, i.e., so that the area of the target scanned is greater than its sensitive area. This procedure during the warm-up period is recommended to prevent burning on the target a raster smaller than that used for on-the-air operation. Note that overscanning the target results in a smaller-than-normal picture on the monitor.

With the lens still capped and the target voltage set at approximately 2 volts negative, adjust the grid-1 voltage until noise or a rough-textured picture of dynode 1 appears on the monitor. Then adjust the alignment-coil current so that the small white dynode spot does not move when the beam-focus (grid 4) control is varied, but simply goes in and out of focus. During alignment of the beam, and also during operation of the tube, always keep the beam current as low as possible to give the best picture quality and to prevent excessive noise and burning of the dynode-1 surface.

Next, uncap the lens and partially open the lens iris, and focus the camera on a test pattern. Advance the target voltage until a reproduction of the test pattern is just discernible on the monitor. This value of target voltage is known as the "target cutoff voltage." The target voltage should then be raised exactly 2 volts above the cutoff-voltage value, and the beam-current control should be adjusted to give just sufficient beam current to discharge the high lights. Then adjust the lens to produce best optical focus, and adjust the voltages on the photocathode and grid 4 to produce the sharpest picture.

At this point, attention should be given to the voltage controls for grids 3 and 5. Grid 5 is used to control the landing of the beam on the target and consequently the uniformity of signal output. The grid-5 control should be adjusted to produce the most uniform picture shading from center to edge with the lens iris opened sufficiently to permit operation with the high lights above the knee of the light-transfer characteristic. The value of grid-5 voltage should be as high as possible consistent with uniform shading. Grid 3 facilitates a more complete collection by dynode 2 of the secondary electrons that are released from dynode 1. The grid-3 control should be adjusted to the position that results in the maximum signal output.

Now, with a test pattern consisting of a straight line centered on the face of the tube, adjust the voltage on grid 6 and the voltage on the photocathode to produce a sharply focused straight line on the monitor. Improper adjustment of the grid-6 control will cause the straight-line pattern to be reproduced with a slight S shape.

The above adjustments constitute a rough setup. Final adjustments necessary to produce the best possible picture from the image orthicon camera are as follows:

With the lens capped, realign the beam. Beam alignment is necessary after each change of the grid-5 control and sometimes after each adjustment of the grid-3 control.

The proper illumination level for camera operation should be determined next. For most 3-inch non-field-mesh tubes, adjust the target voltage accurately to 2 volts above the target-cutoff value. Remove the lens cap and focus the camera on a test pattern. Open the lens iris just to the point at which the high lights of the test pattern do not rise as fast as the low lights when viewed on a video-waveform oscilloscope.

Next, cap the lens and adjust the grid-3 voltage control so that the video signal when viewed on a video-waveform oscilloscope has the flattest possible trace. This represents the black level of the picture.

The lens iris setting then should be noted, and the lens opened not more than one stop beyond this point, unless extreme scene-contrast ranges necessitate further opening of the lens.

The use of a higher value of target voltage than that recommended will shorten the life of the tube. The target-voltage control should not be used as an operating control to match pictures from two different cameras. Matching of cameras should be accomplished by control of the lens-iris openings.

Retention of a scene, sometimes called *sticking picture,* may be experienced if the tube is allowed to remain focused on a stationary bright scene, or if it is focused on a bright scene before it reaches an operating temperature in the range from 35° to 45°C. Often, the retained image disappears in a few seconds, but sometimes it may persist for long periods before it completely disappears. A retained image generally can be removed by focusing the tube on a clear white screen and allowing it to operate for several hours with an illumination of about 1 foot-candle on the photocathode.

To avoid retention of a scene, it is recommended that the tube always be allowed to warm up in the camera for ¼ to ½ hour with the lens iris closed and with a slight amount of beam current. Never allow the tube to remain focused on a stationary bright scene, and never use more illumination than is necessary.

The setup procedure used for the field-mesh type of image orthicon differs in the following respects from that to be used for image orthicons that do not employ field-mesh design: First, because the dynode aperture of the field-mesh tube cannot be brought into focus, different alignment techniques must be used. Second, the field-mesh tube *may not* be operated on certain grid-4 voltage loops because of severe mesh-beat patterns that result. To obtain optimum performance from the field-mesh tube, the following setup procedure should be followed carefully:

Before the proper voltages are applied to the tube, the lens should be uncapped and the lens iris opened. *This is a very important step for this type of image orthicon.* The proper voltages should then be applied, and the grid-1 voltage should be adjusted immediately to produce a small amount of beam current. This procedure will prevent the mesh from being electrostatically pulled into contact with the glass disc. Make certain that the deflection circuits are functioning properly to cause the electron beam to scan the target, and adjust the deflection circuits so that the beam overscans the target. (The purpose of this procedure during the warm-up period is to prevent burning on the target a raster smaller than that used for on-the-air operation.) The lens should then be capped, and the tube should be allowed to warm up for 1/4 to 1/2 hour before use or before other adjustments are made.

Next, uncap the lens and partially open the lens iris. Increase the target voltage until information appears on the monitor. Then adjust the beam focus, image focus, and optical focus until detail can be discerned in the picture. Adjust the controls for alignment-coil current until picture response is maximum. If the picture appears in negative contrast, increase the beam current. Further adjust the alignment-coil current so that the center of the picture does not move when the beam-focus (grid 4) control is varied, but simply goes in and out of focus. During alignment of the beam, and also during operation of the tube, always keep the beam current as low as possible to give the best picture and prevent excessive noise.

Next, focus the camera on a test pattern. The distance from the camera to the test pattern should be set so that the corners of the test-pattern image just touch the inside of the target ring. Next, the deflection circuits are adjusted so that the entire test pattern just fills the TV raster. The target voltage is then advanced or reduced to the point at which a reproduction of the test pattern is just discernible on the monitor. This value of target voltage is known as the target-cutoff voltage. The target voltage should be raised exactly 2 volts above the cutoff-voltage value, and the beam-current control should be adjusted to give just sufficient beam current to discharge the high lights.

Now adjust the lens to produce best optical focus, and adjust the voltages on the photocathode and grid 4 to produce the sharpest picture. To prevent mesh-beat problems, the voltage on grid 4 should be adjusted to a value between 150 and 180 volts.

Proper adjustment for suppression of high-light flare, or "ghosts," and for proper geometry is obtained when the grid-6 voltage is accurately set at 73 percent of the photocathode voltage. This adjustment may be effected by positioning a small, bright spot of light on the edge of the field to be viewed and then adjusting the grid-6 voltage so that the "ghost" seen on the viewing monitor disappears as the image section is brought into sharpest focus. Improper adjustment is evident when a light spot on the right edge of the viewing monitor produces a "ghost" above the spot and when a light spot on the left edge of the monitor produces a "ghost" below the spot.

Following the adjustment of the grid-6 voltage, the voltage on grid 5 should be adjusted to produce the best compromise between high signal output in the picture corners and best geometry. Best geometry is indicated by the absence of S distortion of straight lines and by a rectangular raster.

After adjustment of the image-section voltages, the voltage of grid 3 should be set for maximum signal output. The deflection yoke should be rotated, if necessary, so that the horizontal scanning of the camera is parallel to the horizontal plane of the scene.

Finally, readjust the target voltage so that it is accurately set to 2 volts above target cutoff. With the lens open, the lens iris should be opened to one or two lens stops beyond the point at which the high lights of the scene reach the knee of the light transfer characteristic.

NOTE: The general setup techniques for $4\frac{1}{2}$-inch tubes are similar to those described, except that the target voltage is 2.3 to 4 volts above cutoff, and the field-mesh potential is adjustable for optimum performance.

In color-camera operation of an image orthicon, the tube normally is operated at no more than $\frac{1}{2}$ stop over the knee of the transfer curve.

The return beam in the I.O. scans an area of about $\frac{1}{4}$ inch on the first dynode around the "aperture" of the beam at the gun end of the tube. In a non-field-mesh tube, this beam aperture is apparent as a small white dot near the center of the picture when the lens is capped and the target and beam are adjusted so that the texture of the first dynode can be seen. In beam alignment of this type of tube, the grid-4 voltage is "rocked" manually by means of the beam-focus control, and the beam alignment-coil current is adjusted so that the dot does not swirl but simply "blinks" in and out of focus.

Most monochrome and color cameras now employ the field-mesh type of image orthicon. The mesh defocuses the return beam so that the texture of the first dynode does not appear in the background (dark areas) of the picture. To make beam alignment more accurate and rapid, a *rock*, or *wobbulator*, circuit (Fig. 4-9) for grid 4 usually is provided in the camera. This circuit is an Eccles-Jordan flip-flop with a pulse from the vertical-sweep circuits applied to the emitters as shown. This pulse cuts off the transistor that is conducting, causing its collector to go more negative. Triggering causes no change in the nonconducting transistor, since it is merely driven further into cutoff. Collector coupling causes the transistor that is cut off to drive the other into conduction. This state remains until the next pulse arrives to initiate the reverse action. Thus, one output pulse occurs at the collector of Q1 for every two input pulses. This provides a 30-Hz square wave to grid 4 when the align switch is in the on position. With the camera looking at an alignment test pattern, a "split field," or two images, occurs if the beam is not properly aligned. The cur-

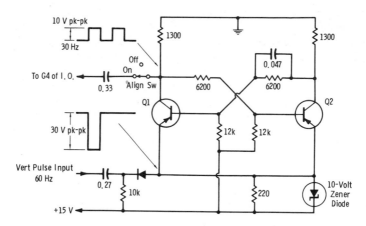

Fig. 4-9. Wobbulator circuit for beam alignment.

rents in the horizontal and vertical alignment coils are then adjusted to superimpose the two fields.

4-2. THE YOKE ASSEMBLY

Electrons from the electron gun in the tube are first focused into a narrow beam, and then this beam is caused to sweep back and forth across the image on the target with a definite time interval and sequence. Such movement of the electron beam is called the *scanning* process.

The beam is focused by a coil that surrounds the outside of the deflection-coil assembly and creates a magnetic cross field that narrows the emitted electrons into a beam of constant diameter. The cross section of the electron beam is termed the *scanning aperture*, probably a carryover from the days of revolving mechanical discs with small holes that traversed the projected area of the scene.

The beam is caused to scan the image by horizontal- and vertical-deflection coils constituting a *yoke* around the neck of the pickup tube. The deflection coils are provided with sawtooth current waves that deflect the electronic beam electromagnetically.

The entire yoke assembly is surrounded with a *Mumetal* wrap (magnetic shield) with removable end caps. The pickup tube mounts within the yoke assembly. Openings in the end caps expose the pickup-tube faceplate and base.

The yoke assembly is mounted on a cradle and base plate that pivots out from one side of the camera housing to facilitate maintenance, adjustments, and I.O. installation and removal (see Fig. 4-10). The entire assembly normally pivots on a large screw located at the connector end of the yoke. A locking thumbscrew is provided to secure the assembly in place.

(A) Normal yoke position.

(B) Assembly pivoted out.

(C) Faceplate coil removed.

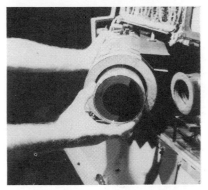

Fig. 4-10. Yoke assembly for 4½-inch I.O. in RCA TK-42 color camera.

NOTE: Older monochrome cameras require the lens turret to be removed to replace the I.O. The yoke assembly is fixed on sliding rails to allow optical focusing by means of the focus handle on the camera head.

A temperature-control plenum with three heating elements is mounted on one side and end of the yoke assembly but is not a part of the yoke-coil assembly. The temperature-control plenum is automatically connected to or disconnected from the plenum of an air-circulating blower when the yoke assembly is pivoted in or out of the operating position.

Adjusting screws are provided on the yoke assembly for optical focusing and target-mask alignment. Note that the I.O. yoke in a color camera, once properly positioned for correct optical focus at the correct image size, is fixed. Optical focusing during normal operation is achieved by means of controls that operate the zoom-lens assembly.

The yoke assembly of Fig. 4-10 performs these functions:

1. Horizontal deflection
2. Vertical deflection

3. Beam alignment
4. Focusing
5. I.O. temperature control

Horizontal- and vertical-deflection coils produce transverse magnetic fields that deflect the I.O. beam in a scan pattern determined by external deflection circuits. Alignment coils produce magnetic fields to effect alignment of the beam. A focus coil provides a magnetic field for the image orthicon. A temperature-control system circulates warm or cool air through the yoke assembly as required to maintain the I.O. temperature at approximately 37° C.

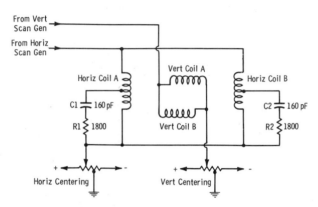

Fig. 4-11. Schematic diagram of monochrome deflection yoke.

Two center-tapped, parallel-connected coils are employed in the yoke assembly for horizontal deflection (Fig. 4-11). The center tap and one side of each coil are connected to RC damping networks (C1-R1 and C2-R2) on a terminal board located within the coil-assembly shield. The coils are driven by the output transformer of the horizontal-deflection circuit. Connections are made through pins of the yoke connector.

Two parallel-connected coils are employed in the yoke assembly for vertical deflection (Fig. 4-11). These coils are driven by the output transformer of the vertical-deflection circuit. As with the horizontal-deflection coils, connections are made through pins of the yoke connector.

Centering of the sweep normally is obtained by passing dc of the required polarity through the coils, as indicated in Fig. 4-11.

Fig. 4-12 shows the basic approach to achieving identical scanning rasters in a three-tube color camera (all pickup tubes alike). Deflection coils are driven in parallel from a common source, as shown in Fig. 4-12 for vertical deflection.

NOTE: Horizontal- and vertical-deflection amplifiers are covered in Chapter 7. We are concerned at this point only with the yoke assembly.

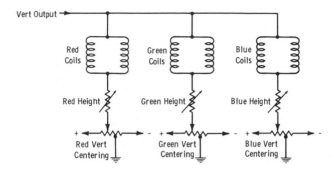

Fig. 4-12. Vertical-deflection coils in color camera.

When different kinds of tubes are employed in a color camera (for example, an I.O. for luminance and vidicons for chrominance), different deflection amplitudes must be obtained. Fig. 4-13 shows a typical arrangement for vertical-deflection circuits. Master vertical size and linearity adjustments affect all four channels. These generally are set for the monochrome tube. Deflection for the color channels then is obtained through a waveshape and attenuation network.

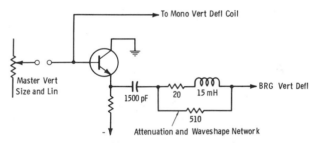

Fig. 4-13. Vertical-deflection circuits for different tube types.

Fig. 4-14 shows in simplified form a typical arrangement for a four-tube camera. Note from this configuration that the green size control becomes the master for all chroma tubes, with individual controls provided for red and blue. Linearity controls for red and blue are variable resistors in series with the deflection coils. Remember that all of this is for the purpose of most readily obtaining registration.

To get satisfactory registration of all three images, a characteristic known as *skew*[4] must be controlled. Compensation for this effect is provided by cross-mixing a small amount of vertical sawtooth into the horizontal sawtooth current, as shown in Fig. 4-14. In a four-tube camera, skew controls

---

[4]Harold E. Ennes, *Television Broadcasting: Equipment, Systems, and Operating Fundamentals* (Indianapolis: Howard W. Sams & Co., Inc., 1971) p 202.

normally are provided only in the three color channels to match the luminance-tube yoke as a standard. In three-tube cameras, the green yoke is taken as standard, and only red and blue skew controls are provided. Some cameras do not use electronic circuitry for skew correction. A mechanical adjustment is provided in the yoke assembly to permit rotation of the horizontal coil relative to the vertical coil.

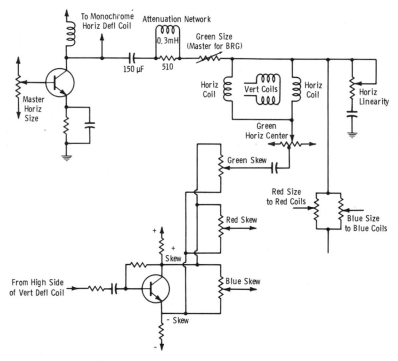

Fig. 4-14. Horizontal-deflection circuits for four-tube camera.

The alignment-coil arrangement for older I.O. cameras was as shown in Fig. 4-4. Only one vertical- and one horizontal-alignment coil was used, and sometimes it was necessary to rotate the alignment-coil assembly in addition to adjusting the current in the coils. In more recent camera designs, two series-connected horizontal-alignment coils and two series-connected vertical-alignment coils are employed (Fig. 4-15). The coils are potted in epoxy and mounted on the inner mandrel of the yoke behind the deflection-coil assembly. The alignment coils are connected in series with the I.O. focus coils.

The focus-coil assembly consists of series-connected, *pi*-wound coils on a mandrel in the yoke assembly. The front end of the focus-coil winding is brought out to a protruding pin on the inside of the yoke mandrel. A second pin, located 120° from the first pin, is connected to the focus-current source through the yoke connector. The rear end of the focus-coil winding is connected to the focus-current source through the same connector. A third protruding pin on the inside of the yoke mandrel has no electrical connection. This pin and the two current-carrying pins serve to position and retain an I.O. faceplate coil.

The faceplate coil fits inside the yoke mandrel and is equipped with contacts that engage the protruding pins when the coil is properly positioned and then rotated counterclockwise. The faceplate coil is thereby connected in series with the focus-coil winding and must be in place to complete the focus-current circuit.

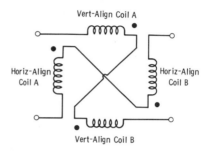

Fig. 4-15. Series-connected alignment coils.

Focus coils for electromagnetically focused tubes are in series across a common regulated focus-current supply so that identical focus fields are obtained. In four-tube color cameras employing three vidicons in the color channels, the vidicons normally are electrostatically focused, eliminating the focus coils. In this case, the focus-current supply is used only for the luminance tube. Otherwise, two supplies would be required because of the different current requirements and the need for individual current adjustment.

4-3. YOKE MAINTENANCE

The deflection yoke and focusing coil used with the I.O. must incorporate means for preventing the magnetic field produced by the yoke from extending into the image section of the tube. Unless proper shielding is provided, cross talk from the yoke into the image section will cause the electron image to "jitter." This jitter produces a loss of picture sharpness. It is common practice to enclose the focusing coil in a cylindrical magnetic shield. Additional shielding can be provided by fitting the inside portion of the focusing coil that is directly over the image section of the I.O. with a copper cylinder having a length of approximately 2¼ inches and a wall thickness of 1/32 inch. If these camera-design considerations are followed, the optical focus, image size, and centering characteristics will be uniform from tube to tube.

Modern yoke assemblies are so designed that very little trouble occurs in this section of the camera head. This is not true of older monochrome cameras, many of which are still in use. The type of distortion produced by yoke problems is termed *geometric distortion*. The five general types of geometric distortion are illustrated in Fig. 4-16.

Fig. 4-16A shows *S distortion*. You are probably familiar with this type of distortion in conjunction with the grid-6 (image accelerator) voltage control on the camera. You pan the camera along horizontal lines and observe the departure from straight lines in the reproduction. This departure is the result of a nonuniform axial field in the pickup tube; such a field causes nonuniform rotation to the electrons in the scanning pattern. In practice, S distortion results from improper adjustment of pickup-tube potentials, from stray magnetic fields, or from a magnetized yoke.

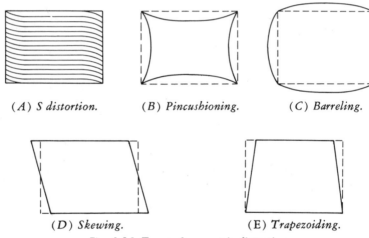

(A) *S distortion.*   (B) *Pincushioning.*   (C) *Barreling.*

(D) *Skewing.*   (E) *Trapezoiding.*

Fig. 4-16. Types of geometric distortion.

Pincushioning (Fig. 4-16B) normally results from improper distribution of windings in a picture-tube (monitor) deflection yoke. It is quite common in low-cost yokes, or it may result from an attempt to substitute a different yoke than that intended for the particular kinescope. *Barreling* (Fig. 4-16C) may result from the same causes.

Skewing (Fig. 4-16D), can occur in either the pickup tube or the monitor kinescope. It results when the horizontal- and vertical-deflection coils are not perpendicular to each other. Color cameras employ skew controls (either mechanical or electrical). In a monochrome camera, skew can be caused by a magnetized yoke, or stray magnetic fields.

Trapezoiding (Fig. 4-16E) can be introduced into either the pickup tube or the display monitor. It occurs when one set of deflection coils is not symmetrically placed with respect to the axis of the other; the axes of the horizontal- and vertical-deflection coils should effectively bisect each other. Trapezoiding also can be caused by a defective capacitor or resistor network used as built-in compensation for the difference in effective capacitance of the two sides of the coil.

If you encounter a bothersome degree of geometric distortion such as that in Fig. 4-16A, 4-16D, or 4-16E, you may have a magnetized yoke. First, note whether the type or shape of distortion changes with change of location of the camera or monitor. If it *does* change with location, you have a stray magnetic field. If it *does not* change with location, you have a magnetized yoke, or something in the camera or monitor is strongly magnetized.

1. Disconnect the *focus coil* (*not* the deflection coil) leads from the terminal board. If no terminal board is used for the focus-coil leads, simply disconnect the camera cable, and locate where these leads go on the camera-cable receptacle.

2. Attach the output of a variable autotransformer (with switch off) across the focus-coil leads.

3. Set the autotransformer arm on 115 volts, and plug the cord into an outlet.

4. Turn the variable autotransformer on. Reduce the voltage to zero in about 5 seconds.

5. Turn the autotransformer off, remove the autotransformer leads, and restore the focus coil to normal operation. The yoke should be demagnetized.

If the above method does not work, it will be necessary to use the longer procedure of removing the entire yoke from the surrounding shield, and demagnetizing it with a degaussing coil of the type used for color picture tubes and receivers. Use the same coil on the entire camera. The small hand-type degaussers used for magnetic audio-recording heads are not effective in this procedure.

4-4. SPECIAL NOTES ON THE 3-INCH FIELD-MESH I.O.

A problem that may be encountered is changing over from the non-field-mesh I.O. (such as the 5820) to a field-mesh tube (such as the 7293). For the field-mesh tube, *never have the lens capped when you first apply dc voltages to the camera.* The lens should be uncapped with some light falling on the photocathode, and the beam control should be turned up so that there is some beam current. This prevents formation of static charges between the field mesh and the target; such charges can cause "sagging" of elements and consequent damage or shortened tube life.

The field-mesh tube is more critical in beam alignment than is the non-field-mesh type. It usually will operate at top performance only at one particular mode of focus. For the 3-inch field-mesh tube, the grid-4 voltage normally is in the range of 140 to 170 volts. Now look at Fig. 4-17, which is the circuit associated with the ORTH FOCUS control in the RCA TK-11 camera. This particular circuit has maximum voltage occuring at maximum counterclockwise rotation of the control. As the control is turned clockwise, the voltage on grid 4 is reduced; the lowest voltage obtained is about 130 volts. Thus, in this particular camera, the first mode encountered when starting from the maximum-clockwise position of the ORTH FOCUS control is usually the optimum mode of focus. On any other mode of focus, there may be a coarse-mesh background in picture low lights, and noticeable with the lens capped.

The field-mesh I.O. has no "white dynode spot" by which beam alignment may be judged. Therefore, the tube can be aligned with the same procedure as that used for the vidicon. The alignment current is adjusted so that the center of the picture does not change position as the grid-4 voltage is varied.

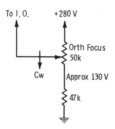

Fig. 4-17. Circuit for orth-focus control in TK-11 camera.

A much quicker and more accurate method of beam alignment for these tubes is the wobbulator method. If you feed a synchronized 30-Hz square wave to grid 4, beam misalignment will result in a split image of the entire pattern. It is then only necessary to adjust the alignment controls to converge the pattern into one well-defined image. Older cameras did not provide this facility, but it is particularly helpful in aligning the field-mesh I.O. (Section 4-1).

Fig. 4-18A is the schematic of a practical device for this purpose. It has only two tubes, since you can "borrow" +280 volts from a regulated supply in your existing installation. The 30-Hz square-wave output can be looped through all camera control units as shown in Fig. 4-18B; thus only one wobbulator is capable of handling all cameras. Note that this line should not be terminated. This system requires adding the necessary loop-through coaxial connectors to each control unit, and the addition of a switch as shown.

The reader may have built other wobbulator circuits that have appeared in various publications in the past, and found they did not work. The trouble may not have been in the wobbulator, but simply a result of the ignoring of other aspects. For example, see Fig. 4-18C. If your camera is similar to the RCA TK-11, the camera focus circuit has a resistor and by-

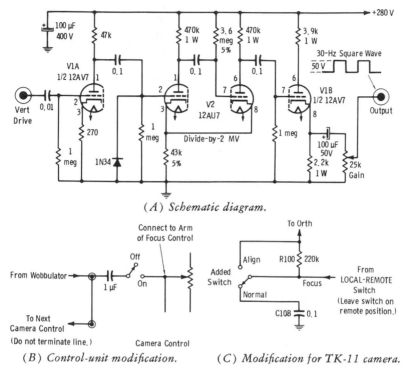

(A) Schematic diagram.

(B) Control-unit modification.   (C) Modification for TK-11 camera.

Fig. 4-18. Tube-type wobbulator circuit.

pass capacitor incorporated as shown. As a result, very little of the wobbulator signal is transferred to the I.O. Fig. 4-18C shows the modification required for use of the wobbulator signal.

In practice, the switches in both the camera control and camera are placed in the align-on position, and the beam is quickly aligned. (Remember to have grid 4 on the optimum mode of focus.) Then both switches are thrown to the off, or normal, position.

4-5. THE IMAGE-ORTHICON COOLING SYSTEM

Fig. 4-19 illustrates the cooling system of the Visual 3-inch I.O. zoom camera. This automatic temperature-control system is designed for fast warm-up of the image orthicon and accurate control of its temperature as well as the inside temperature of the camera. Fig. 4-19 shows a side view of the camera with the lens on the left separated by a plenum chamber from the image section of the I.O. tube. The zoom lens has a relatively long back focal length, so there is room in front of the I.O. photocathode for this chamber, which contains the light filter wheel, heating elements, and the outside-air intake.

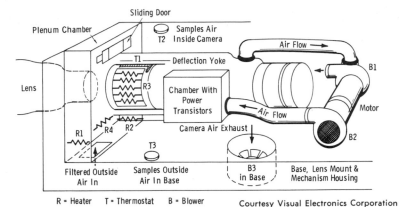

R = Heater   T = Thermostat   B = Blower   Courtesy Visual Electronics Corporation

Fig. 4-19. Camera temperature-control system.

A mercury-contact thermometer is located inside the focus coil, adjacent to the image section, with its bulb near the target. The thermometer has two contacts; one operates at 86° F, and the other operates at 104° F. When the camera is first turned on, three heaters are turned on for fast warm-up; these are R3 surrounding the image section, and R1 and R2 in the plenum chamber. Until the thermometer reaches 86° F, the I.O. beam is held off by control of grid 2. When the temperature reaches 104° F, the second contact in the thermometer turns off the three heaters (R1, R2, and R3). Air is constantly drawn in through the filters in the bottom of the camera, through the plenum chamber, through the deflection yoke, and into the camera interior by blower B1. The motor for this blower is on at all times. The same motor also drives blower B2, which forces air through a chamber surrounding the power transistors. Thermostat T2 samples the air temperature at the top of the camera, and at 86° F it turns on blower B3, located in the camera base, which exhausts air from the camera.

Thermostat T3 samples the outside air, and if the ambient temperature is below 14° F T3 turns on an additional heater, R4, in the intake plenum chamber for extra fast warm-up under very cold field conditions. Also, a door is provided between the intake plenum chamber and the camera interior; this door can be opened to recirculate camera air to the I.O. tube for extra cold field conditions.

4-6. THE VIDICON

There are several basic differences in types of vidicons,[5] as follows: Physical size:

---

[5]For additional information on vidicon tubes, see Harold E. Ennes, *Television Broadcasting: Equipment, Systems, and Operating Fundamentals* (Indianapolis: Howard W. Sams & Co., Inc., 1971).

1. 1-inch diameter
2. 1½-inch diameter. The larger tube diameter provides more TV lines per picture height (resolution).

Focus and deflection method:

1. Magnetic focus and magnetic deflection. This method normally provides highest center and corner resolution but requires the most electrical power and results in cameras having the most weight.
2. Electrostatic focus and magnetic deflection. This type of tube has intermediate resolution capability and requires intermediate electrical power. It finds use in compact, lightweight cameras and in four-tube color cameras where a larger tube (such as a 1½-inch vidicon or large image orthicon) is used for luminance information.
3. Electrostatic focus and electrostatic deflection. This type of tube requires the least electrical power and permits a considerable reduction in necessary camera size and weight. Its resolution capability usually is less than for other types of vidicons when operating voltages are the same. Normally, it is found only in portable and mobile TV cameras.

Conventional mesh or separate mesh:

1. The conventional-mesh type has the mesh (grid 5) and wall (grid 4) electrodes internally connected.
2. In the separate-mesh type, wall and mesh electrodes are not internally connected and are brought out to separate pins on the tube base.

Grid 3 and grid 4 connections:

1. Grid 3 and grid 4 internally connected
2. Grid 3 and grid 4 brought to separate terminals

Figs. 4-20, 4-22, and 4-23 illustrate the basic differences in vidicons. The structural arrangement of the vidicon shown in Fig. 4-20A consists of: a target composed of a transparent conducting film (the signal electrode) on the inner surface of the faceplate and a thin photoconductive layer deposited on the film; a fine-mesh screen (grid 4) located adjacent to the photoconductive layer; a beam-focusing electrode (grid 3) connected to grid 4; and an electron gun.

Each element of the photoconductive layer is an insulator in the dark but becomes slightly conductive when it is illuminated; it then acts like a leaky capacitor having one plate at the positive potential of the signal electrode and the other floating. When light from the scene being televised is focused on the photoconductive-layer surface next to the faceplate, each illuminated layer element conducts slightly, depending on the amount of illumination on the element. The potential of the opposite surface (on the gun side) is then caused to rise in less than the time of one frame

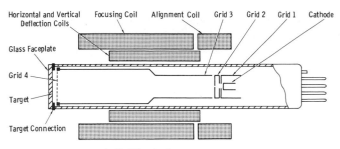

(A) Physical arrangement.

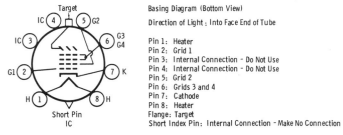

Basing Diagram (Bottom View)

Direction of Light : Into Face End of Tube

Pin 1: Heater
Pin 2: Grid 1
Pin 3: Internal Connection – Do Not Use
Pin 4: Internal Connection – Do Not Use
Pin 5: Grid 2
Pin 6: Grids 3 and 4
Pin 7: Cathode
Pin 8: Heater
Flange: Target
Short Index Pin: Internal Connection – Make No Connection

(B) Basing diagram.

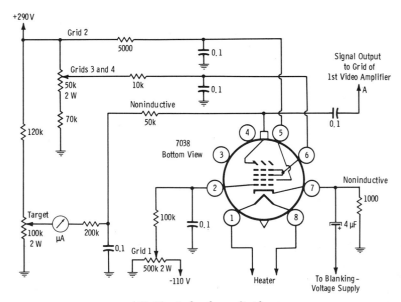

(C) Typical voltage dividers.

Courtesy RCA

Fig. 4-20. Vidicon with grids 3 and 4 connected internally.

toward the signal-electrode potential. Hence, there appears on the gun side of the entire layer surface a positive-potential pattern, composed of the various element potentials, corresponding to the pattern of light that falls on the layer.

The gun side of the photoconductive layer is scanned by a low-velocity electron beam produced by the electron gun. This gun contains a thermionic cathode, a control grid (grid 1), and an accelerating grid (grid 2). The beam is focused at the surface of the photoconductive layer by the combined action of the uniform magnetic field of an external coil and the electrostatic field of grid 3. Grid 4 serves to provide a uniform decelerating field between itself and the photoconductive layer so that the electron beam tends to approach the layer in a direction perpendicular to it—a condition necessary for driving the surface to cathode potential. The beam electrons approach the layer at low velocity because of the low operating potential of the signal electrode.

When the gun side of the photoconductive layer with its positive-potential pattern is scanned by the electron beam, electrons are deposited from the beam until the surface potential is reduced to that of the cathode; thereafter, the electrons are turned back to form a return beam, which is not utilized. Deposition of electrons on the scanned surface of any particular element of the layer causes a change in the difference of potential between the two surfaces of the element. When the two surfaces of the element, which in effect is a charged capacitor, are connected through the external target (signal-electrode) circuit and the scanning beam, a capacitive current is produced and constitutes the video signal. The magnitude of the current is proportional to the surface potential of the element being scanned and to the rate of scan. The video-signal current is used to develop a signal-output voltage across a load resistor. The signal polarity is such that for high lights in the image, the grid of the first video-amplifier tube swings in a negative direction.

As with the image orthicon, alignment of the beam is accomplished by a transverse magnetic field produced by external coils located at the base end of the focusing coil. Deflection of the beam is accomplished by transverse magnetic fields produced by external deflecting coils.

The basing diagram for this type of vidicon appears in Fig. 4-20B. The diagram shown is a bottom view.

The focusing-electrode (grid 3) voltage may be fixed at a value of about 280 volts when focusing control is obtained by adjusting the current through the focusing coil. In general, resolution decreases with decreasing grid-3 voltage. The necessary range of current adjustment depends on the design of the coil, but should be such as to provide a field-strength range of 36 to 44 gausses. When a fixed value of focusing-coil current capable of providing a fixed strength of 40 gausses at the center of the focusing device is used, the grid-3 voltage is made adjustable over a range between 250 and 300 volts.

Definition, focus uniformity, and picture quality decrease with decreasing grid-4 and grid-3 voltage. In general, grid 4 and grid 3 should be operated above 250 volts. The grid-1 supply voltage should be adjustable from 0 to −110 volts. The dc voltages required by the vidicon can be provided by the circuit shown in Fig. 4-20C. The Type 7038 vidicon is a typical tube of the kind being discussed.

A blanking signal is applied to grid 1 or to the cathode to prevent the electron beam from striking the photoconductive layer during the return portions of the horizontal- and vertical-deflecting cycles. Unless this is done, the camera-tube return lines will appear in the reproduced picture. The blanking signal is a series of negative voltage pulses applied to grid 1, or a series of positive voltage pulses applied to the cathode.

Beam intensity is controlled by the amount of negative voltage on grid 1. The beam must have adequate intensity to drive the high-light elements of the surface of the photoconductive layer to cathode potential on each scan. When this does not occur, and only the low-light elements are driven to cathode potential, the picture high lights all have the same brightness and show no detail. Also, when the beam has insufficient intensity, the photo-conductive-layer surface that normally rises in potential by only a small fraction of the signal-electrode potential during each scan, gradually rises to a value approaching the full signal-electrode potential in the high lights. Under this condition, many scans are required to drive to cathode potential any element that has changed from a high light to a low light because of movement of the image. As a result, the high lights tend to "stick." The loss of high-light detail and sticking of the high lights is referred to as "bloom." At the other extreme, a beam with excessively high intensity should not be used because the size of the scanning spot increases with resultant decrease in resolution.

Uniform signal output over the scanned area can be obtained if the vidicon is operated with a deflecting-yoke and focusing-coil system designed so that no beam-landing errors are produced in the vidicon. If the tube is to be utilized with focusing and deflecting systems that introduce such errors (as was common in older cameras, many of which are still in use), uniform sensitivity over the scanned area can be achieved by compensating for the beam-landing errors.

Sensitivity variations resulting from beam-landing errors are in the form of lower signal from the edges of the scanned area than from the center. However, because of the uniformity of the photoconductive layer, these variations in sensitivity are the same from tube to tube. Compensation for the beam-landing errors to achieve uniform sensitivity can be obtained by supplying a modulating voltage of a suitable waveform to the cathode. The desired waveform is parabolic in shape and of such polarity that the cathode voltage is lowered as the beam approaches the edges of the scanned area. The modulating waveform contains parabolic components of both the horizontal- and vertical-scanning frequencies. The horizontal component should have the greater amplitude and is the most effective in obtaining uniform sensitivity.

Fig. 4-21 shows the amount of parabolic-waveform voltage required and the method of applying the waveform to the cathode, grid 1, and grid 2 of a vidicon. The modulating voltage is applied to grid 1 and grid 2 as well as to the cathode to prevent modulation of the scanning beam.

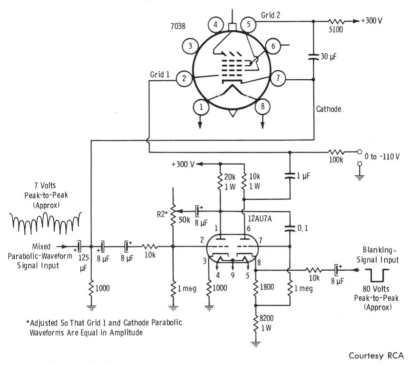

Courtesy RCA

Fig. 4-21. Circuit for compensation for beam-landing errors.

The use of this modulating waveform also improves the center-to-edge focus of the vidicon and assures that sensitivity over the scanned area will be uniform for the recommended dark current for any specified service. Care must be taken that identical waveforms are applied to the electrodes of the three tubes in a three-vidicon color camera to insure good registration of all signals over the entire scanned area.

Fig. 4-22A is a diagram of a vidicon (such as the Type 6326) with separate grids 3 and 4. Note that although the screen mesh adjacent to the photoconductive layer is connected to grid 4, it is given the separate designation of grid 5. Grid 4 is the beam-focusing electrode, and grid 3 can be used as a vernier focusing electrode. Fig. 4-22B illustrates the base-pin designations for this type of tube.

Fig. 4-22C shows a typical voltage-divider arrangement for this type of vidicon when grid 3 is externally connected to grids 4 and 5. Fig. 4-22D illustrates how grid 3 can be individually adjusted for optimum electrical focus of the image.

Fig. 4-23 illustrates a more recent vidicon of the electrostatic-focus, magnetic-deflection type, with a separate-mesh arrangement. Grid 5 is the additional electrostatic focus grid. Note that grid 5 is brought to a separate base pin, and that the screen adjacent to the photo-conductive layer is now designated grid 6.

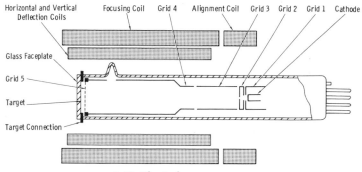

(A) Physical arrangement.

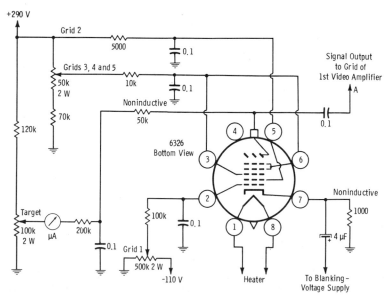

(C) Grid 3 connected to grids 4 and 5.

Fig. 4-22. Vidicon with

The weight, size, and power requirements of TV cameras employing this tube are substantially less than the requirements of cameras using conventional magnetic-focus, magnetic-deflection vidicons of comparable size (1-inch diameter). Camera size and weight are automatically reduced by elimination of the magnetic focusing coil. Negligible power is required for electrostatic focusing. Deflection power is one-fourth of that required for vidicons using magnetic focusing and deflection, having a separate mesh connection, and operating at equivalent mesh potential; it is one-sixth of that required when the mesh and wall electrodes are connected

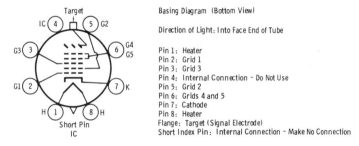

Basing Diagram (Bottom View)

Direction of Light: Into Face End of Tube

Pin 1: Heater
Pin 2: Grid 1
Pin 3: Grid 3
Pin 4: Internal Connection – Do Not Use
Pin 5: Grid 2
Pin 6: Grids 4 and 5
Pin 7: Cathode
Pin 8: Heater
Flange: Target (Signal Electrode)
Short Index Pin: Internal Connection – Make No Connection

(B) Basing diagram.

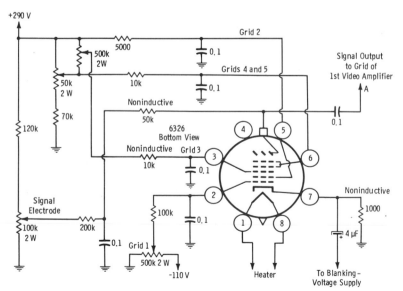

(D) Grid 3 as vernier focus electrode.

Courtesy RCA

separate grids 3 and 4.

internally or operated at the same potential. In addition, the precision outer diameter of the bulb permits the use of low-power, close-fitting deflection yokes of small size and low impedance.

This type of vidicon can be operated with "low" electrode potentials, as described for Figs. 4-20 and 4-22, or at "high" potentials, as illustrated in Fig. 4-23A. Note the greatly improved resolution at higher voltages shown by Fig. 4-23C. This is the uncompensated response. With aperture compensation, excellent resolving power at 800 TV lines can be obtained.

Control over beam landing is obtained by adjustment of the voltages on grids 6 and 3 and grid 5. This voltage relationship is determined by the

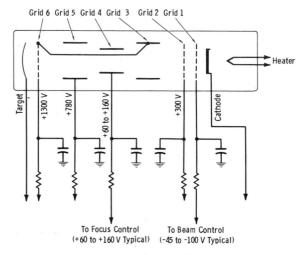

(A) Typical one-inch tube.

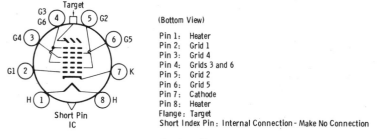

(B) Basing diagram.

Fig. 4-23. Vidicon with

camera designer and is not recommended as an operational control. In general, best geometrical accuracy is obtained with a ratio of grid 6 and 3 voltage to grid 5 voltage of 1.67, and most uniform signal output is obtained with a ratio of 2.

Definition, focus uniformity, and picture quality decrease with decreasing voltages on grids 6 and 3 and grid 5. In general, grids 6 and 3 should be operated at or above 300 volts, and grid 5 should be operated at or above 180 volts. A substantial increase in both limiting resolution and amplitude response is obtained by operating the tube with higher voltages applied to these electrodes. It should be noted that deflection-current re-

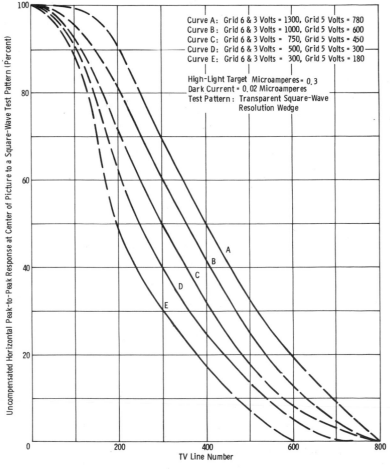

(C) Resolution curves.

Courtesy RCA

additional focusing grid.

quirements of the tube on grids 6 and 3 are increased when the voltages are increased.

A composite of manufacturers' instructions concerning operation of the vidicon in studio and film cameras follows. We will explore advanced maintenance techniques for the vidicon further in Chapter 11. The target connection is made by a suitable spring contact that bears against the edge of the metal ring at the face end of the tube. This spring contact may be provided as part of the focusing coil.

The deflection yoke and focusing coil used with the vidicon are designed to cause the scanning beam to land perpendicularly to the target at all points of the scanned area. This is done to obtain minimum beam-landing error and resultant uniformity of sensitivity and focus over the scanned area. The recommended location of these components for monochrome cameras is shown in Fig. 4-20A.

The deflection yoke and focusing coil in color cameras should extend $\frac{1}{4}$ to $\frac{1}{2}$ inch beyond the faceplate of the tube, as shown in Fig. 4-22A. The yoke must have a minimum inside diameter of $1\frac{5}{8}$ inch to provide clearance for the side tip. A long yoke, in comparison with a short yoke, deflects the beam through a narrower angle, which effectively gives better center-to-edge focus and reduces geometric distortion of the image. Freedom from such distortion is particularly important in color cameras utilizing the method of simultaneous pickup, in which three images must be identical for proper registration.

The polarity of the focusing coil is such that a north-seeking pole is attracted to the image end of the focusing coil, with the indicator located outside of and at the image end of the focusing coil.

The scanning speed must be constant in order to obtain good black-level reproduction when the vidicon is operated at high dark current with resultant higher effective sensitivity. (Dark current is the current with the lens capped and target voltage applied.) The dark-current signal is proportional to the scanning speed. Therefore, any change in scanning speed will produce a nonuniformity in black level in direct proportion to the change in scanning speed.

NOTE: A constant scanning speed means that the current through the deflection coils describes a pure sawtooth waveform. Any departure results in picture nonlinearity. This emphasizes the importance of proper sweep-linearity adjustments in the operation of vidicon cameras.

The alignment coil should be located on the tube so that its center is at a distance of $3\frac{11}{16}$ inches from the face of the tube. This coil should be positioned so that its axis is coincident with the axis of the tube, the deflection yoke, and the focusing coil.

Electrostatic shielding of the target from external fields is required to prevent interference effects in the picture. Effective shielding from the fields produced by the deflection components ordinarily is provided by grounding a shield on the inside of the faceplate end of the focusing coil and by grounding a shield on the inside of the deflection yoke at a point near the input of the video amplifier.

The temperature of the faceplate should not exceed 60°C (140°F) during either operation or storage of the vidicon. Operation with a faceplate temperature in the range from about 25 to 35°C (77 to 95°F) is recommended. The temperature of the faceplate is determined by the combined heating effects of the incident illumination on the faceplate, the associated components, and the tube itself. To reduce these heating effects in film-pickup cameras and permit operation in the preferred temperature range with a high value of illumination, it is normal to use an infrared filter between the projector and faceplate and to provide a blast of cooling air across the faceplate.

The dark current is doubled for every 10°C rise in the temperature of the faceplate, and halved for every 10°C decrease in the temperature of the faceplate. To obtain optimum performance, it is desirable to operate the vidicon at a pre-established value of dark current. Therefore, if the temperature of the faceplate is allowed to vary, it will be necessary to adjust the target voltage to maintain the desired dark current, as shown in Fig. 4-24. Since the sensitivity of the tube decreases with increasing tempera-

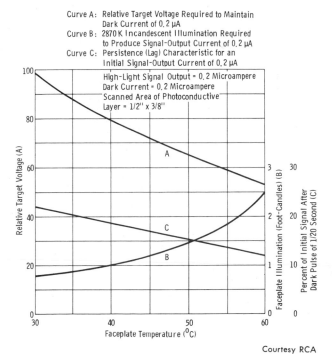

Curve A: Relative Target Voltage Required to Maintain Dark Current of 0.2 μA
Curve B: 2870 K Incandescent Illumination Required to Produce Signal-Output Current of 0.2 μA
Curve C: Persistence (Lag) Characteristic for an Initial Signal-Output Current of 0.2 μA

High-Light Signal Output = 0.2 Microampere
Dark Current = 0.2 Microampere
Scanned Area of Photoconductive Layer = 1/2" x 3/8"

Fig. 4-24. Characteristics of typical vidicon.

ture, the amount of faceplate illumination necessary to produce a given signal as a function of faceplate temperature also is shown in Fig. 4-24. In addition, the lag decreases with increasing temperature, as shown by curve C in Fig. 4-24. For live pickup, it is desirable to select an operating temperature that provides the best balance between lag and sensitivity. The faceplate should be held close to this temperature to assure stability of black level and signal-output level.

The target voltage is obtained from an adjustable dc source. As the target voltage is increased, the dark current increases (Fig. 4-25). The target voltage must be adjusted to produce the desired value of dark current for the type of operation. The target-voltage range of the vidicon for a given value of dark current is small, as shown in Fig. 4-25. This feature permits utilization of simplified circuits in cameras where automatic change in target voltage is desired to compensate for varying light levels.

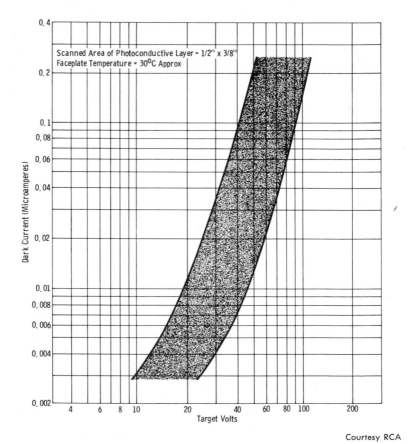

Courtesy RCA

Fig. 4-25. Range of dark current for typical vidicon.

The illumination incident on the faceplate ranges from relatively high values for film pickup to relatively low values for direct pickup. For satisfactory operation of the vidicon at these extremely different light levels, it is essential that the target voltage be properly adjusted with reference to the curves in Figs. 4-26A, 4-26B, and 4-26C to give the proper value of dark current for the desired service. (Adjustment of the target voltage to obtain the desired dark current is covered later.)

For practical purposes, the illumination on the tube faceplate can be calculated from the following relationship:

$$E = \frac{E_s R T}{4f^2}$$

where,

E is the tube-face illumination in foot-candles,
E_s is the scene illumination in foot-candles,
R is the reflectance of the scene,
T is the transmission of the lens,
f is the f-stop number of the lens.

Assuming a lens transmission of 80 percent and a scenic high-light reflectance of 60 percent, with a scene illumination of 300 foot-candles and the lens stopped to $f/2.8$:

$$E = \frac{(300)(0.6)(0.8)}{4\ (2.8)^2}$$

$$= \frac{144}{31.4}$$

$$= 4.6 \text{ foot-candles on tube faceplate}$$
$$\text{for high lights}$$

For live pickup involving low illumination levels, a good picture can be obtained with a high-light illumination of 1 to 3 foot-candles on the faceplate of the vidicon. Such a low illumination level, however, requires maximum-sensitivity operation of the tube. For this type of operation, a dark current of 0.2 microampere is required. This value is obtained with a target voltage in the range of 60 to 100 volts. Under such low-level illumination conditions, the lag will be greater and the black-level uniformity will be poorer than for live-pickup conditions with higher faceplate illumination and lower dark current.

When the vidicon is used for live pickup with illumination levels of 10 to 20 foot-candles on the faceplate, a dark current of 0.02 microampere is required. This value is obtained with a target voltage in the range of 30 to 50 volts.

For film pickup, an average high-light illumination of 50 to 200 foot-candles is required on the faceplate of the average vidicon for minimum lag and best black-level uniformity. For this range of illumination, a dark

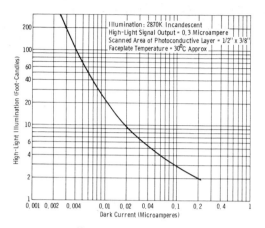

(A) *Illumination vs dark current.*

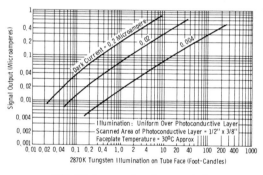

(B) *Light transfer characteristic.*

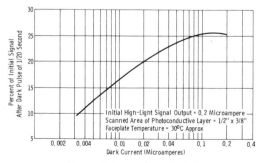

(C) *Persistence characteristic.*

Courtesy RCA

Fig. 4-26. Typical vidicon characteristics.

current of about 0.004 microampere is required, and the target voltage is between 10 and 20 volts.

The exact value of target voltage to give the required dark current depends on the individual tube and on the temperature at which its faceplate is operated. It is important that the tube be allowed to reach a stable operating temperature before the operating dark current is determined; otherwise, the dark current will change as the temperature of the tube changes.

In all cases, the illumination level and/or dark current must be limited or adjusted so that the peak signal-output current does not exceed 0.3 microampere for the 1-inch tube, or 0.6 microampere for the 1½-inch vidicon. In order that the signal-output current and dark current will be known at all times, the camera sometimes is provided with a microammeter in the target circuit of each vidicon to read average target current, or a calibration pulse of the proper magnitude is fed into the input of the video preamplifier to indicate peak target currents.

The maximum amount of illumination on the photoconductive layer is limited primarily by the temperature of the faceplate, which should never exceed 60°C and should preferably be maintained within the operating range from 25°C to 35°C for most satisfactory performance.

Signal output as a function of uniform 2870K tungsten illumination on the photoconductive layer for different values of dark current is shown in Fig. 4-26B. Note that these curves are for a typical tube under the conditions indicated. Because the target voltage needed to give maximum sensitivity at a dark current of 0.2 microampere may range between 60 and 100 volts, it is essential that the best operating target voltage be determined for each vidicon. From these curves, it also should be noted that the illumination must be increased about 30 times to produce an increase of 10 times in signal-output current for any given value of dark current.

The average gamma, or slope, of the light-transfer characteristic curves in Fig. 4-26B is approximately 0.65. This value is relatively constant over an adjustment range of 4 to 1 in target voltage, or 50 to 1 in dark current, for a signal-output current range between 0.01 and 0.3 microampere. Close uniformity in the value of gamma between individual tubes is maintained to insure satisfactory operation of color cameras in which the signal-output currents of the three vidicons must match closely over a wide range of scene illumination. Because its transfer characteristic is approximately the complement of the transfer characteristic of a picture tube, the vidicon can produce a picture having proper tone rendition.

The spectral response of the typical vidicon is shown by curves A and C in Fig. 4-27. Curve A is on the basis of equal values of signal-output current at all wavelengths, whereas curve C is on the basis of equal values of signal-output current with radiant flux from a tungsten source at 2870K. For comparison purposes, the response of the eye is shown in curve B.

Full-size scanning of the ½″ × ⅜″ area of the photoconductive layer should always be used. This condition can be assured by first adjusting the deflection circuits to overscan the photoconductive layer so that the edges of the sensitive area can be seen on the monitor. Then, after centering the image on the sensitive area, reduce the scanning until the edges of the image just disappear. In this way, the maximum signal-to-noise ratio and maximum resolution can be obtained. It should be noted that overscanning the photoconductive layer produces a smaller-than-normal picture on the monitor.

Underscanning of the photoconductive layer, i.e., scanning an area of the layer less than $\frac{1}{2}'' \times \frac{3}{8}''$, should never be permitted. This condition (which produces a larger-than-normal picture on the monitor) not only causes sacrifice in signal-to-noise ratio and resolution, but also may cause permanent change in sensitivity and dark current of the underscanned area.

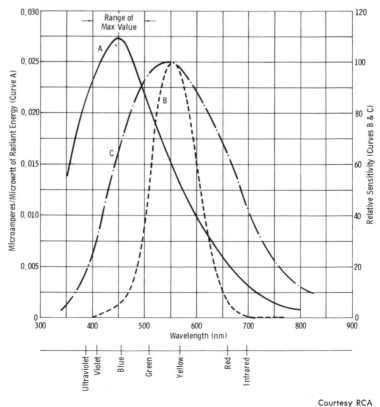

Curve A: For Equal Values of Signal-Output Current at All Wavelengths
Signal Output From Scanned Area of 1/2'' x 3/8'' = 0.02 Microampere
Dark Current = 0.02 Microampere
Curve B: Spectral Characteristics of Average Human Eye
Curve C: For Equal Values of Signal-Output Current With Radiant Flux From Tungsten Source at 2870K

Courtesy RCA

Fig. 4-27. Spectral sensitivity of typical vidicon.

An underscanned area showing such a change will be visible in the picture when full-size scanning is restored.

Failure of scanning even for a few seconds may permanently damage the photoconductive layer. The damaged area shows up as a spot or line in the picture during subsequent operation. To avoid damaging the vidicon during scanning failure, it is necessary to prevent the electron beam from reaching the layer. This can be accomplished conveniently by increasing the grid-1 voltage to cutoff.

The sequence of adjustments in operating the vidicon for live pickup is as follows: With the grid-1 voltage control set for maximum negative bias (beam cutoff), the target voltage control set for minimum voltage (about 10 volts), and the deflection controls set for maximum overscan, apply other voltages to the tube. Next, with a $\frac{1}{2}'' \times \frac{3}{8}''$ mask centered on the face of the tube, and with the iris set for minimum opening, decrease the grid-1 bias to just bring out the high-light details of the picture on the monitor. Adjust the beam-focus voltage control, the lens stop, and the optical focus to obtain the best picture. Reduce horizontal and vertical scanning so that the edges of the image extend just outside the scanned area on the monitor. Then adjust the alignment field so that the center of the picture does not move as the beam-focus voltage is varied. Some readjustment of horizontal and vertical centering may be necessary after alignment.

For maximum-sensitivity operation of the vidicon in live-pickup service, proceed as follows: With no illumination on the face of the tube, increase the target voltage until a dark current of 0.2 microampere is measured. The current should be measured with a sensitive microammeter, or by another procedure, as outlined in Chapter 11 of this book. Next, open the lens and adjust the aperture to give a peak signal-output current of 0.2 to 0.3 microampere. A good procedure for doing this is to focus the camera on a uniform white area having the same brightness as the high lights in the scene to be televised. The image of this white area must at least cover the scanned area of the tube face. The current read on the microammeter will be the dark current plus the peak signal-output current, i.e., high-light target current.

A waveform monitor can be used to compare the peak signal-output current produced by any scene to the peak value measured with the microammeter when the camera is focused on a uniformly bright scene. When a camera is adjusted in this manner, video gain should be kept constant, and the light level on the tube face should be controlled to maintain the constant predetermined value of peak signal as observed on the oscilloscope.

After the light level is adjusted to obtain the correct signal-output current, the grid-1 bias voltage should be adjusted to just discharge the high lights. Too much current will result in poor resolution and poor picture quality. After the grid-1 bias is adjusted properly, it will be necessary to check and readjust the dark current and the peak signal-output current. Proper adjustment of the dark current, the peak signal-output current, and the grid-1 bias will result in a picture of good quality with minimum smearing of moving objects.

For average-sensitivity operation of the vidicon in live-pickup service, the adjustments are similar to those for maximum-sensitivity operation, except that the target voltage should be adjusted to produce a dark current of 0.02 microampere. When sufficient light is available, decreased lag can be obtained by operating with this lower value of dark current.

For film-pickup operation of the vidicon, the adjustments are similar to those for live pickup, except that the target voltage should be adjusted to produce a dark current of 0.004 microampere, and the peak signal-output current should be adjusted to the desired value by controlling the light level on the faceplate of the tube.

In setting up three vidicons in a color camera, particular attention must be given to proper alignment, best obtainable focus, and identical centering of scanned areas on the photoconductive layers. For best color balance and color tracking over a wide range of light levels, the light level in each color channel should be controlled so that each vidicon develops the same value of peak signal output for white portions of a scene. Observation of these operating conditions should assure good registration and good color balance.

4-7. THE LEAD-OXIDE VIDICON

The lead-oxide vidicon (*Plumbicon*) is a photoconductive tube with higher sensitivity than the conventional vidicon, virtually zero dark current, and a gamma of close to unity. Being the most recently developed pickup tube, it is still more "experimental" than either the image orthicon or the vidicon described in the previous section. It is, however, rapidly becoming a widely used pickup tube in television cameras. The term *Plumbicon* is a registered trade mark of N. V. Philips of Holland. "Lead oxide vidicon" is the RCA terminology for the same type of pickup tube.

The lead-oxide tube is similar in electrode arrangement to the conventional vidicon, uses the same general type of focusing and deflection coils, and has similar grid characteristics. The major difference is in the properties and construction of the photoconductive layer.

The *Plumbicon* element consists of three layers. The middle layer is a relatively thick lead oxide (PbO) acting as an intrinsic (designated i) semiconductor. The outer and inner layers are very thin. The layer toward the electron gun is doped into a p-type semiconductor. The layer on the signal-electrode side is doped into an n-type semiconductor. Thus, the three layers form a p-i-n diode (positive-intrinsic-negative diode). This diode is connected in the "reverse" direction; the p-type material is toward the tube cathode, and the n-type material is biased by the positive signal-electrode potential.

Although the basic action of the *Plumbicon* is identical to that of the vidicon, the intrinsic region of the p-i-n structure contributes largely to the

Table 4-3. Comparison of *Plumbicons* and *Vidicons*

Tube Size	Typical Plumbicons			Typical Vidicons	
Diameter	30 mm (1.2")	1"	5/8"	1"	1½"
Length	8.7"	6.25"	5"	6.25"	10.25"
Scanned Area (Diagonal)	21 mm (0.8")	16 mm (0.62")	10 mm (0.4")	0.62"	1.0"
Luminous Sensitivity (μA/Lumen)	400	400	350	225 (At Dark Current = 20 nA)	300 (At Dark Current = 20 nA)
Resolution (Modulation Depth at 400 TV Lines, Percent)	50	30	15	50 (High-Voltage Operation)	60
Limiting Resolution (TV Lines For 10-Percent Min Depth)	800	600	400	800	1200
Output Capacitance (pF)	5.0	4.5	1.5	4.6	11.0
Gamma	0.95	0.95	0.95	0.65	0.65
Dark Current (nA)	0.8	0.5	1.0	20 (Target Volts = 30)	20 (Target Volts = 20 to 60)
Lag 3rd Field (Percent)	3.5	3.5	5.0	32*	20*
Lag 12th Field (Percent)	1.1	1.2	1.2	10*	5*

*Dependent on Tube Illumination and Target Voltage

relatively high sensitivity of the *Plumbicon*. In this region, the conductivity is low while the electrical field strength is high. Thus, all the liberated carriers in this region contribute to the photocurrent when the target potential is sufficiently high. A much higher ratio of signal current to dark current is obtained than is the case for the vidicon described previously. Table 4-3 compares typical characteristics of three available sizes of *Plumbicons* with typical characteristics of 1-inch and 1½-inch conventional vidicons.

Note that in Table 4-3 luminous sensitivity is given in microamperes per lumen ($\mu A/lm$). Sometimes this type of information in tube specification sheets is given in terms of microamperes per watt ($\mu A/W$). Usually the light source is specified as a tungsten light of 2870K color temperature. This light source emits 20.4 lumens per watt of total radiant flux. Thus, you can convert microamperes per lumen to microamperes per watt by using the 20.4 multiplication factor. For example, to convert 300 $\mu A/lm$ to $\mu A/W$:

$$300 \ \mu A/lm \times 20.4 = 6120 \ \mu A/W$$
$$= 6.12 \ mA/W$$

Conversely, you can convert microamperes per watt to microamperes per lumen with the relationship:

$$\mu A/lm = \frac{\mu A/W}{20.4}$$

4-8. CAMERA OPTICS

It is important that the reader already be familiar with the fundamentals of lenses, including lens angles, field of view and depth of field for a given lens and pickup-tube useful scanned area, the differences between the turret-mounted lens and the zoom lens, etc.[6] It now remains to examine the more advanced applications of camera optics, and servo control of lens-iris functions.

Fig. 4-28 illustrates the problem of bringing three images into optical focus at exactly the same time and with exactly the same orientation with respect to the scanning rasters. The lens in Fig. 4-28 may be either an objective lens or a field relay lens. It would be possible to mount a number of these lenses of different focal lengths on a turret, but the back focal distance must be rather large, limiting the installation to lenses that are of very large focal length and therefore are unsuitable for average studio pickups. So consider the lens to be a field relay lens. This means that an objective lens is ahead of the relay lens, or a film-projector lens is throwing a real image on the relay lens. In this case, each pickup tube would have its

---

[6]See, for example, Harold E. Ennes, *Television Broadcasting: Equipment, Systems, and Operating Fundamentals* (Indianapolis: Howard W. Sams & Co., Inc., 1971), Sections 4-1 and 4-2.

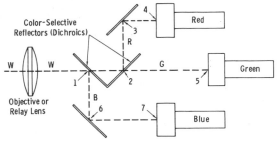

Fig. 4-28. Basic color-separating optical system.

own objective lens (normally termed *reimage* lens) focused on the real-image plane of the relay lens. This is the basic type of color-camera optical system for both studio and film applications.

Since the images on all three pickup-tube photocathodes must have identical size and must come into focus at one common adjustment, paths 1-6-7 (blue), 1-2-3-4 (red), and 1-2-5 (green) must be of identical length. This is the reason for the physical placement shown.

Dichroic mirrors divide the incoming light roughly into blue, red, and green components. The mirror at point 1 reflects blue and passes red and green. The mirror at point 2 reflects red and passes the remaining light, which is primarily green. Front-surface mirrors are located at points 3 and 6.

Fig. 4-29 shows typical transmission and reflection characteristics of a pair of dichroic mirrors. These curves reveal why it was stated that dichroic mirrors split the light "roughly" into the three primary color components. It is obvious that additional *trim filters* in the light paths following the dichroic mirrors are necessary. The characteristics of dichroic surfaces are always given for a fixed angle of incidence.

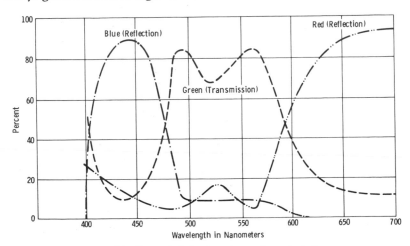

Fig. 4-29. Typical dichroic-mirror characteristics.

It is now time to examine the basic problems involved in color separation, and the reasons behind color optics. This will make it easier to understand which problems are under our control and which problems are not.
Conventional dichroic mirrors have the following drawbacks:

1. The dichroic interference layers are evaporated onto a plane-parallel
   glass plate. Certain aberrations (coma and astigmatism) are introduced and make compensating optical elements necessary.
2. Reflections from the rear surface of the glass plate result in a faint
   ghost image. This image is particularly noticeable under high-contrast
   conditions.

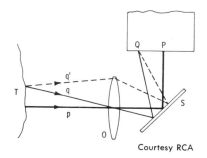

**Fig. 4-30. Influence of angle
of incidence.**

Courtesy RCA

3. Properties of the interference layer depend on the angle of incidence,
   as shown by Fig. 4-30. Rays p and q from two points of scene T meet
   dichroic mirror S at different angles. Since the spectral reflection
   characteristic depends on the angle of incidence, color rendition at
   points P and Q is different, resulting in a spurious shift of color across
   the image. Rays q and q', originating at the same point of the scene,
   likewise strike the mirror at different angles. Consequently, their con-
   tributions to Q are governed by slightly differing spectral reflection
   characteristics. The result is a loss of color discrimination.
4. In practice, light can become polarized by reflection from polished
   objects, specular surfaces, or even perspiring foreheads of persons in
   the scene. Red and blue reflecting surfaces of color splitters in the
   camera are sensitive to the plane of polarization, since they are at
   angles with the incoming light. (Green passes straight through and
   therefore is not affected by polarization). Fig. 4-31 shows the result
   of a red-reflecting dichroic mirror under polarized-light conditions.
   This shows in essence that the perpendicular component (dash line)
   reflects at a shorter wavelength than does the horizontal component
   (solid line).

NOTE: Both the color discrimination of Fig. 4-30 and the difference-of-
wavelength effect of Fig. 4-31 become worse with increasing angles of
incidence.

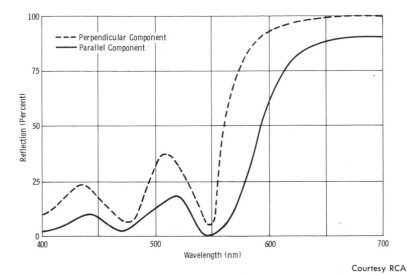

Courtesy RCA

Fig. 4-31. Influence of light polarization.

The foregoing drawbacks are minimized by prism-type optics (Fig. 4-32). Note that the pickup tubes in this example are not parallel to each other, but are "fanned." This arrangement minimizes the required angle of incidence of all optics. The front part of the prism optical assembly contains a ¼-wavelength plate, in diagonal position, such that the characteristics of perpendicularly polarized and parallel polarized light are averaged. The effectiveness is dependent on having minimum angles of incidence of the following reflective surfaces.

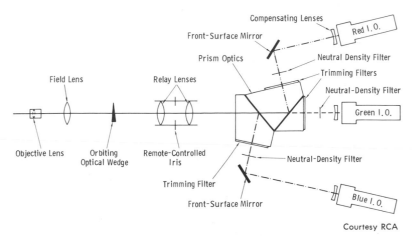

Courtesy RCA

Fig. 4-32. Optical system of RCA TK-41 camera.

Without this correction for polarization, the red and blue outputs are reduced at certain critical angles of view and with variations in scenic reflectance or light levels, leaving a predominantly green high light. This is sometimes noticed as a greenish flesh tone of a person at (for example) the left of the screen, while flesh tones in the rest of the scene are reproduced properly.

The primary advantage of the parallel pickup-tube assembly of Fig. 4-28 is that the influence of any magnetic field will affect all tubes by the same amount; hence compensation can be made easily with centering adjustments. The primary disadvantage is that this is not the optimum arrangement for an optical system employing a prism block, because of the long back focal length required and the relatively large angles of incidence for the red and blue reflecting surfaces.

One type of prism-block assembly is illustrated in Fig. 4-33. With this arrangement, the vertex angle of the central prism is reduced, decreasing the angle of incidence to the red and blue tubes to approximately 30°. The tubes are fanned as in Fig. 4-32. Fig. 4-34 is a view of the Marconi Mark VII color camera; note the fanned yoke assemblies.

Another type of prism-block assembly is shown in Fig. 4-35. To gain efficiency in the separation of red from green, the angle of the ray incident on the red reflecting surface at the central point is only 13°. The angle of incidence onto the blue reflector is approximately 25°. An air gap is provided between the blue-reflecting and red-reflecting prisms to reflect the red component internally. The back focal length required with this assembly is less than half that of the methods previously described. The fanning angle of the pickup-tube yoke assemblies is greater than for the former methods. Fig. 4-36 illustrates the yoke configuration in the RCA TK-44 color camera.

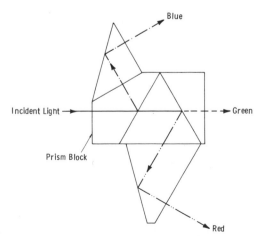

Fig. 4-33. Prism-block assembly with 30° angle of incidence.

Fig. 4-34. Interior view of color camera.

Fig. 4-37 illustrates the zoom and focus controls normally mounted on the RCA TK-42 color camera. These controls are supplied by Albion Optical Co. and are adaptable to all makes of cameras. In the RCA TK-42, the zoom lens assembly is internal. The focus and zoom control cables feed through two holes in the side of the casting below the camera as shown.

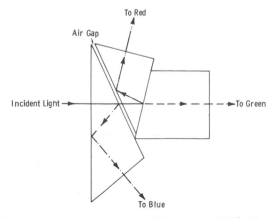

Fig. 4-35. Prism-block assembly with 13° (red) and 25° (blue) angles of incidence and reduced back focal length.

Courtesy RCA

Fig. 4-36. Fanning of color-camera yoke assemblies.

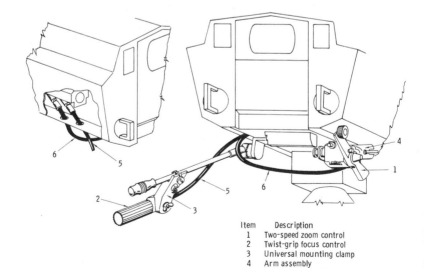

Item	Description
1	Two-speed zoom control
2	Twist-grip focus control
3	Universal mounting clamp
4	Arm assembly
5	Focus control cable
6	Zoom control cable

Courtesy RCA

Fig. 4-37. Zoom and focus controls.

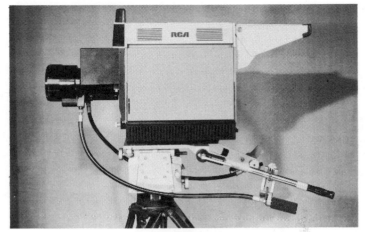

(A) Side view.

(B) Front view.

Courtesy RCA

Fig. 4-38. Camera with external zoom lens.

A two-speed lever permits fast or slow zoom control. In the fast position, one turn of the handle zooms through the entire range. In the slow position, slightly more than two turns of the handle are required for full zoom range.

The zoom lens assembly on the RCA TK-44 color camera is external (Fig. 4-38). Thus, the focus and zoom control cables connect directly into the zoom assembly.

The Philips PC-100 color camera (Fig. 4-39) employs a lens-mounting design that automatically couples internal mechanical drive shafts as well as completing the electrical connections. There are no external lens cables.

Courtesy Philips Broadcast Equipment Corp.

Fig. 4-39. Camera with quick-connect lens assembly.

4-9. THE IRIS-CONTROL SERVO

Most modern cameras employ remote control of the lens iris from the operating panel. Servo functions for this purpose generally are located in a servo module of the camera head.

Fig. 4-40A represents the basic servo function. The motor is geared to a follower potentiometer which turns with the motor rotation. The voltage applied to the follower also is applied to a control potentiometer. When the net voltage at the amplifier input is zero (equal and opposite voltages from the two potentiometers), the motor cannot receive power. Assume the control is turned toward a positive voltage. The amplifier now has an input that is amplified and applied to the motor. When the motor rotates, the follower rotates in the opposite voltage direction, in this case toward a negative voltage. When the amplifier input again receives equal and opposite voltages (net zero), the motor stops. Bridge circuits normally are involved in practice.

Fig. 4-40B is a functional block diagram of an iris servo. When either of the silicon controlled rectifiers is "fired," its associated diode bridge circuit is able to pass ac. When this happens, the circuit of the winding of the two-phase motor is completed to ground directly through the bridge, and

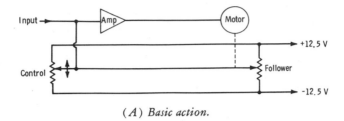

(A) Basic action.

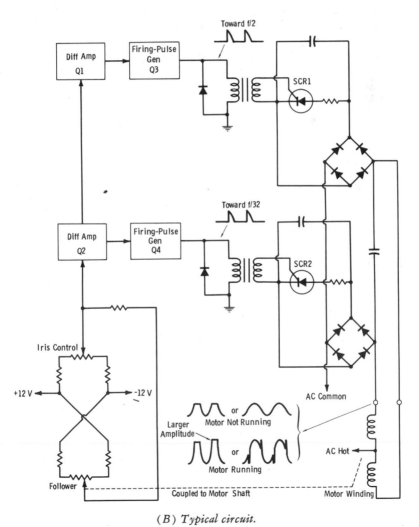

(B) Typical circuit.

Fig. 4-40. Principle of lens-iris servo.

the circuit of the other winding is completed to ground through the bridge and a capacitor. The direction in which the motor runs depends on which winding is in series with the capacitor, and this in turn depends on which bridge is conducting (and therefore which SCR has fired).

When the inputs to the differential amplifier are equal, neither bridge conducts. A change at the Q2 base causes an opposite reaction on Q1. The change occurs because of operation of the iris control; one extreme of this control is the maximum stop opening, and the opposite extreme is the minimum stop opening. An opposite rotation of the iris control causes the opposite SCR to fire, and the motor turns in the reverse direction.

The waveforms on the drawing give a clue as to how to signal trace in probing for troubles. The two types of signals depend on the circuitry involved; these are the most prevalent. Remember, you will always see the 60-Hz power waveform on the scope at the point indicated whether the circuit is completed to ground or not. The clue to completion of the circuit for the motor is the increase in the amplitude clipping point or the presence of the SCR "spikes" on the waveform.

EXERCISES

Q4-1.   What is the purpose of the "pancake" coil in front of the 4½-inch image orthicon?

Q4-2.   How is electrical focus obtained in the image section of the image orthicon?

Q4-3.   How is focusing achieved in the scanning section of the image orthicon?

Q4-4.   Does the multiplier section of the image orthicon contain any beam-focusing element?

Q4-5.   What is the purpose of grid 5 in the image orthicon, and how is the proper voltage adjustment for this grid determined?

Q4-6.   Compare I.O. curve B in Fig. 4-6 with vidicon curve A in Fig. 4-23C. The I.O. amplitude response at 350 lines is approximately 70 percent. The vidicon amplitude response at 350 lines is slightly less than 60 percent. Does this mean the resolution capability of the vidicon camera is less than that of the image-orthicon camera?

Q4-7.   What adjustment in a vidicon camera generally assures good uniformity in black level (lack of picture shading)?

Q4-8.   In the vidicon, what is the relationship between dark current and target voltage?

Q4-9.   In the *Plumbicon*, what is the relationship between dark current and target voltage?

Q4-10. What is the difference between the *Plumbicon* and the lead-oxide vidicon?

CHAPTER 5

Video Preamplifiers

Shunt circuit capacitances, both stray wiring capacitance and tube or transistor interelectrode capacitance, inevitably result in a signal loss that increases as the frequency increases. Therefore, if a video amplifier is to have a "flat" frequency response, "peaking" circuitry or heavy amounts of inverse signal feedback must be used. The coupling between the pickup tube and the preamplifier input, since it occurs at high impedances, is also a cause of high-frequency losses and phase shift.

Video-peaking circuitry to obtain a flat frequency response results in a certain amount of phase shift across the video passband. Therefore, additional compensation circuitry is required for phase correction. Adjustable controls for this type of signal correction are termed *high-peaker* or *phase-correction* controls. Sometimes both terms are used for separate controls, depending on the stage in which the correction is applied, and on the time constant of the correcting network. High-peaking and phase-correction controls normally are found in the video preamplifier, which receives the signal output from the pickup tube for further amplification.

5-1. THE VACUUM-TUBE VIDEO AMPLIFIER

Vacuum-tube circuitry depends more heavily on peaking and phase correction than is true of the most recent solid-state amplifiers. Since there are many tube-type amplifiers still in daily service, we would be remiss in our duty if we omitted them from our studies.

Vacuum-Tube Peaking Circuits

Fig. 5-1A illustrates the shunt-peaking method of compensating for the usual high-frequency losses. In the equivalent circuit (Fig. 5-1B), C_T is the total of the output capacitance of V1, the stray capacitance of the circuit, and the input capacitance of V2. Electrically, the circuit amounts to a parallel resonant circuit, designed so that the resonant frequency is approximately 1.41 times the highest frequency to be amplified. Thus, a boosting of the high frequencies occurs, but there is no effect on the lower frequencies. In practice, peaking coils may have values between one and several hundred microhenries. Ten to fifty microhenries is the average range found in commercial equipment.

Fig. 5-2A illustrates the series type of peaking circuit. The series coil, in combination with the effective circuit capacitances, forms a low-pass filter network (Fig. 5-2B). At first thought, it might appear that such a circuit defeats the purpose intended, an increase in the efficiency of amplification at higher frequencies. A basic analysis is therefore necessary.

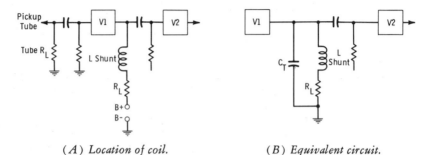

(A) Location of coil.   (B) Equivalent circuit.

Fig. 5-1. Shunt peaking.

It is necessary that C1 have twice the capacitance of C2. Load resistor R_L is always connected to the low-capacitance side of the circuit. In practice, therefore, a small physical capacitor may be found in the circuit where effective capacitance C1 would appear; this physical capacitor is considered to be in parallel with C1, effecting a 2-to-1 ratio in effective capacitances.

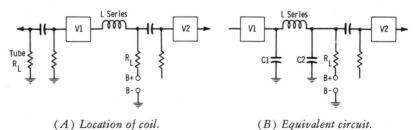

(A) Location of coil.   (B) Equivalent circuit.

Fig. 5-2. Series peaking.

Inductor L and capacitance C1 form a series resonant circuit with an effective increase in current as the frequency increases. Capacitance C2 is separated from C1 by inductance L, with a resultant reduction in shunting effect across L at high frequencies in the passband. Therefore, since the voltage drop in the series resonant circuit increases with an increase in frequency, and this voltage is applied across load resistor R_L shunted by C2, the voltage developed in the load likewise increases for high frequencies in the passband. The increase of high-frequency voltages as a result of the

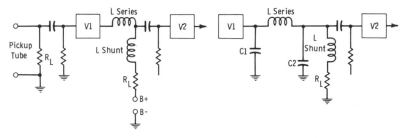

(A) Locations of coils.   (B) Equivalent circuit.

Fig. 5-3. Shunt-series peaking.

resonant effect of L and C1 more than offsets the effect of decreased reactance of C2 with increasing frequencies. The resulting video voltage is
coupled to V2 in the usual manner.

Fig. 5-3 illustrates the most efficient design used in video amplifiers to
increase the high-pass range. This is the shunt-series peaking circuit, which
can be seen to be a combination of the two methods just discussed. The
increased voltage gain from such a circuit is aided materially by the fact
that it is possible to use a load resistor of around 80 percent greater value
than can be used with a simple shunt-peaked circuit. Since the gain of a
stage is equal to the transconductance of the tube times the value of R_L, it
may be seen that the gain is increased appreciably by this means alone. It
should be remembered that the value of plate load resistance in ordinary
amplifiers is limited by the bandpass required; too great a value of load
resistance reduces the bandpass capabilities of the stage.

The relative gains of video amplifier circuits may be tabulated as follows:

Uncompensated 0.707
Shunt Peaked 1.0
Series Peaked 1.5
Shunt-Series Peaked 1.8

Low frequencies are of just as much importance in the video amplifier
as are high frequencies. Fig. 5-4 illustrates the addition of a low-frequency

Fig. 5-4. Low-frequency compensation.

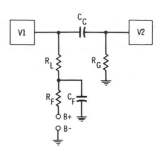

filter circuit to an amplifier stage. At low frequencies, the voltage across R_G, the grid resistor of V2, decreases because of the increased reactance of coupling capacitor C_C. However, the reactance of C_F also increases at the lower frequencies, and the combined impedance of R_F and C_F is added to that of load resistor R_L. Thus, as low-frequency signals become somewhat attenuated by the limited value of coupling capacitance, the total plate load impedance becomes greater, and the resulting boosting effect helps maintain constant gain across the entire passband.

NOTE: The reader should be cautioned that the frequency-compensation circuitry presently being discussed is for normal losses that result from circuit characteristics. The aperture distortion previously discussed is a loss of high-frequency definition without phase distortion. Correction for this effect is made by aperture-correction circuitry in processing video amplifiers (Chapter 6).

Typical Vacuum-Tube Preamplifiers

The video preamplifier shown schematically in Fig. 5-5 consists of eight stages (six tubes). The first four stages, V1 to V4, are video amplifiers using Type 6AH6 tubes with shunt peaking in the plate circuit of each amplifier. The cathodes of the first and fourth stages are unbypassed. RC high-peaking networks in the cathode circuits of the second and third stages, with a variable resistor in each cathode circuit, permit adjustment of the response in two distinct steps. This peaks the higher frequencies to compensate for high-frequency attenuation associated with the image-orthicon output.

The four feedback and output stages are built around two dual tubes, a Type 6U8 pentode-triode (V5) and a Type 5687 dual triode (V6). Feedback from one cathode (pin 6) of V6 to one cathode (pin 7) of V5 is adjustable by means of a 7-35 pF variable capacitor (C_T) to obtain flat overall response for the feedback pair. The feedback circuit also provides sending-end termination and isolation for the viewfinder output.

Horizontal shading (Chapter 6) in the viewfinder picture is corrected by feeding a positive or negative sawtooth voltage to the cathode of V1.

The video preamplifier in Fig. 5-5 has no gain control. The image-orthicon output is controlled by varying the dynode gain by means of the ORTH GAIN control. This allows adjustment for variations in image orthicons in order to keep the video input to the preamplifier constant.

Note that the preamplifier input signal from the image orthicon is about 0.05 volt peak-to-peak. Two outputs are provided: a 1.0-volt peak-to-peak signal for the viewfinder and a 0.4-volt peak-to-peak signal to the 50-ohm coaxial section of the camera cable. The ORTH GAIN (dynode gain) control in the camera head is adjusted so that these output levels exist for high lights in the scene.

Observe in Fig. 5-5 the small values of the cathode-bypass capacitors in the second and third stages. The greater the value of cathode-return resis-

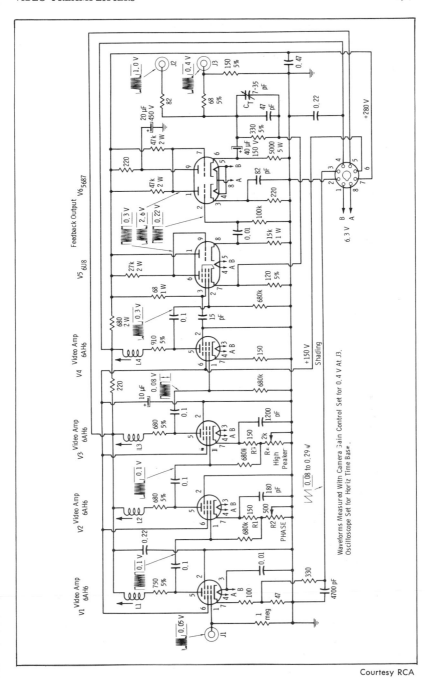

Fig. 5-5. Preamplifier in RCA TK-11 camera.

tance (variable), the greater is the degree of low-frequency degeneration relative to higher frequencies. Sometimes a fixed cathode-return resistance is used with variable bypass trimmer capacitors. Fig. 5-6 is a simplified schematic diagram of a more recently developed tube-type video preamp. The signal voltage to feed the preamplifier input is developed across R1, the I.O. load resistor. Note that this load resistor is returned to ground for ac through C1. This is normal in most I.O. output coupling circuitry.

Shading or calibration signals generated in rack-mounted processors in the control room are terminated by R4 (51 ohms) and fed into the bottom of the I.O. load (R1) through C1. These signals may be disconnected and a test signal inserted across R2 (10k).

The video signal is amplified in the first cascode amplifier and then coupled into the second cascode amplifier, which incorporates the high-peaker circuit. The signal then is amplified further by two feedback pairs as drivers for the output line driver.

Most modern plate-coupled, low-impedance output stages take the form of the single-ended push-pull circuit, as shown for V7 and V8 in Fig. 5-6. This circuit provides the advantages of push-pull amplification while supplying a single-ended output for the coaxial cable. Typical values of output coupling capacitors range from 200 to 400 μF.

The cascode input stage for video preamplifiers has become quite common in both tube and solid-state circuitry. The preamplifier and frequency-phase compensation methods are similar for all pickup tubes whether image orthicon, vidicon, or lead-oxide vidicon.

5-2. SOLID-STATE VIDEO PREAMPLIFIERS

A solid-state video preamplifier frequently used in vidicon channels of RCA color cameras is illustrated in Fig. 5-7. The input amplifier unit and preamplifier module provide current amplification for low-level signals from a vidicon tube to drive the 93-ohm input of the following video amplifier stage.

The input amplifier, consisting of a field-effect transistor (FET) and a resistor, is a separate unit that is designed for mounting on a vidicon yoke assembly, thereby minimizing input circuit capacitance. The output from the FET amplifier provides the input signal for the preamplifier module.

When used together, the input amplifier and the preamplifier function as a signal amplifier with selectable current gain of 50, 100, or 150. The video output is passed through a peaking network in the preamplifier module to compensate for input source capacitance that causes high-frequency attenuation of the video input signal from the vidicon. Ac feedback from the preamplifier to the input amplifier is employed to cancel the capacitive loading effect on the transistor elements; dc feedback is employed for bias stabilization.

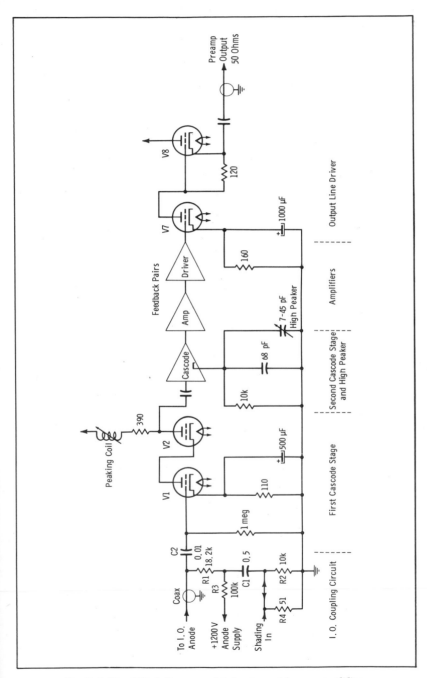

Fig. 5-6. Simplified diagram of tube-type video preamplifier.

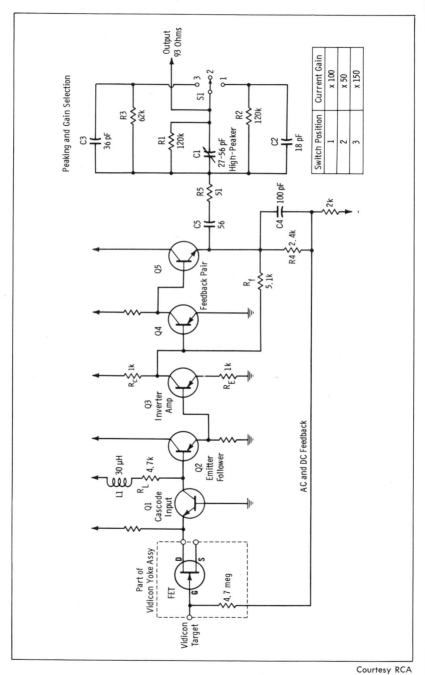

Fig. 5-7. Simplified diagram of video preamplifier for vidicons.

A field-effect transistor (FET), which has characteristic high input impedance and low noise figure, operates as a current amplifier in the input amplifier. A 4.7-megohm resistor in the input gate circuit provides the path for degenerative video-signal and bias-stabilization feedback. The signal current from a vidicon target (constant-current device) is passed through the gate-biasing resistor. The signal voltage developed across the resistor is applied to the gate (G) terminal of the FET, setting up within the transistor space charges that vary with the signal. These varying space charges modulate the current in the path between the drain (D) and source (S) terminals. The result is amplification of the input signal current without polarity inversion.

The output from the field-effect transistor is coupled to cascode amplifier Q1 through the input amplifier cable. Transistor Q1, a grounded-base amplifier with low input impedance and high output impedance, provides neutralization of the degenerative capacitance effects between the drain and gate materials of the FET. The video output signal from Q1, developed across resistor R_L and high-frequency peaking coil L1, is applied to the base of Q2, an emitter follower with a high input impedance and a low output impedance. The video signal at the emitter of Q2 is amplified and inverted by Q3, and is applied to a feedback amplifier pair, Q4 and Q5.

One output from the emitter of Q5, the degenerative video feedback signal, is coupled through C4 and R4 to the gate terminal of the FET. A second output from Q5 is coupled through C5 and R5 to a peaking and gain-selection network. High-frequency peaking trim is adjusted by means of capacitor C1 to compensate for losses caused by vidicon-target capacitive loading at the input to the input amplifier. Gain-selection switch S1 sets the current gain at 50 when in position 2, at 100 when in position 1, and at 150 when in position 3. The high-frequency-peaking trim requires adjustment when the gain is changed.

Note that when S1 is in position 2, the output is through R1 and high-peaker trimmer C1. When S1 is in position 1, R2 is paralleled with R1. Since the resistances of R1 and R2 are equal, the current gain is doubled. When the switch is in position 3, R3 is paralleled with R1; the value of R3 is such that the current gain is triple that for position 2.

Capacitors C2 (across R2) and C3 (across R3) are compensating capacitors to help maintain the proper frequency-phase response for the various switch positions. However, as stated before, the high peaker (C1) normally must be readjusted when the switch position is changed.

The feedback pair (such as Q4 and Q5 in Fig. 5-7) is a common circuit in modern cameras, where it finds use particularly in drivers for 50- or 75-ohm lines. The voltage amplification is approximately equal to the ratio of feedback resistance to signal-source resistance (Fig. 5-8). In Fig. 5-7, the source resistance for the feedback pair is the effective output impedance of inverter stage Q3. Since the emitter resistance (R_E) is large and degenerative (unbypassed), this source resistance (R_S) is essentially the value of

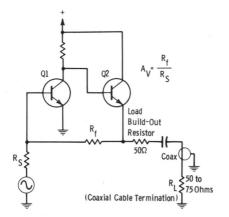

$$A_V = \frac{R_f}{R_S}$$

Fig. 5-8. Typical feedback pair.

R_C, or 1000 ohms. Thus, the voltage gain expected in the feedback pair is 5100/1000, or 5 (approximately). Both the input and output resistances are low so that capacitive effect at high frequencies is negligible.

Observe that in both Fig. 5-7 and Fig. 5-8 a resistor is placed in series with the load feed. All modern video-amplifier outputs provide a source impedance equivalent to the load impedance. Let us examine now why we normally find a *build-out resistor* in the output stage of a coaxal line driver.

Recall that the emitter-follower circuit is characterized by a relatively high input impedance and a relatively low output impedance. Therefore, it is a "natural" for feeding low-impedance video lines. The input and output impedances of an emitter follower (feeding a low impedance load) are somewhat interdependent. This is because the input load is part of the output, and the outut load is part of the input. If the reader will observe Figs. 5-7 and 5-8, he will note that the same is true of the feedback-pair circuit.

Observe Fig. 5-9A. The approximate internal output impedance of this circuit, as given on the drawing, is the source impedance divided by the sum of beta of the transistor plus 1. Assuming a beta of 50, we can see that this output impedance becomes 680/51, or 13, ohms (approximately). For the coaxial-line termination of 50 ohms to "see" a sending-end impedance of 50 ohms, a 37-ohm resistor must be used as shown in Fig. 5-9B.

Now see Fig. 5-9C. In this circuit, emitter follower Q2 is driven by emitter follower Q1. Assume that Q1 sees a source resistance of 1000 ohms and that beta is 50 for both transistors. The output resistance of Q1 is:

$$Z_{OUT} = \frac{1000}{50 + 1} = 20 \text{ ohms (approx)}$$

Then the output resistance of Q2 is:

$$Z_{OUT} = \frac{20}{50 + 1} = 0.4 \text{ ohm (approx)}$$

The equivalent circuit of the output stage is shown in Fig. 5-9D. To feed a 50-ohm load, a build-out resistance of 50 ohms is needed so that the effective internal output impedance (looking back from the load) becomes 50 ohms. With such a low-impedance transistor output, it is simply necessary to use a 75-ohm build-out to feed a 75-ohm coaxial line. In either case, the build-out and load combine to form a 2-to-1 voltage divider. Therefore, the signal voltage at the emitter of Q2 is twice as great as the signal voltage across the coax-line termination (input to following video amplifier). This is a normal condition for modern video line drivers, and it should be understood in particular by the maintenance personnel.

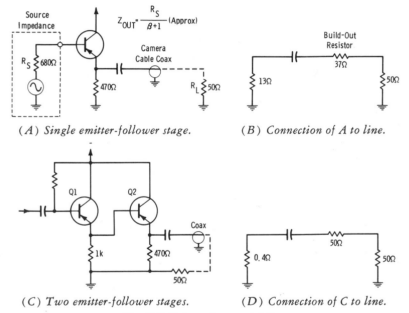

(A) Single emitter-follower stage.   (B) Connection of A to line.

(C) Two emitter-follower stages.   (D) Connection of C to line.

Fig. 5-9. Drivers for coaxial lines.

The ideal phase-frequency characteristic is a straight line that rises from a value of pi radians (180°) at zero frequency (frequency plotted along the x axis and phase angle plotted along the y axis). The scale is linear, which indicates that the phase angle increases linearly with frequency, maintaining the same time delay for all frequencies. Departure from this characteristic results in phase distortion that is most noticeable as trailing white edges following black edges on a gray or white background.

If the phase characteristic departs from the ideal straight slope and bows upward away from the frequency (x) axis, the increasing slope with frequency is an indication that the time delay is increasing with frequency. The shifted components making up the pulse now add in such a manner that a leading-edge overshoot and a trailing-edge undershoot on the passed pulse occur. Conversely, if the phase characteristic bows toward the x axis, the decreasing slope with frequency indicates a decrease in time delay with frequency. The shifted pulse components now add to produce a leading-edge undershoot and trailing-edge overshoot. Note that one type of phase distortion may be compensated by an equal and opposite phase correction. A lagging phase shift is corrected by an equal leading phase shift, and vice versa.

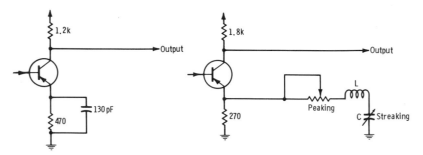

(A) Conventional emitter peaking.　　　(B) Peaking and phase controls.

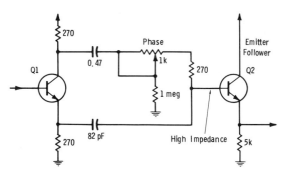

(C) Phase-split correction method.

Fig. 5-10. Representative high-peaking and phase circuitry.

In Fig. 5-10A, the emitter resistor forms an inverse-feedback path that is incompletely bypassed by the small value of capacitance. Signals of higher frequencies are bypassed, but signals at lower frequencies must pass through the inverse feedback path. Thus, gain is reduced at lower frequencies but not at higher frequencies.

In Fig. 5-10B, the LC circuit is resonant near the top of the intended passband. The variable resistance (peaking control) determines the magnitude of the resonant peak introduced. The variable trimmer capacitor (streaking control) adjusts the resonant frequency to introduce the required lead or lag in phase.

In Fig. 5-10C, Q2 is fed from both the emitter and collector of Q1. Thus, Q2 is driven from signals 180° apart at high frequencies. Phase changes in Q1 itself are prevented by the low values of the emitter and collector resistors. Transistor Q2 presents a high impedance to Q1 and the coupling networks. Note from the relative values of the coupling capacitors that low-frequency gain is higher in the collector circuit than in the emitter circuit, to maintain amplitude response independent of frequency under adjustment of the phase control. The relative phase shift at high and low frequencies is adjustable by means of this control.

5-3. MAINTENANCE OF VIDEO PREAMPLIFIERS

We are now ready to study procedures for maintenance and adjustment of the video preamplifier. Older tube-type preamps require a more consistent maintenance schedule than do the latest solid-state versions. In fact, some of the most recent solid-state amplifiers have few, if any, adjustable peaking coils in the circuitry. This is because of the very high gain-bandwidth product of transistors and integrated circuits in the latest designs. Also, the stability of performance has been improved greatly over that possible in vacuum-tube circuitry.

Preparation for Video Sweep

Older tube-type video preamps normally require a video-sweep alignment on regular preventive-maintenance schedules about every 90 to 120 days. In any case, the maintenance technician should be familiar with video-sweep techniques. Even newer solid-state amplifiers require such testing whenever components are replaced.

Preliminary steps are necessary before video sweeping is attempted. This is particularly true for those stations still using the older TV oscilloscopes, such as the Tektronix Model 524. First, we will review briefly the practical use of the oscilloscope with normal probes and with the video-sweep detector probe.

Because of capacitive loading effects, the direct scope probe is severely limited in its application to TV equipment maintenance, even when it is applied directly across 50- or 75-ohm terminations. A direct probe should never be used where frequency response or transient response is a factor; therefore, use of this type of probe is limited to certain applications in which the IRE response is employed.

For most applications, the 10-to-1 capacitance divider probe should be used. For a scope with 1-megohm input shunted by a 40-pF capacitance, the simplest 10-to-1 probe consists of a series-connected 9-megohm resistor shunted by a trimmer capacitor of 3 to 12 pF. When the probe is connected to the scope, the input impedance from the probe tip becomes 10 megohms shunted by approximately 12 pF. The trimmer capacitor is adjusted so that the RC product of the probe is equal to the RC product of the scope input, thus making the voltage division independent of frequency. This is done by touching the probe to the scope calibration-pulse output (or the output of a square-wave generator set to about 1 kHz) and adjusting the trimmer so that the leading edge of the pulse is not rounded on the top (undercompensated) and does not overshoot (overcompensated). This adjustment should be checked often, and it always must be checked when the probe is used with a different scope, even one of the same make and model. The frequency response and transient response of the scope itself should be checked with this probe so that all variables are calibrated.

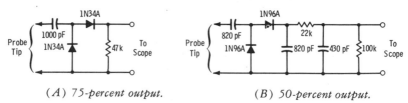

(A) 75-percent output.   (B) 50-percent output.

Fig. 5-11. Video-sweep detector probes.

A logical step-by-step initial calibration of the scope can be outlined as follows:

1. *Video Sweep (Detected).* Terminate the video-sweep generator in 75 ohms directly at the generator output connector. Use a video-detector probe (Fig. 5-11). The probe in Fig. 5-11A will read approximately 75 percent of the actual peak-to-peak output signal, whereas the higher-isolation probe in Fig. 5-11B will read about 50 percent. Adjust the output amplitude for 1 volt, which will read approximately 0.75 volt with the probe of Fig. 5-11A or 0.5 volt with the probe of Fig. 5-11B. Also, adjust the scope gain to provide a convenient display (Fig. 5-12). This makes possible a check of the flatness of the sweep generator itself, since the detected sweep envelope does not depend on the high-frequency response of the scope. The Tektronix scope may be used on any response position; or a scope with limited response can be used, provided it has reasonably good low-frequency square-wave response. If the video-sweep generator cannot be made perfectly flat, as observed on the scope, the

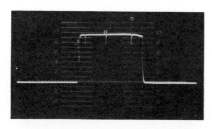

Fig. 5-12. Detected video-sweep signal waveform.

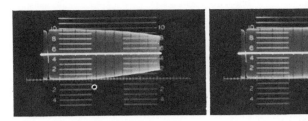

(A) Scope on normal response.　　　(B) Scope on flat response.

Fig. 5-13. Display of undetected video-sweep signal.

deviations must be plotted as a correction factor for equipment checks and scope calibration.

2. *Video Sweep (Wideband).* (This should be observed only after determining the flatness of the sweep generator as in Step 1.) Although this method is used only in very special cases (and with extreme care), the rf envelope may be observed directly without detection as a "quickie" check on scope-amplifier response (Fig. 5-13). This check, however, is valid only if the probe (10 to 1) to be used for equipment checks is used on the scope and a signal of the same amplitude is employed so that the scope compensated attenuator is at the same setting as that to be used. It is good engineering practice to run these checks with all probes in stock, and through the scope preamp as well as directly to the vertical-amplifier output. Use varying levels from the sweep generator to permit use of convenient scales on the scope with different attenuator positions that might be incorrectly compensated. An attempt to employ correction factors for different attenuator settings becomes both cumbersome and inaccurate in system measurements. Normally there will be some correction factor when using the preamp and when feeding the vertical amplifier directly. Plot these responses either on a graph or by tabulation in peak-to-peak values. Normally the detector probe is employed when video sweep is used. The wideband display provides a quick check of scope response to single-frequency sine waves or in similar applications.

3. *Single-Frequency Sine-Wave Checks.* The most accurate method of checking scope-amplifier frequency response is to run single-frequency sine-wave checks over the range of 100 kHz to 10 MHz. The same generator and probes that will be used for system checks should be used for scope calibration. Commercial sine-wave generators such as the Hewlett-Packard 650-A incorporate a frequency-compensated metering circuit at the output to aid in maintaining a constant input to the scope or equipment at all frequencies. If a generator of this type is not available, a VTVM with good response to 10 MHz can be used across the terminated generator output. As

in Step 2, it is good practice to check all probes and all attenuator settings that are likely to be used in system checks. When the calibration is posted on the scope, the particular generator, meter, and probe should be identified, unless all such items have been found to be directly interchangeable. Such checks normally should be made about twice a year, or at any time that considerable maintenance (tube or component changes, etc.) on the scope or signal generators has been required.

4. *Low-Frequency and Transient Response.* Determine the rise time and percent of overshoot of the square wave as read on the scope, both through the preamp and directly into the main amplifier, at the frequencies normally used. Unless a generator with short rise time is available, higher-frequency square waves (above 75 kHz) are not particularly useful, because for response checks at the higher frequencies the rise time of the pulse must be shorter than the rise time of the amplifiers to be checked. A 60-Hz square wave fed to the Tektronix 524 AD (dc position) should have an absolutely flat top, as shown in Fig. 5-14A. Fig. 5-14B shows the normal amount of tilt introduced by the input coupling capacitor when the scope selector is on the ac position. Remember that the last two (highest-gain) positions of this particular scope are ac only, since the preamp is used on these positions. An adjustable grid time constant (low-frequency compensation control) is used in the preamp; this control should be adjusted according to the manufacturer's instructions. When a scope employs either external or plug-in preamps, always include these units in all scope-calibration procedures.

NOTE: More recent Tektronix television oscilloscopes, such as the Type 545 (30-MHz bandwidth) or Type 547 (50-MHz bandwidth), are dc coupled even at the highest gains. These scopes require checking only about once a year unless obvious performance deterioration is noted.

It is possible for a wideband scope amplifier to exhibit a leading-edge overshoot caused by a vacuum-tube defect known as *cathode interface.* This low-frequency phase shift results from series-resistance and capacitive-

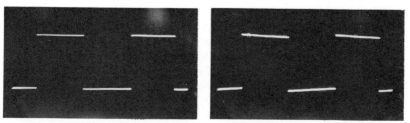

(A) 60 Hz, scope on dc position.   (B) 60 Hz, scope on ac position.

Fig. 5-14. Square-wave response patterns.

bypassing effects of a chemical interface layer that forms between the sleeve and oxide coating of the cathode. Since some tubes have been known to develop cathode interface in less than 500 hours of operation, the scope should be checked about every two months for this type of tube defect. The following procedure may be used:

1. Adjust the frequency of the square-wave generator to 500 kHz. The waveform should have a rise time of 0.2 microsecond or less.

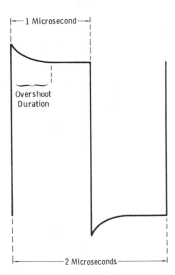

Fig. 5-15. Interface distortion of square wave.

2. Adjust the time base so that several cycles of the square wave are displayed. If an overshoot with a duration of 0.2 to 0.6 microsecond appears (Fig. 5-15), one or more tubes in the vertical amplifier may have cathode interface. (Overshoot duration, or time constant, is the time required for the overshoot to decay to the final flat-top value.) A 500-kHz square wave completes one cycle in 2 microseconds; thus it has a pulse width of 1 microsecond, as shown in Fig. 5-15. The overshoot duration normally is between 20 and 60 percent of the total pulse width when cathode interface is present. As a double check, plug the scope into a variable autotransformer and increase the line voltage to the upper limit allowed. If cathode interface is present, the increased tube-heater voltage will reduce the overshoot; a decrease in the voltage will increase the overshoot. When this occurs, it is best to replace all tubes in the vertical amplifier with new ones; then substitute the old tubes one at a time while observing the square wave. Discard any tube that tends to show this effect. Leave the line voltage at the lower limit (105-108 volts) to emphasize the effect of cathode interface.

Video Sweeping the Preamp

A sweep generator consists of a fixed-frequency oscillator the output of which is beat with the output of a sweep oscillator that is frequency modulated at 60 Hz. The frequency modulation is such as to cause the beat frequency to swing over a usable range from about 100 kHz to approximately 10 or 20 MHz. The frequency swing may be produced by a reactance-tube circuit with 60-Hz excitation from the power line, or, in many cases, by a motor-drive capacitor in the oscillator tank circuit. A 3600-rpm motor provides a 60-Hz sweep of the oscillator frequency. Such a sweep generator usually incorporates an absorption marker generator that places a notch (or series of notches) at any reference frequency (or frequencies) over the usable range.

The fundamentals of checking the high-frequency characteristics of a video amplifier are illustrated in Fig. 5-16. The output frequency of the sweep generator is swept over a range of 100 kHz to 10 or 20 MHz, with a tunable frequency maker (notch) placed at any desired frequency. The sweep is repeated 60 times per second. This test signal is applied to the amplifier to be tested. A detector of the type shown in Fig. 5-11 is connected to the output of the amplifier. This detector rectifies the signal output as shown (in this case the amplifier is considered to be theoretically ideal: no distortion has occurred), and the output of the detector is fed to the vertical input of the oscilloscope. By this means, the oscilloscope traces a graph of output voltage versus frequency over the passband above 100 kHz. The scope for this test should have excellent low-frequency response so that no distortion of the 60-Hz square wave takes place. High-frequency response need extend no farther than 50 kHz.

It is very important not to overload the amplifier(s) when using video sweep. If the normal output is a 0.7-volt (peak to peak) signal, feed just enough input sweep level to result in a 0.5-volt (peak to peak) amplifier output. Remember to calibrate the detector probe so that you know how much loss occurs in the probe. For example, if the detector-probe gain is 50 percent, a 0.5-volt (peak to peak) actual output level reads 0.25 volt (peak to peak) through the probe.

Some engineers prefer to check the output level by using the regular 10-to-1 probe with the scope in the wideband position (Fig. 5-13). When using this method, adjust the sweep-generator output so that the 2-to-3 MHz region of the sweep signal is at reference level at the output.

When the more modern oscilloscopes with 30- to 50-MHz bandwidths (flat response across the passband of interest) are used, the detector probe need not be employed if care is taken. Even here, however, it is advisable to use the detector probe to minimize hookup and grounding problems.

Most modern video-sweep generators for broadcast service have an internal output impedance (sending-end impedance) of 75 ohms. In making response adjustments, which may require feeding an interstage circuit, the

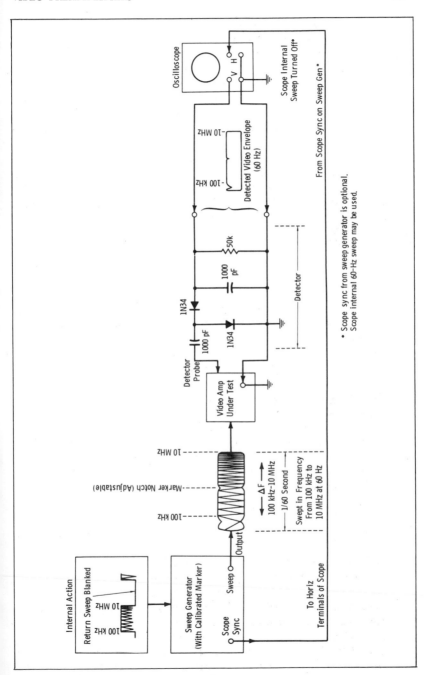

Fig. 5-16. Method of testing video amplifier for high-frequency characteristics.

video-sweep generator feeds a high impedance unless the particular instruction manual for the unit specifies otherwise.

In checking individual units, the coaxial output cables should be disconnected and replaced with 50-ohm terminations (for preamplifiers). The detector probe is placed directly across the termination and retained in this position for response alignment. NOTE: When the video preamplifier feeds a following amplifier in the camera head (rather than the camera cable), the output impedance may be something other than 50 ohms. A 93-ohm coaxial cable sometimes is used.

Let us assume we are going to sweep and align the preamplifier of Fig. 5-5. The output coaxial cable at J3 is removed, and J3 is terminated with a 50-ohm resistor (actual value is 51 ohms). The scope probe or detector probe is connected across this termination, where it remains throughout the alignment procedure. The alignment procedure may be outlined as follows:

1. Feed the sweep-generator signal to the grid (pin 2) of V5. Adjust the sweep generator (in this example) to obtain a signal of no more than 0.4 volt at the termination feeding the scope.
2. Note that the only adjustable component in the last two stages is the trimmer capacitor (C_T). Adjust this capacitor for the flattest response possible.
3. Feed the sweep signal to the grid (pin 1) of V4. Reduce the generator output to obtain the reference level at the output. Adjust L4 for flattest response.
4. Note now that the grid of V3 (where the sweep-generator signal is to be fed next) has a dc component from the return to the high side of the high-peaker control. Therefore, it is necessary to feed the sweep signal to this grid through a capacitor of about $0.1\mu F$. This capacitor is needed because most video-sweep generators do not employ a blocking capacitor at the output, since the actual output normally is through a built-in pad to obtain the desired level.

   Also in this stage is the frequency-phase compensation network associated with the high-peaker. This circuit must be bypassed temporarily with a capacitor of 0.25 to 0.47 μF. The final adjustment of high-peaker and phase controls *must* be made only with pickup-tube signals to eliminate smear or streaking in the picture. The purpose of video alignment is to make the amplifier response flat, by using a video-sweep signal or single-frequency sine-wave signals, with any special phase-correction circuitry bypassed. Peaking coil L3 is now adjusted for flattest sweep.
5. Note that the conditions for the circuit of V3 also apply to the circuit of V2. Bypass the cathode of V2 with another $0.47-\mu F$ capacitor, leaving the bypass for the V3 cathode in place. Feed the grid of V2 through a $0.1-\mu F$ capacitor, and adjust L2 for flattest response. Re-

member to reduce the gain of the video-sweep generator in all steps
to maintain the reference output level.

6. This will be a rough adjustment of the phase and high-peaker con-
trols, which will be adjusted finally with a pickup-tube picture as
mentioned above. Remove the temporary bypass capacitors at the
V2 and V3 cathodes. Feed the sweep-generator signal to the grid of
V1 through the circuit of Fig. 5-17B (described below). Adjust
L1, R2, and R4 for proper frequency response.

If circuits such as the cathode circuits of V2 and V3 (Fig. 5-5) are not
bypassed with a capacitance of around 0.47μF, the video sweep through
the amplifier is distorted (Fig. 5-17A) and is meaningless. There is practically no response below about 2 MHz. This is where the term "high-
peaker" originates, but it is a misleading term. To see the actual effect on

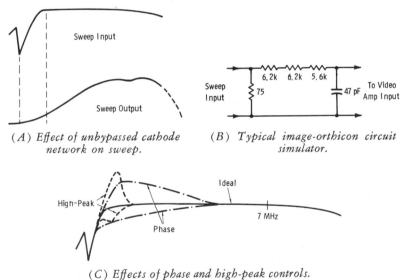

(A) Effect of unbypassed cathode
network on sweep.

(B) Typical image-orthicon circuit
simulator.

(C) Effects of phase and high-peak controls.

Fig. 5-17. Modifications in sweep response.

frequency response, you need to feed the video-sweep signal through an
image-orthicon circuit simulator, an example of which is illustrated in Fig.
5-17B. Note that the so-called "high peaker" affects only the very low end
of the video sweep (see Fig. 5-17C, which shows the amplifier output
with the video-sweep signal fed through the I.O. circuit simulator). The
"phase" control in this instance affects a higher-frequency range up to
around 5 MHz. The main purpose of such controls is phase correction
made necessary by the image-orthicon input resistance-capacitance network. Obviously, the same type of correction is required for vidicons or
any other type of pickup tube.

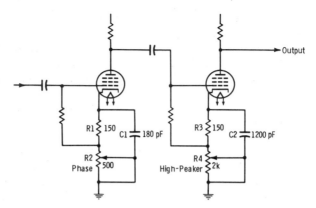

Fig. 5-18. Phase and high-peaker controls.

You can get a rough idea of the intended range of effect on the length of a smear (or streaking) by noting the maximum time constant of the correction networks (Fig. 5-18). For example, the maximum time constant of the phase network (R1, R2, C1) is 0.12 microsecond. Consider the active line interval to be 53 μs (63.5 μs minus the horizontal-blanking time). Note that 0.12 μs is approximately equal to 0.00226 of 53 μs. Assume the raster is 18 inches wide. Then:

$$0.00226 \times 18 = 0.04 \text{ inch, or about } 3/64 \text{ inch}$$

This type of circuit with this time constant corrects short trailing smears.

The maximum time constant of the high-peaker network (R3, R4, C2) is about 2.5μs. This is approximately 0.047 of the raster width. Then on a raster 18 inches wide:

$$0.047 \times 18 = 0.846 \text{ inch, or about } 13/16 \text{ inch}$$

Therefore, this time constant corrects long streaking of trailing edges.

All pickup-tube preamps do not have the same specified frequency response. Usually, the luminance-channel preamp will be specified as "flat" within 1 dB to around 7 MHz. In 4-channel cameras, the chariance tubes have preamps with a much narrower bandwidth, such as to 4 MHz, whereas the luminance channel has response to 7 or 8 MHz. Remember that in the 4-channel camera, the luminance channel carries the high-resolution information, and the color channels provide a relatively broad "paint brush" that requires less high-frequency response.

Figs. 5-19 through 5-22 illustrate various detected sweep curves. Fig. 5-23 identifies the components discussed in connection with these curves in the following paragraphs.

If plate-load resistor R_L (Fig. 5-23) should increase from the normal value, the trace obtained would appear similar to curve 2 in Fig. 5-19. Remember that a higher value of coupling resistance causes a departure

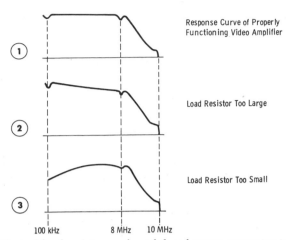

Response Curve of Properly
Functioning Video Amplifier

Load Resistor Too Large

Load Resistor Too Small

100 kHz   8 MHz   10 MHz

Fig. 5-19. Effect of load resistor on demodulated sweep-generator test signal.

from flat response at both high and low frequencies. In this case, we are observing the high passband from 100 kHz to 8 MHz, and the droop toward the upper end of the band is noticeable. If the slope is very pronounced, phase distortion is bound to occur, and loss of resolution is apparent in the picture. Although this effect might be caused by an actual change in value of R_L, this is not necessarily the sole cause. Anything that would affect the dynamic plate load so as to increase its effective impedance over the passband would have the same result. For example, observe curve 2 in Fig. 5-22. This is essentially the same trace as curve 2 of Fig. 5-19, and it is caused by reduced inductance of the shunt peaking coil. The use of the proper value of shunt peaking coil allows a higher value of plate load resistor to be used than when peaking is not employed. Thus reduction of the inductance of this coil results in a condition similar to that of curve 2.

Understanding of these basic circuit relationships materially aids the maintenance engineer in interpreting resultant scope traces. Curve 3 of Fig. 5-19 is a typical trace when R_L has decreased from its normal value. Observation of curve 1 in Fig. 5-22, for which L1 is larger than the optimum value, reveals the effect on plate-load impedance at the higher frequencies, effectively decreasing the value of R_L. The traces are therefore similar in appearance.

The effect of the series-peaking coil is shown in Fig. 5-20. When the series coil is larger than the optimum value (curve 1), a gradual upward slope occurs from 100 kHz (low end of sweep) toward the mid-range of sweep. The larger the value of inductance, the farther the hump is shifted to the left. Compare this with curve 3 in Fig. 5-19, which indicates R_L is too low in value. The major difference in the resulting traces is the extremely reduced cutoff level (at start of maximum downward slope of

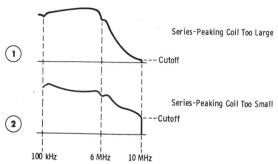

Series-Peaking Coil Too Large

--Cutoff

Series-Peaking Coil Too Small

---Cutoff

100 kHz   6 MHz   10 MHz

Fig. 5-20. Effect of series-peaking coil on response curve.

curve) indicated in Fig. 5-20 (series coil too large). This causes the notch to "slide down" on the sloping portion of the curve at the high end. From the slope of this curve, it may be inferred that the effective load is reduced as the series-peaking coil is increased in value, just as in the case of the shunt-peaking coil. Increasing the value of the series coil lowers the resonant frequency (greater LC ratio) at which the series coil performs. This causes the hump in amplitude response to move to the left (lower in frequency), and the effective load at the highest frequencies in the desired passband is reduced.

If the series peaking coil is too small in value (curve 2 in Fig. 5-20), the response from 100 kHz to midrange of the sweep is too large, and the amplitude at cutoff is increased. This condition also may be caused by a reduced value of damping resistor shunted across the series coil.

The effects of increased values of damping resistance are shown in Fig. 5-21. Curve 1 is displayed when the resistance value has increased to the point at which insufficient damping of the resonant peak occurs. This

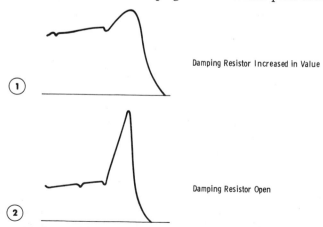

Damping Resistor Increased in Value

Damping Resistor Open

Fig. 5-21. Effect of damping resistor on response curve.

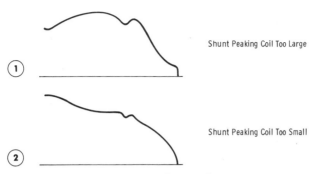

Shunt Peaking Coil Too Large

Shunt Peaking Coil Too Small

Fig. 5-22. Effect of shunt-peaking coil on response curve.

is one possible cause of transient oscillation. Curve 2 indicates an open damping resistor that allows the resonant peak to appear.

The peaking circuits of most video amplifiers are adjustable, as indicated by the variable inductances in Fig. 5-23. The proper alignment of these stages constitutes an important function of the maintenance engineer both in initial setup of amplifiers and in routine and priority checks of equipment. With the sweep generator connected at point 1 in Fig. 5-23, the effects of varying the adjustments of L1 and L2 may be noted. It will be observed that varying series-peaking coil L2 mostly affects the trace at the right of the pattern, and varying L1 mostly affects the trace through the center of the pattern. The marker-notch frequency should be set so that it appears at approximately the assumed limit of the flat portion of the curve, such as 7 or 8 MHz. If, on varying L2, the peak starts moving to the left, the adjustment should be made in the opposite direction to obtain as flat response as possible. Similarly, while the scope pattern is observed, L1 is adjusted to obtain the ideal response characteristic. No more than approximately 2 percent variation should occur from 100 kHz to the limit of the flat portion of the curve. Always compare results with manufacturer's specifications.

The typical traces shown in Figs. 5-19 through 5-22 assume only one component fault, as is usually the case in preventive maintenance or in

Fig. 5-23. Components for shunt and series peaking (Figs. 5-19) through 5-22).

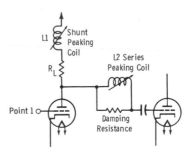

trouble occcurring during operation. If a number of amplitude variations show up in the pattern, several defects may exist simultaneously. In this case, the engineer familiar with the effect of any given adjustment on the corresponding pattern will establish a basis from which to proceed. It is important that the setup of test equipment and test leads produce no spurious response on the screen. Experience with any particular installation is necessary before the engineer can readily determine whether a trace is normal or abnormal.

If a great number of pronounced "humps" or "wiggles" is observed on the scope screen, there probably is a poor ground connection. Always use the shortest possible ground leads in video-sweep measurements. If changing a ground connection to a different point changes the pattern, the grounding arrangement is faulty.

Adjustment of High-Peaker and Phase Controls

As emphasized earlier, final adjustments of the high-peaker and phase controls are made with the camera looking at a test pattern. The horizontal bars in the standard test pattern serve as a reference for adjusting preamp controls. Fig. 5-24A shows slight *positive streaking* (black after black or white after white), and Fig. 5-24B shows *negative streaking* (white after black or black after white). Fig. 5-24C illustrates the appearance of

(A) *Slight positive streaking.*

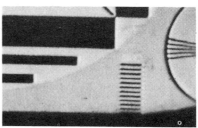

(B) *Negative streaking.*

(C) *Severe negative streaking.*

Fig. 5-24. Examples of streaking.

lettering under severe negative-streaking conditions. A condition this severe may originate in the preamp or in following video amplifiers that incorporate clamping circuitry (Chapter 6).

It may be asked how "anticipatory" streaking can occur before (for example) a white area in the scene, causing streaks all the way from left to right on the raster as in Fig. 5-24C. In Fig. 5-25A, the build-up of low-frequency response causes a gradual decline toward black after the white

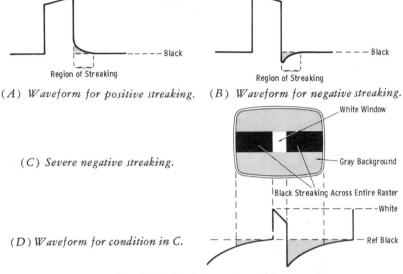

(A) Waveform for positive streaking.   (B) Waveform for negative streaking.

(C) Severe negative streaking.

(D) Waveform for condition in C.

Fig. 5-25. Conditions of streaking.

signal, resulting in white streaking of the white image. In Fig. 5-25B, the loss of low-frequency response causes an overshoot into the black region on the trailing edge of the signal, causing black streaking after the white image. These waveforms are representative of "short streaking" (as in Figs. 5-24A and 5-24B). Control of such streaking normally is within the range of high-peaker and phase controls in the preamplifier.

Fig. 5-25C illustrates a white window on a gray background (analogous to the lettering of Fig. 5-24C) with severe negative streaking. The reason for this may be made clear if we consider more than one line of information, as in Fig. 5-25D. A severe loss of low-frequency response not only affects the shape of the white signal (ideally a flat-topped pulse), but also affects the base line as shown. Thus, the black streaking occurs all the way across the raster, not just following the white signal.

Obviously, the best way to check the low-frequency response of the pre-amplifier is with a square-wave signal. Always be certain of the back-to-back response of your square-wave generator and scope (as described earlier in this chapter) before measurement is attempted. Also, since the input amplitude of a preamplifier normally is very small, 75-ohm pads for the input will be needed.

Fig. 5-26A illustrates a video pad suitable for test-signal hookups to preamplifiers. With all the switches in the "up" position, the signal is passed without attenuation. Any pad or combination of pads may be selected to obtain a number of attenuation values from 3 to 53 dB. The resistors are carbon film, 1 percent, ½ watt, and should be mounted in a shielding box with a spacing of at least two inches from the shield to avoid capacitive effects. The proper coaxial connectors, either BNC or UHF, should be used for input and output.

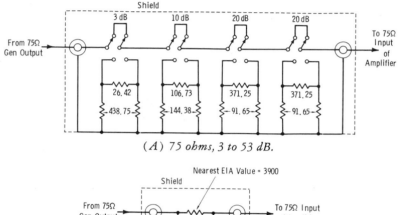

(A) 75 ohms, 3 to 53 dB.

(B) 75 ohms, 40 dB.

Fig. 5-26. Video attenuators.

Fig. 5-26B illustrates a simple 40-dB pad. The series resistor may be either the 1-percent value of 3750 ohms or (when absolute gain measurements are not required) the nearest EIA value of 3900 ohms. In the latter case, the attenuation is slightly more than 40 dB.

In running square-wave response checks, small-value cathode (or emitter) bypass capacitors, such as in Fig. 5-18, must be bypassed with a large-value capacitor as described previously.

It is extremely important that the square-wave input amplitude be sufficiently attenuated that no overload occurs. Overloading obviously would flatten the waveshape and lead to an erroneous interpretation. The 40-dB pad is valuable for this purpose. For example, a 1-volt signal into the pad will give a 10-millivolt signal into the preamp. When observing the output of the preamplifier with the scope, always reduce the input level and check for any effect on the output waveform, so that compression can be avoided.

Video-sweep and square-wave response checks on amplifiers employing clamping circuits are covered in the next chapter.

EXERCISES

Q5-1. Why are phase controls necessary?

Q5-2. Does the output of a video preamplifier always feed a 50-ohm (actually 51-ohm) coaxial cable?

Q5-3. Do peaking circuits in the preamplifier compensate for scanning-beam aperture distortion?

Q5-4. When the video preamplifier does not incorporate a gain control (usually the case), what adjustment is set to give the normal output level of the preamplifier?

Q5-5. Why is it important to have a 40-dB video pad available to the maintenance department?

Q5-6. What could cause white compression in the pickup tube even though the associated amplifiers are linear?

Q5-7. What is the result of using excessive beam current in the pickup tube?

CHAPTER 6

Video Processing

In this chapter, we will consider the special kinds of video-signal processing in amplifiers following the video preamplifier. This includes:

1. Resolution and transient response (bandwidth)
2. Detail contrast (aperture compensation)
3. Cable-equalizing amplifiers
4. The dc component (clamping)
5. The transfer curve (gamma correction)
6. Blanking and sync insertion onto the video signal
7. Testing and maintenance

6-1. THE TRUE MEANING OF BANDWIDTH

The *bandwidth* of a video amplifier normally is defined in terms of the frequency span between the -3-dB amplitude points of the response curve. However, there is an additional characteristic of prime importance to the television engineer; this is the shape of the rolloff of the response curve for a given gain-bandwidth product.

Fig. 6-1 illustrates the point of interest at this time. The solid line indicates a gradual rolloff starting at some frequency below the -3-dB point. The dash line indicates a rolloff starting at a frequency closer to the -3-dB point (extended flat response) but with a much more rapid fall in response. We will now examine the effects of this most important characteristic.

NOTE: It is important that the reader understand the following relationships. (See for example, Harold E. Ennes, *Television Broadcasting: Equipment, Systems, and Operating Fundamentals* [Indianapolis: Howard W. Sams & Co., Inc., 1971].)

1. Horizontal resolution (ability to resolve sharp vertical transitions in the picture) is dependent on bandwidth, and 1 MHz corresponds to approximately 80 TV lines.

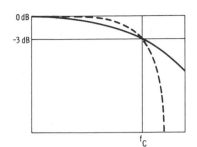

**Fig. 6-1. Rolloff of
frequency response.**

2. Vertical resolution is fixed by the FCC standards and is limited by
the number of active scanning lines, not bandwidth. Because of
the *scanning factor,* vertical resolution becomes about 320 lines,
somewhat less than the actual number of active scanning lines.
3. One cycle of video signal (whether pulse or sine wave) equals two
picture elements, one black and one white.

The High-Frequency Spectrum of Bandwidth

We know that high definition of sharp vertical transitions in the image
requires a well focused scanning beam and sufficient amplifier bandwidth.
A perfect reproduction would require a waveform with zero time for the
vertical transition; it would require an infinitely small scanning beam and
infinite amplifier bandwidth, both impossible in practice. This is to emphasize that high horizontal resolution requires a pulse concept of effective
bandwidth rather than a sine-wave concept. This applies just as well to
the low-frequency spectrum of bandwidth, which we will cover later.

Since the video signal must be considered a "pulse" signal, we know that
the pulse rise time versus amplifier bandwidth is important. Fourier
analysis tells us that for an amplifier to pass a pulse with the same rise
time and shape as the input pulse, the amplifier bandwidth must be:

$$BW = \frac{1}{2RT}$$

where,
BW is the bandwidth,
RT is the rise time of the pulse.

This says that the bandwidth must be equal to half the reciprocal of the
rise time. For a pulse with $RT = 0.02$ μs:

$$BW = \frac{1}{2(0.02)} = \frac{1}{0.04} = 25 \text{ MHz}$$

(Since the rise time is in microseconds, the result is in megahertz.)

Since the limit of bandwidth normally is taken to be the −3-dB point (Fig. 6-2A), the frequency at this point is designated the cutoff frequency (f_C). Assuming the input pulse does not have a faster rise time than the amplifier, phase shift is proportional to frequency (Fig. 6-2B), with uniform time delay at all frequencies in the passband. Thus, the reproduced waveform (Fig. 6-2C) is the same as that applied, with no overshoot or undershoot. Bear in mind that the curve in Fig. 6-2A has a gradual rolloff beyond the −3-dB amplitude-response point. This is normally specified as a 6-dB/octave rolloff.

For the wideband curve of Fig. 6-2A, since:

$$BW = \frac{1}{2RT}$$

it follows that:

$$RT = \frac{1}{2BW}$$

Thus, if the wideband response is 10 MHz, the rise time of the amplifier is:

$$RT = \frac{1}{20} = 0.05 \ \mu s$$

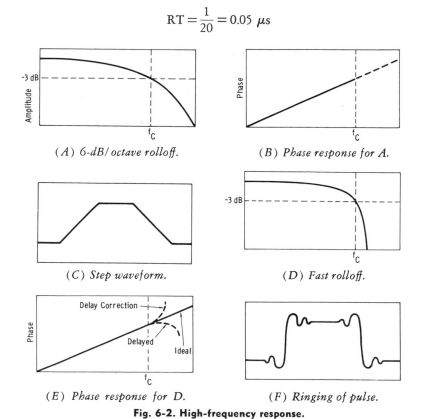

(A) 6-dB/octave rolloff.

(B) Phase response for A.

(C) Step waveform.

(D) Fast rolloff.

(E) Phase response for D.

(F) Ringing of pulse.

Fig. 6-2. High-frequency response.

If the applied pulse waveform (Fig. 6-2C) has a rise time of no more than 0.05 μs, the output will be free of overshoot or undershoot. If the wideband response with the rolloff of Fig. 6-2A is 25 MHz, the amplifier will pass a pulse with a 0.02 μs rise time without distortion.

In practice, designers make a compromise in wideband amplifiers, and the shape of the rolloff is made only sufficient to allow no more than a 2 to 3 percent overshoot for a pulse rise time equivalent to the amplifier rise time. For the purpose of visualizing this relationship, a k factor is defined as follows:

$$(BW) \ (RT) = k$$

where,

BW is the bandwidth in megahertz (to the -3-dB point),
RT is the rise time in microseconds (measured between 10 and 90 percent of peak value),
k is a factor between 0.3 and 0.5, depending on the type and amount of high-frequency compensation.

The limit on factor k is that the overshoot on the leading edge of a pulse must be less than 3 percent. In fact, a system has an equivalent bandwidth and rise time only within the limits of 3-percent overshoot.

The most typical value for k is 0.35, and the equation may be expressed in three possible ways:

$$(BW) \ (RT) = 0.35$$

$$(BW) = \frac{0.35}{(RT)}$$

$$(RT) = \frac{0.35}{(BW)}$$

Table 6-1 is based on this relationship and shows the equivalent TV lines of horizontal resolution for different values of rise time and bandwidth.

Table 6-1. Resolution and Amplifier Response

BW (MHz)	RT (μs)	Equivalent TV Lines
1	0.35	80
2	0.175	160
3	0.1166	240
4	0.0875	320
5	0.07	400
6	0.058	480
7	0.05	560
8	0.0437	640
9	0.039	720
10	0.035	800

Notice that a pulse with 0.035 μs rise time will be passed with no more than 3 percent overshoot by an amplifier that has a 10-MHz bandwidth. For example, an oscilloscope ideally must show all TV waveforms, including sync pulses, exactly as they are, without introducing any distortion in the scope amplifier. The "wideband" position of the Tektronix 524 scope has a bandwidth to 10 MHz (-3-dB point) with a *gaussian* rolloff curve. With this characteristic, the -3-dB point occurs at 10 MHz, and the -12-dB point occurs at 20 MHz. This is a 9-dB/octave rolloff. (When a frequency is doubled, it has been increased by one octave.) This type of rolloff is suitable for scope amplifiers used to display television waveforms. More recent oscilloscopes, such as the Tektronix 545 (30 MHz) and 547 (50 MHz), have much shorter rise times (11 nanoseconds and 7 nanoseconds, respectively).

A gaussian response curve, while essential in oscilloscope vertical amplifiers, is not found in television camera chains or in video distribution amplifiers. This is because of the limitation of rise time in a series of amplifiers forming a cascaded system. The rise time of the original waveform is reduced by the square root of the sum of the squares of the individual amplifier rise times.

For example, suppose we pass a signal through two identical amplifiers with 10-MHz bandwidth and gaussian response. The combined rise time of the amplifiers is:

$$RT = \sqrt{0.035^2 + 0.035^2}$$
$$= \sqrt{0.00245} = 0.05 \ \mu s \ (\text{approx})$$

This is a 40-percent increase in rise time as a result of passing the signal through just two cascaded 10-MHz gaussian amplifiers. In practice, many video amplifiers are cascaded in forming a complete system.

Thus, the practical video amplifier must have a flat frequency response up to and including the highest anticipated frequency, with a relatively rapid rolloff beyond this frequency. It can be shown from pulse theory that rise time is proportional to the *area* under the amplitude-frequency response curve; hence cascading such amplifiers does not appreciably affect the rise time. However, such an amplifier will *not* reproduce a step transition without overshoot, ringing, or other transient distortions at the output.

It is actually the shape of the curve that is being changed when video-peaking coils are adjusted. Leading and trailing transients of a rapid transition in picture content must be adequately controlled. Hence it is necessary for maintenance personnel to have complete familiarity with the scope-amplifier and video-amplifier characteristics.

Fig. 6-2D illustrates a frequency-response curve which more nearly approaches that of the average video amplifier. The resultant phase response is shown in Fig. 6-2E. Remember that a "pulse" (step transition) requires transmission of the higher-order harmonics (which may actually be above the passband intended) to be free of waveshape distortion. Fig. 6-2F shows the resultant ringing that occurs. The amplitude of such ringing depends on the step-transition rise time for a given amplifier bandwidth and rolloff characteristic. The distribution of ringing (leading and trailing) is an indication of direction and degree of phase shift. Actually, Fig. 6-2F shows a "phase corrected" signal, as indicated by the even distribution of ringing at the leading and trailing edges. Late arrival of high-frequency components causes most of the ringing to occur on the trailing edge, whereas early arrival of these components causes most of the ringing to occur at the leading edge.

The Sine-Squared Pulse

Thus far, we have learned that the sine-wave response of a video amplifier does not provide a complete story of the amplifier performance for a video signal. Likewise, a step transition (or a square-wave signal) is not a particularly useful signal for evaluation unless the exact rise time of the pulse is correlated properly with the intended passband of the amplifier. The very important *transient response,* which accounts for the degree of picture ringing, smearing, or streaking, requires a rather precise analysis method to assure valid tests in practice.

It is pertinent to recall here the actual pickup-tube output when a "square-wave" test pattern is scanned. Remember that, because of the round shape and finite size of the scanning beam, the actual output more nearly approaches a sine wave (actually a sine-squared wave, which we will develop) rather than a square wave. Therefore, the practical video amplifier is not required to handle square-wave video information (discounting blanking and synchronizing pulses in a composite picture).

The problem of pulse rise times that are not related to the actual picture transmission spectrum is the reason why, sooner or later, we will be concerned with the sine-squared (sin^2) pulse. The basic usefulness of this test signal is in complete system testing, and therefore a complete study of the uses of the sin^2 pulse is more appropriate for a text on system maintenance. However, we will go through an introductory examination here, because the sin^2 pulse provides an excellent means for revealing the practical requirements of a video amplifier, and because it has certain limited applications in camera-chain maintenance, as covered later in this chapter.

First of all, be sure to understand what a "picture element" is. A picture element is determined by the available bandwidth. Our complete TV system is fixed by FCC standards that allow the video transmitter only a 4-MHz bandwidth for the picture signal.

One cycle occurs in a time equal to the reciprocal of its frequency; for example:

$$1 \text{ cycle at 4 MHz} = \frac{1}{4 \times 10^6} = 0.25 \text{ microsecond}$$

Therefore, a black-to-white transition with a width representing 4 MHz will occur in 0.25 microsecond. But black is one picture element, and white is one picture element. Therefore, a picture element in a 4-MHz system represents 0.125 microsecond (one alternation of the complete cycle). In the sin^2 technique, a time duration equivalent to one picture element is given the symbol T, whereas a time duration equivalent to two picture elements (for the system bandwidth under test) is symbolized by 2T.

An explanation of the sin^2 pulse is shown in Fig. 6-3. In Fig. 6-3A, note the conventional continuous sine wave with a frequency of 4 MHz

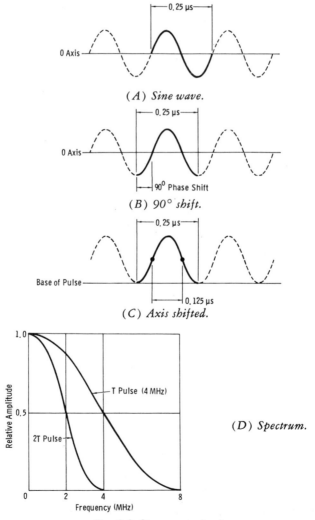

(A) Sine wave.

(B) 90° shift.

(C) Axis shifted.

(D) Spectrum.

Fig. 6-3. Sine-squared pulse.

(one cycle occurs in 0.25 μs). We realize from fundamental theory that any phase shift of a continuous sine wave can be measured only by laborious methods not suitable for routine testing of transmission facilities. Also, the amplitude-frequency characteristic of a system simply shows the amplitude of the continuous sine wave relative to the amplitude at a reference frequency, unless we are equipped to measure the phase relative to a known reference.

Observe Fig. 6-3B. If we shift the waveform 90°, we have one complete cycle of an inverted 4-MHz cosine wave (starting and finishing at negative peaks). Now with an added dc component of such value as to raise the negative peaks to the zero line, we have the T pulse for a 4-MHz system (Fig. 6-3C). Note that the half-amplitude duration (h.a.d.) is 0.125 μs, equivalent to one picture element for a 4-MHz bandwidth. Fig. 6-3D shows that the significant energy of the T pulse is down 50 percent (6 dB) at 4 MHz, and there is practically no energy beyond 8 MHz. The energy of the 2T pulse (h.a.d. of 0.25 μs) is 50 percent (6 dB) down at 2 MHz, with no significant energy beyond 4 MHz. Thus, the system can be checked with a pulse that essentially duplicates actual picture conditions and that provides known frequency content upon which to base judgement of system performance. Please note that any similarity to the sine wave no longer exists; a *pure* sine wave has no harmonic content at all.

Fig. 6-4A shows the preceding definition in terms of T and system bandwidth. Fig. 6-4B shows the terminology used with a pulse that has been passed through an amplifier or (more usually) a complete system. The *first lobe* is a negative overshoot, and the *second lobe* is a positive overshoot, either preceding or following the pulse.

The sin^2 pulse generator normally also generates a window signal following the pulse (Fig. 6-4C) so that an amplitude reference for low frequencies is established. The sin^2 pulse appears as a thin, vertical, white line on the left of the raster immediately following blanking, and the white window is on the right side, centered vertically on the raster as indicated by the field display. This signal is used frequently for system and line checks. The window signal has a rise time equivalent to that of the sin^2 pulse.

The amplitude-frequency and amplitude-phase response at higher frequencies (beyond about 100 kHz) will be most evident in the measurement of the sin^2 pulse. Amplitude-phase response is most evident in the measurement of the window signal.

NOTE: Some generators place the sin^2 pulse following rather than preceding the window. This has no effect on a basic understanding of the measurement principles.

Distortions at low frequencies produce waveform changes with a long time constant as, for example, in streaking. This is most evident by window measurement. Distortions at higher frequencies produce waveform changes with shorter time constants as, for example, in smearing, loss of resolution, or "edge effects" from bad transient response. This is most evident by sin² pulse measurement.

High-frequency rolloff results in loss of amplitude. Loss of amplitude results in a *widening* of the pulse, since the area of the pulse represents a

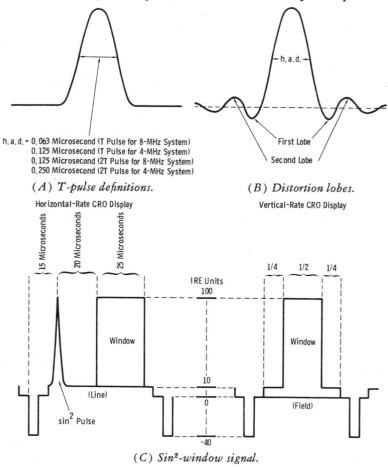

h.a.d. = 0.063 Microsecond (T Pulse for 8-MHz System)
0.125 Microsecond (T Pulse for 4-MHz System)
0.125 Microsecond (2T Pulse for 8-MHz System)
0.250 Microsecond (2T Pulse for 4-MHz System)

(*A*) *T-pulse definitions.*   (*B*) *Distortion lobes.*

Fig. 6-4. Television test signals.

(*C*) *Sin²-window signal.*

constant dc component. A slow rolloff within the video band produces large reduction in amplitude (and pulse-width increase) with little or no ringing. A rapid rolloff close to the top of the band but still within the desired video bandwidth produces both ringing and a reduction (perhaps slight) in amplitude. A rapid rolloff (almost a cutoff) just above the video bandwidth concerned results in practically no effect on amplitude, but does produce ringing. The shape of the rolloff and whether the resulting phase shift is leading or lagging is revealed by the distribution of ringing before and after the pulse.

The window detects low-frequency distortion which has practically no effect on the sin² pulse. The window shows undershoot, overshoot, and horizontal tilt depending on the time constant of the impairment. The window when used with the sin² pulse has the same rise time as the pulse so that no frequencies beyond the system test reference are introduced.

The Low-Frequency Spectrum of Bandwidth

See Fig. 6-5A. The rise and decay times of a pulse depend on the system high-frequency response and the shape of the passband response curve. We are concerned now with the duration response (t_d), which depends on t/RC, or time divided by the RC product. This, of course, is the low-frequency characteristic in practice.

The output voltage as a function of t/RC is shown in Fig. 6-5B. As t increases from 0, the factor t/RC increases and the output voltage decreases until, at $t/RC = 1$, the output voltage drops to 0.37 of the initial voltage.

Since the pulse durations required in a TV system are known, it is most convenient to use the reciprocal of the above relationship in thinking of practical RC-coupled circuits. Fig. 6-5C shows a plot of the output voltage during a pulse in relation to the RC/t_d ratio. Note that it is necessary to have an RC product of 10 times the pulse duration (t_d) to avoid more than a 10-percent tilt over the duration of the pulse. It is obvious that the time-constant problem becomes severe in a practical circuit when the duration of a field is 16,666 μs (the reciprocal of 1/60 second). The time constant (TC) is given in seconds when R is in ohms and C is in farads or when R is in megohms and C is in microfarads. It is given in microseconds when R is in ohms and C is in microfarads or when R is in megohms and C is in picofarads.

For example, the combination of a 0.1-μF coupling capacitor and a 1-megohm grid resistor results in a time constant of 100,000 μs. You can see that this value is not 10 times the field duration. The TC value in practical circuits is limited by the stability factor (motorboating, large capacitances to ground, etc.); this is the reason why either negative feedback to flatten the lows (as well as the highs) is used, or the low-frequency boost circuit is employed (previous chapter). In amplifiers incorporating clamping circuits, the low-frequency characteristics are almost entirely dependent on proper operation of the clamp-pulse former and clamping circuit (discussed later).

See Fig. 6-5D. This cathode-follower circuit has a sine-wave frequency response that is only 3 dB down (relative to 1 MHz) at 5 Hz. But, for all practical purposes, the time constant of this circuit is 1000 × 12, or 12,000 μs. This is an RC/t_d ratio of less than 1 for the field duration in a television signal. The circuit in Fig. 6-5D is that of an oscilloscope probe.

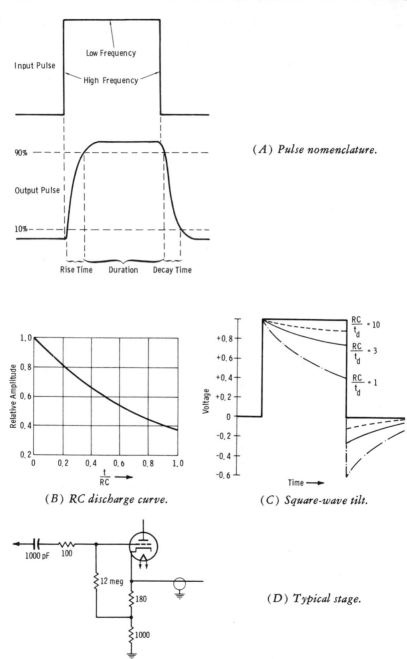

(A) Pulse nomenclature.

(B) RC discharge curve.

(C) Square-wave tilt.

(D) Typical stage.

Fig. 6-5. Factors in low-frequency response.

Since capacitive loading of the circuit by the probe must be minimized, a small coupling capacitor must be used. This probe permits making circuit checks at medium and high frequencies without a sacrifice in gain such as occurs when a 10-to-1 capacitance-divider probe is used. It is not intended to be used in those instances in which low-frequency duration checks are important.

This example is intended to form a sharp demarcation line in your mind between "sine-wave response" and "transient response." An amplifier that checks "flat" down to 5 Hz may produce excessive "tilt" in a low-frequency step-response waveform.

6-2. THE TELEVISION-CAMERA RESOLUTION CHART

The resolution chart was designed to provide a standard reference for measuring resolution of television cameras and as an aid in testing for streaking, ringing, interlace, shading, scanning linearity, aspect ratio, and gray-scale reproduction. It contains low-frequency, midfrequency, and high-frequency information.

Resolution

The horizontal resolution that may be obtained from many camera chains is limited by the resolving capabilities of the camera tube, and not by the bandwidth of the video amplifiers employed. Therefore, much useful information concerning the limiting resolution, percent response at various line numbers, and degradation of resolution with aging of camera tubes and components can be obtained from a test chart containing a high number of lines. Fig. 6-6 illustrates a typical resolution chart with high-frequency resolution detail up to 1600 TV lines.

The types of resolution charts vary with the intended purpose. For example, the usual station-identification resolution chart often broadcast just prior to sign-on has vertical- and horizontal-resolution wedges going only to 320 lines. This is the limiting vertical resolution set by FCC standards, and the limiting horizontal resolution (also from FCC standards) corresponding to the bandwidth limitation of 4 MHz at the visual transmitter.

The chart shown in Fig. 6-6 is more useful at the studio, where the bandwidth normally extends to 10 MHz, for the purpose of evaluating and adjusting camera chains. The center horizontal and vertical wedges are composed of four black lines separated by three white lines of equal width. The numbers printed alongside the wedges correspond to the total number of lines (black and white) of the indicated thickness that may be placed adjacent to one another in the *height* of the chart. For example, if black and white lines having the same thickness as those indicated at the 300 position were placed adjacent to one another, a total of 300 (black *and* white) lines could be fitted into the height of the chart. Since the aspect ratio of the chart is 4 to 3, a total of 300 × 4/3, or 400, lines of this thickness could be placed in the width of the chart.

The fundamental frequency (based on FCC television standards) developed in scanning through the 300 position of one vertical wedge may be calculated as follows:

Horizontal scannning frequency (nominal) = 15,750 Hz
H = time for active scan + horizontal blanking = 63.5 μs
Assume horizontal blanking = 0.17H
Therefore active scan = 0.83H
Active scanning time = 52.7 μs

The total number of vertical black and white lines, having the thickness indicated at the 300-line position on the chart, that could be placed adjacent to each other in the width of the chart is 400 (see preceding paragraph). Since a complete cycle includes one black line and one white line, there should be 400/2, or 200, cyclic variations in scanning this pattern (200 cycles in 52.7 μs). Thus, the time to scan (horizontally) one black and one white line at the 300-line position is 52.7/200, or 0.26 μs. The fundamental video frequency is:

$$f = \frac{1}{t} = \frac{1}{0.26}$$

$$= 3.8 \text{ MHz}$$

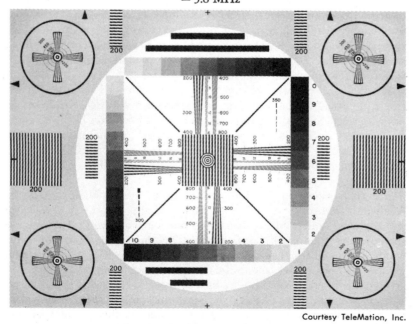

Fig. 6-6. Typical resolution chart.

The limiting resolution as a result of the TV-transmitter bandwidth (4 MHz) is 320 TV lines (from Table 6-1). Therefore, ideally, sufficient aperture correction should be applied to the pickup-tube response to obtain 100-percent response at 300 to 320 lines on the resolution pattern. Circuitry and problems of aperture correction are covered later in this chapter.

The fundamental video frequencies generated by scanning through different parts of the vertical wedges may be determined from the formula:

$$\frac{N}{f} = K$$

where,

N is the indicated line number on the chart,
f is the fundamental video frequency in MHz,
K is 80 (320 lines/4 MHz = 80),

When solved for frequency, the formula becomes:

$$f = \frac{N}{80}$$

Table 6-1 lists the TV line number for whole multiples of 1 MHz.

Shading

Shading may be checked by visual inspection of the picture monitor to determine if the background is an even gray, and if the same number of steps is discernible on all four gray scales. Also, a waveform monitor may be used to determine if the average picture-signal axis is parallel with the black-level line at both line and field frequencies.

Streaking

Streaking of the horizontal black bars at the top or bottom of the large circle is an indication of low-frequency phase shift or of poor dc restoration. The black bars are also useful in adjusting the high-peaking circuits that are used in camera chains to compensate for the high-frequency rolloff of the coupling network between the camera tube and first video amplifier.

Interlace

The four diagonal black lines inside the square formed by the gray scales may be used to check interlace. A jagged line indicates pairing of the interlaced lines.

Gray Scale Reproduction

The transfer characteristic of the camera, for given operating conditions, may be determined by using an oscilloscope with a line detector (discussed later in this chapter). The gray-scale reproduction achieved depends on the amount of gamma correction employed, the manner in which the camera tube is operated, and the adjustment of the picture monitor. The user will have to standardize these operating conditions if comparative subjective measurements are to be made. (Gamma correction is covered later in this chapter.)

The four ten-step gray scales cover a contrast range of approximately 30 to 1. The reflectance of step 1 is determined by the reflection density of the chart material forming the center circle. The nine-step "paste-on" gray scales cover a nominal contrast range of 20 to 1, step 2 having a reflectance of 60 percent and step 10 a reflectance of 3 percent. The steps are arranged in logarithmically decreasing values of reflectance such that the difference in reflection density between adjacent steps is 0.16. Table 6-2 gives the reflectance and reflection density of the steps on the gray scales. The background reflectance of the outer useful area of the chart is 40 percent ± 5 percent.

Table 6-2. Specifications for Gray Scales

Gray Scale Number	Nominal Reflectance Relative to MgO (Percent)	Nominal Reflectance Density
1*	>60.0	<0.22
2	60.0	0.22
3	41.7	0.38
4	28.2	0.55
5	19.5	0.71
6	13.5	0.87
7	9.3	1.03
8	6.3	1.20
9	4.4	1.36
10	3.0	1.52

*Center circle

6-3. CLAMPING CIRCUITRY

This section assumes the reader has a background in the basic reasons for and theory of clamping circuitry.[1] It remains to explore the more advanced features of various types of clampers found in both tube-type and solid-state equipment.

It is important to understand the difference between "clamping" and the broader term "dc restoration." The clamper restores the dc component by means of a keying signal such that a line-to-line reference is established

---

[1]See, for example, Harold E. Ennes, *Television Broadcasting: Equipment, Systems, and Operating Fundamentals* (Indianapolis: Howard W. Sams & Co., Inc., 1971).

at either sync-tip level or back-porch blanking level. Other, more simple forms, such as unkeyed diodes, are grouped under the general classifica-cation "dc restoration."

There are two basic types of clampers: the "fast" clamp that practically eliminates any 60-Hz hum component, and the "slow" clamp that prevents a change in baseline reference under varying picture levels (APL's), but is sufficiently slow to allow 60-Hz hum components to pass. The latter form is used in modern camera-waveform monitors to prevent the blank-ing level from shifting under varying APL's, but to allow a hum component to pass with very little attenuation so that this trouble is apparent to the operator. In this section, we will consider only the fast clamp as used in camera chains. The slow clamp as used in camera-waveform monitors is covered in Chapter 8.

Clamping is performed in any circuit in which a video reference black must be established. Examples are blanking- or sync-insertion stages, gam-ma-correction circuits, sync-stretching circuits, etc.

Fig. 6-7A shows the basic form of keyed clamping. The charge time of coupling capacitor C_C is limited only by the value of source resistance R_S. If a small-value coupling capacitor is used together with a low source

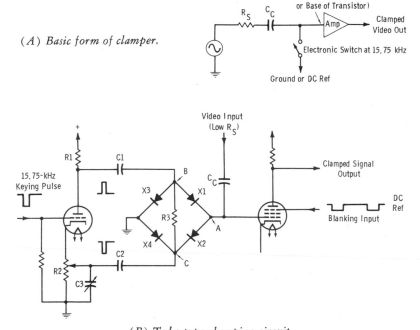

(A) Basic form of clamper.

(B) Tube-type clamping circuit.

Fig. 6-7. Clamping circuits.

resistance (such as from a cathode follower or emitter follower), C_C is charged rapidly when the electronic switch is closed. The switch is closed only for about 1.5 μs either at sync-tip time or during the back-porch blanking times. Discharge time is limited only by the impedance in the switch.

Fig. 6-7B illustrates a basic tube-type clamping arrangement. In this example, blanking is inserted on the suppressor grid of the clamped tube. A dc reference of picture black (or camera blanking) must be established at the control grid for the start of composite blanking so that this reference does not change with average picture level.

Negative pulses from the cathode circuit and positive pulses from the plate circuit of the clamp-driver tube supply clamping pulses at line frequency for the four-diode clamper. Point A is the point to be clamped; that is, point A must be maintained at a predetermined reference level on successive pulses of the video signal. The diodes function as an electronic switch triggered by the line-frequency pulses. When pulses are applied, point B becomes positive and point C becomes negative, creating the condition for diode conduction to occur. This is equivalent to closing the switch of the simple circuit in Fig. 6-7A. When the diodes conduct, a low-resistance charge (or discharge) path for capacitor C_C results. During the active line-scanning interval, the diodes do not conduct. During this time, should point A develop a negative dc potential, conduction will occur through X1 to point B. If point A should develop a positive dc potential during this time, conduction will occur through X2 to point C. In either case, the clamped point is maintained at a certain reference level that determines the dc level of the video signal. Resistor R3 prevents point B from drifting in the positive direction and point C from drifting in the negative direction, which would render the clamp inoperative. Thus, the clamped point (A) is maintained at reference level during the line-scanning interval, and it is discharged (or charged) at the conclusion of each scanning line. By this means, interaction on the video signal information is prevented, since the time constant (product of C_C and the forward clamp resistance) is long compared with the line interval.

Potentiometer R2 and capacitor C3 are used for balancing the clamp circuit. The need for this may be realized by noting that point A is at high impedance, and therefore a small component of the pulse appearing at point B will appear at point A through the cathode-to-anode capacitance of X1. From the same principle, a pulse from point C will also appear at point A. If these pulses are effectively balanced in phase, amplitude, and rise time, their effect at point A is nullified, since they are equal and opposite in polarity. Without R2 and C3, however, a slight unbalance would occur, since the cathode clamping pulse tends to rise more rapidly than the plate clamping pulse. By adjusting R2 to balance the amplitudes, and C3 to delay the cathode pulse slightly, an exact balance may be achieved in practice.

The preferred pulse polarity for the input of the clamp-driver tube is such that the tube conducts during active line-scan intervals (diodes held open) and is cut off for the pulse duration. The resultant high source resistance for the pulses gives a more balanced drive condition for the pulse-coupling capacitors, C1 and C2. Time constants R1C1 and R2C2 must be long compared to the pulse time. The clamping pulses are large compared to the video signal amplitude so that the video signal does not cause diode conduction during the active line interval. As a rule of thumb, the video amplitude at the clamped point is one-third to one-fourth the amplitude of the clamping pulses.

Fig. 6-8A illustrates the usual transfer through an RC-coupled circuit of video waveforms with different APL's. This illustration also indicates exact proportionments of white to black durations for 50, 10, and 90 percent APL. Thus, if the active line-scan interval is 0.83H, a white pulse of half this duration (0.415H) simulates an average picture level of 50 percent (average scene). With the same active line-scan interval, a white pulse of 0.083H simulates a 10 percent APL (dark picture), and two

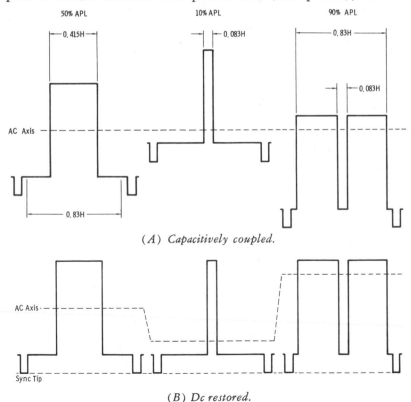

(A) Capacitively coupled.

(B) Dc restored.

Fig. 6-8. Effect of clamping on transferred waveform.

white pulses with a black interval of 0.083H between them simulate a 90 percent APL (almost all-white picture). Note the marked difference in sync levels.

Fig. 6-8B shows the transfer of the same waveforms through a clamped RC-coupled stage. Since the active line-scan signal always starts from the reference dc potential at the beginning of each line, sync and blanking levels remain the same regardless of APL; thus the video-signal ac axis is effectively changed.

Fig. 6-9 illustrates the pulse-transformer type of clamping circuit. The base of clamp driver Q1 receives a sharp negative pulse to drive it from cutoff to saturation. The sudden collector current through the primary of the pulse transformer rings this circuit because of inductive kickback. But after the first positive alternation, diode X5 clamps the negative portion of the ringing waveform by shorting out the primary when the collector attempts to swing negative. Note that this polarity would be reversed if an npn transistor were used. Note also the polarity of the resulting pulses on the secondary and how these pulses result in forward biasing of the quad diode circuit, closing the "switch" and applying −12 volts to the Q2 clamped base. Coupling capacitor C_C is always small, since it must be charged or discharged quickly (during the approximately 1.5-μs duration of clamping) to the reference −12 volts. Time constant R1C1

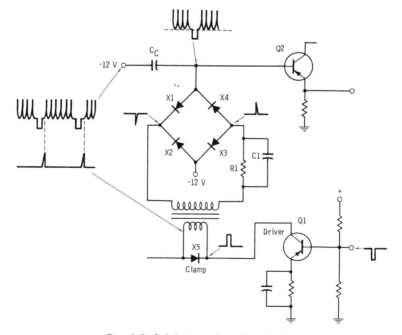

Fig. 6-9. Solid-state clamping circuit.

must be long compared to a line interval so that the charge on C1 will hold the switch open (nonconducting) between pulses during the active line (video) interval.

Back-Porch-Clamp Timing Circuitry

It has been mentioned that clamping may occur either at sync-tip level or immediately following sync at the blanking level of the back porch. We will now examine how the clamping pulses are delayed to the back-porch interval.

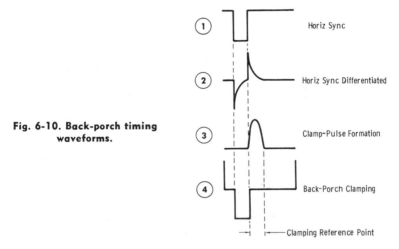

Fig. 6-10. Back-porch timing waveforms.

Horizontal-sync pulses normally are used to form clamping pulses for either sync-tip or back-porch clamp timing. See Fig. 6-10. Horizontal sync (waveform 1) is differentiated (waveform 2). The following circuit is sensitive only to a positive-going pulse so that a clamp pulse is formed at the time of trailing edge of sync (waveform 3). Waveform 4 shows the resultant timing of the clamping action.

See Fig. 6-11A. Since the ratio of R_B to R_L is only ten to one, we should recognize the familiar "boxcar" circuit (transistor saturated in quiescent operating condition). The value of coupling capacitance C is small so that the time constant results in differentiation of the input pulse.

At the base of the transistor, the waveform received through the coupling capacitor would be a negative-going excursion followed by a positive-going excursion if it were not for the base clamping action; all negative-going excursions are clipped because of the low forward impedance of the junction. Since the transistor is already saturated, negative-going excursions have no effect. At the end of the input pulse, the positive-going excursion drives the transistor toward cutoff, and C begins to charge through R_B toward -10 volts. Thus, the output pulse occurs at the trailing

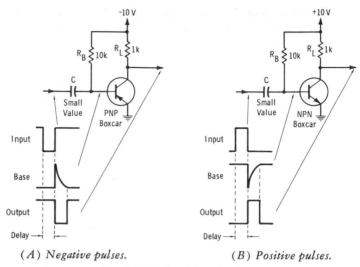

(A) Negative pulses.   (B) Positive pulses.

Fig. 6-11. Pulse-delay circuits.

edge of the input pulse. The width of the output pulse is approximately $0.7R_BC$. Obviously, if R_B is made variable, the pulse width can be adjusted as desired.

Thus, although the input sync pulse is about 5 μs in duration, the width of the output pulse is dependent on time constant R_BC. If C is 170 pF, the output pulse width is:

$$(0.7)\ (0.01)\ (170) = 1.2\ \mu s\ (approx)$$

Fig. 6-11B illustrates the corresponding circuit for a positive-going input pulse. Since the npn transistor is already saturated, the positive-going input excursion has no effect. The trailing-edge negative-going pulse (produced by differentiating action) cuts the transistor off, resulting in a positive-going pulse. As in the previous case, the width of the output pulse is determined by the input time constant, which determines the discharge time of the coupling capacitor.

We will go through the circuit of Fig. 6-12 to show that it is possible to analyze a chain of pulse circuitry rapidly by applying fundamental knowledge. This analysis will show how to determine what waveforms to expect if the circuit is functioning as intended.

Since the circuit in Fig. 6-12 is a back-porch clamper, we know that clamping pulses must occur after the trailing edge of horizontal sync. Comparison of the circuit of Q1 with Fig. 6-11B reveals how the delay is obtained.

Note that Q1 is actually a form of boxcar, since the emitter of the npn transistor is returned directly to −20 volts and the base is grounded through its resistor. The delayed negative-going base pulse drives Q1 to-

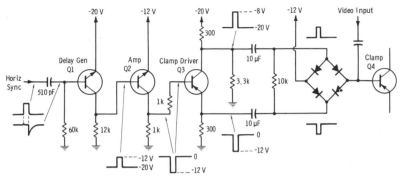

Fig. 6-12. Fundamentals of back-porch clamper.

ward cutoff, but the Q1 collector is directly coupled to the Q2 base. The positive-going pulse excursion at this point is "caught" at −12 volts by the base-emitter clamp of Q2, and can go no further in the positive direction.

We know that prior to the pulse, Q1 is saturated, and its collector is at −20 volts (switch closed). At this same time, since the Q2 emitter is at −12 volts and its base is at −20 volts, Q2 is cut off (the base is negative relative to the emitter). The positive-going pulse at the Q2 base then drives this transistor on, and since there are no limiting diodes and no emitter resistance for self-bias, Q2 saturates. So we should expect a Q2 collector pulse from 0 volts (cutoff) to −12 volts (saturation).

The unbypassed emitter resistance of phase-splitter Q3 tells us that the input impedance of Q3 is quite high (very low base current), so we should expect about the same pulse amplitude directly at the Q3 base as at the collector of Q2.

Now analyze Q3, first the interval between pulses, then the pulse interval. Between pulses, the Q3 base is essentially at zero voltage, so the transistor is cut off. During the pulse interval, since the Q3 emitter and collector loads are identical, we expect essentially unity gain. Therefore, we expect the Q3 emitter pulse to extend from about 0 (cutoff) to −12 volts (unity gain). The emitter current at the −12-volt peak is:

$$I_E = \frac{V}{R_E} = \frac{12}{300} = 40 \text{ mA}$$

For a quick analysis, assume the collector current is equal to the emitter current. Then the signal-voltage swing during the pulse is:

$$V_C = (0.04)(300) = 12 \text{ volts}$$

Therefore, the collector pulse swings 12 volts from −20 volts (cutoff), or up to −8 volts.

In practice, the collector current is slightly less than the emitter current ($I_C = \alpha I_e$), and for this reason you will normally find the collector load resistor of a clamp driver slightly higher in value than the emitter resistor. For a 300-ohm emitter resistance, the collector load usually is about 330 ohms. Thus, the slightly smaller collector-current swing develops a voltage swing the same as that at the emitter.

The above analysis should serve to emphasize an important servicing technique: Always use the dc-amplifier position when making scope checks. This tells you considerably more about circuit operation than does the ac-coupled position. For large-signal operation, this is very convenient.

In case of suspected clamping problems, always check the clamp pulses themselves at the anode and cathode of the diode being driven by the pulses. In some circuits, the pulses are particularly critical in rise and fall times and the shape of the pulse tops. Any difference in rise and fall times between the two opposite-polarity pulses sometimes can result in the appearance of spikes at the clamped grid. A marked difference in pulse-top shapes will appear as a signal voltage at the clamped grid. In practice, a significant difference in amplitude can exist between the pulses *without* causing trouble. This is true up to the point at which the amplitude is not sufficient to drive the diode into conduction.

Loss of clamping normally results in very long streaking of horizontal lines in the image. This type of signal impairment is much more severe than misadjustment of high-peaker stages in video preamplifiers.

6-4. APERTURE-CORRECTION CIRCUITRY

"Phaseless" aperture correction normally is achieved by a delay-line technique in which one end of the line is terminated and the other end is effectively open. The "open" end actually is a high-impedance point in tube or transistor circuitry. A solid-state aperture-correction circuit based on this principle is shown in Fig. 6-13. The problem with transistors is that, because of the finite input impedance, an effective open circuit is hard to obtain if the transistor is used as in a conventional tube circuit. Fig. 6-13A illustrates how this problem may be solved. The collector of Q1 provides a sufficiently high impedance to serve as the open circuit for the delay line, in this case at the sending end. The line is terminated in its characteristic impedance at the collector of Q2.

The signal is divided into two parts: the delayed signal from the Q1 collector at the Q2 collector, and the undelayed signal from the Q1 emitter. Since the collector signal is delayed, the undelayed signal of reverse polarity at the output is, in effect, anticipatory. Now note that the undelayed component appearing at the collector of Q2 goes back down the delay line and is reflected from the relatively high (unterminated) impedance at the Q1 collector. The reflected signal returns to the Q2 collector as a second component of reverse polarity. Thus, this single delay line supplies both an *anticipatory* component and a *following* overshoot on transitions, resulting in symmetrical aperture correction. Fig. 6-13B shows how the aperture correction is made to compensate the scanning-aperture rolloff. Fig. 6-13C shows the sine² pulse response indicating phaseless aperture correction. Recall that conventional peaking circuits introduce phase distortion, which must be compensated by a phase-correction stage.

In Fig. 6-13A, R3 is the correction amplitude adjustment. It sets the ratio of main signal current to aperture-correction current.

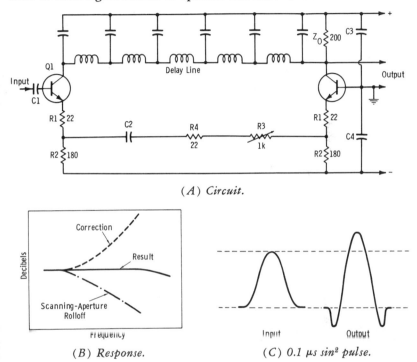

(A) Circuit.

(B) Response.

(C) 0.1 µs sin² pulse.

Fig. 6-13. Aperture correction.

The gain of this circuit at low frequencies is essentially unity. It is best checked with a sine² pulse; equal preshoots and overshoots indicate phaseless response. It also can be checked with the conventional video-sweep technique. The frequency at which the boost is peaked varies between 2.4 MHz and 6 MHz, depending on the specifications of the delay line used.

Measuring Detail Contrast

The fact that one person can "see" 600 lines of horizontal resolution but someone else can "see" only 500 lines on the same test-pattern reproduction is not meaningful. There are too many variables involved, including eyesight, psychology, and the condition of the display device. There is only one way to put resolution on an absolute and measurable basis: the use of the line-selected sweep.

The conventional horizontal-rate scope sweep results in a pattern that contains all the lines of the field. It is, however, possible to observe only a single line of the field on, for example, the Tektronix 524 scope by turning the TRIGGER SELECTOR switch to the delayed-sweep sector. By rotation of the SWEEP DELAY control, any line or lines may be observed. This sweep is obtained internally from the composite television signal by establishing a coarse time delay from a vertical-sync pulse (from the sync-separator circuit) and then actually triggering the sweep from a selected horizontal-sync pulse. Since the scanned line interval is 63.5 microseconds, if the time base is adjusted to 6.35 microseconds per centimeter, a single line will occur in the full-scale 10 centimeters of the scope graticule.

When the SWEEP DELAY control is used in this manner, the sweep is triggered only 30 times per second. The resulting display is correspondingly dim, and, when much ambient light exists, the screen should be viewed through the hood provided for this purpose.

The particular line being observed on the CRO may be determined by connecting a spare video monitor to the LINE INDICATING VIDEO output jack at the rear of the scope. The picture on the monitor is brightened during the time of the sweep gate. The SWEEP DELAY control is rotated until the desired line of the picture is selected (brightened on the monitor). Thus, the amplitude of test-chart bandwidth wedges may be measured relative to gray (100-percent) areas.

Fig. 6-14A shows the picture-monitor display of a multiburst test pattern with the strobe line indicating the point of observation on the waveform monitor. (The scope time base in this instance was adjusted to five TV lines to make the indicating line apparent in the photo.) The highest burst

(A) Multiburst test pattern.

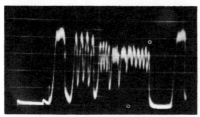 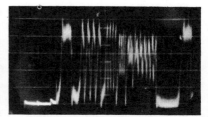

(B) Uncompensated waveform.   *(C) Compensated waveform.*

Fig. 6-14. Strobe-line technique of measuring detail contrast.

frequency in this particular test pattern is 8 MHz, which corresponds to 640 TV lines (Table 6-1). Fig. 6-14B shows the resulting waveform presentation with the aperture-correction circuitry switched off. Fig. 6-14C shows the waveform with the aperture-correction circuitry switched on; note the increased response at mid and high frequencies.

The amount of aperture correction that can be employed is limited because high-frequency noise is increased along with high-frequency picture-signal content. In the image orthicon, the greatest amount of noise is toward the black-level picture information. For example, note the increase in noise at the black level of Fig. 6-14C compared to that in Fig. 6-14B. Where amplitude controls are incorporated for the amount of aperture correction inserted, the noise amplification becomes the limiting factor.

Amplitude-Limited Aperture Correction

Since the human eye is most sensitive to fine detail in picture high lights and less sensitive to detail contrast in shadows, an amplitude-limited aperture correction sometimes is employed. This technique makes aperture correction of the signal effective only over the top 25 or 50 percent of the total video-signal amplitude, thus effectively equalizing only the high lights of the signal. This is sometimes termed a *high-light equalizer.*

Fig. 6-15 is a simplified schematic diagram of such an aperture-compensation circuit. The video-signal amplifier feeds two separate circuits, a white-clipper stage and one input of a difference amplifier. The white clipper is clamped so that the video level does not vary through the RC-coupled stage with changes in APL. The clipping level for the diode is adjusted by the clipper-level control. The output of this stage in turn feeds two paths, a delay stage and the other input of the difference amplifier.

Thus, the difference amplifier has two inputs, the full video signal and the high-light-clipped portion of the signal. Note that the full video signal applied to the lower section of the difference amplifier goes through a polarity reversal at the plate. The clipped signal is applied in effect to a cathode follower; hence this signal does not go through a polarity reversal. Therefore, at the output of the difference amplifier, the clipped signal (which contains no high lights over the amplitude set by the clipping level) is subtracted from the full signal. Thus the remainder is the same as the high-light portion of the signal that was cut off in the clipper stage. This high-light portion is aperture compensated and delivered to one input of a summing amplifier.

Since the aperture equalizer presents a delay to the signal, an equal amount of delay must be presented to the unequalized low-light portion of the signal. This delayed low-light signal then becomes the second input to the summing amplifier.

The summing amplifier has the same configuration as the difference amplifier. Note, however, that the two input signals are now of opposite polarity; hence the output is the sum of the two inputs. The variable resis-

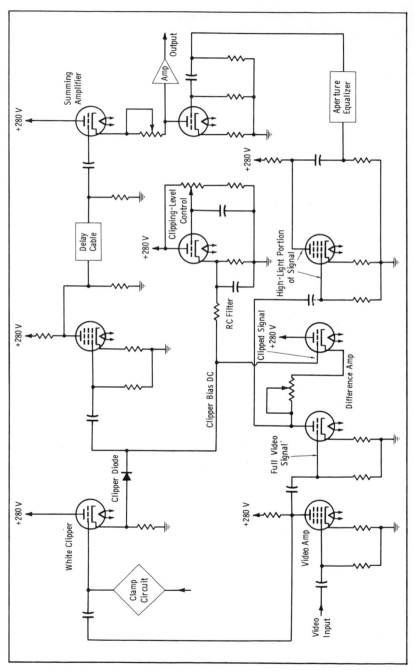

Fig. 6-15. Simplified schematic of amplitude-limited aperture equalizer.

tors between plate and cathode of the difference and summing amplifiers allow exact equalization of gains between the two sections. The gain is slightly less than unity.

If the clipping level is adjusted so that no clipping occurs, complete cancellation results at the difference amplifier, and nothing is equalized. At the other extreme, if the entire signal is clipped, no subtraction occurs at the difference amplifier, and the entire signal is equalized. The clipping-level control normally is adjusted so that about 50 percent of the total signal amplitude is clipped at this input of the difference amplifier.

Contours out of Green

The *contours-out-of-green* technique is employed in some three-*Plumbicon* color cameras to provide "crispening" of the luminance signal similar to that obtained from aperture compensation. Sharpened edges (both vertical and horizontal) produced from the green channel of a three-tube camera are fed into the red, green, and blue channels. This provides the

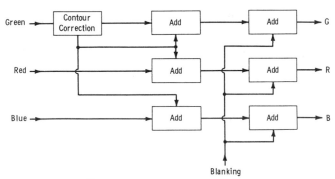

Fig. 6-16. Contours out of green.

same tolerance to misregistration that a fourth tube provides. Remember, however, that this applies only to a monochrome receiver reproducing a color program; misregistration will not result in a loss of sharpness on the monochrome receiver, but it will cause color fringing, and therefore loss of sharpness, on color receivers.

See Fig. 6-16. Enhanced contours in the image from the green tube correct all three channels and become the derived luminance signal. Since the enhanced contours come from a single tube, these contours cannot be degraded by registration errors. The delay of the green signal and the contour signal relative to uncompensated red and blue signals is corrected automatically in the normal registration procedures.

Vertical Aperture Correction

The contour signal is enhanced by employing both vertical and horizontal aperture correction. We know that horizontal aperture correction is limited by the bandwidth of the system, including the home receiver. Vertical aperture correction (being a low-frequency correction) produces a noticeable increase in sharpness not limited by bandwidth. For example, enhancement of a vertical transition from black to white is obtained by slightly darkening one line and whitening the succeeding line in the transition.

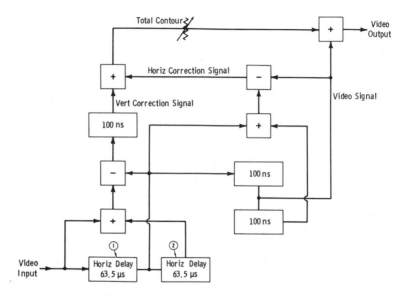

Fig. 6-17. Vertical and horizontal aperture correction.

To accomplish this result some means of delaying a line of video information is necessary. Fig. 6-17 shows that the video is delayed 1 line (1H) in block 1, and another line in block 2. The signal that is delayed one line is the main uncorrected signal. This signal goes to a subtractor, an adder, and a 100-nanosecond (one picture element) delay line. Thus, it serves as the input to the horizontal aperture corrector. The same unit results in both horizontal and vertical aperture correction.

The second line delay (block 2) is used to achieve the functions described as follows by the Philips laboratories:

For large areas in which each line is like every other, subtracting signals from adjacent lines is like subtracting a signal from itself. To get more equalization, more adjacent-line signal is subtracted, and less overall large-area signal remains. This means changing equalization would also change gain. To avoid this problem, a detail signal that has no large-area information is made. This is done by subtracting enough adjacent signal information to cancel the signal completely when adjacent lines are alike. The contour signal is a signal containing only vertical- and horizontal-transition detail. This contour signal is then added in any desired amount to make the main signal without changing gain.

6-5. CABLE EQUALIZATION

See Fig. 6-18. In the camera on the studio floor, amplifiers are made to have a flat frequency response by means of peaking circuitry and negative feedback—that is, by design. Because of the scanning-aperture effect, aperture compensation is employed to maintain this "flat" amplifier response.

The cable that connects the camera with the studio rack or console equipment may be of any length from 50 to 2000 feet. Since the coaxial element within the camera cable exhibits a high-frequency loss approximately proportional to the square root of the frequency, an equalizer must be used to maintain flat response. Since the equalizer attenuates the signal by the amount of equalization required, it is generally followed by an amplifier that compensates for this amplitude loss.

6-6. GAMMA CORRECTION

In the ideal situation, the light output of the kinescope would be directly proportional to the light input from the televised scene. But the kinescope is nonlinear in the direction of compressing blacks and stretching whites. The I.O. tube is nonlinear in the direction of stretching blacks and compressing whites. Also, the transfer characteristic of the I.O. is dependent in a complex manner on whether the scene is high-key or low-key, which is another way of saying that the dynamic transfer curve varies somewhat with APL.

The resultant overall characteristic of an uncompensated system is black compression, because the kinescope is more nonlinear than the I.O. tube. The same is true for the vidicon. The *Plumbicon* has an almost linear transfer characteristic, having a gamma of about 0.95, or close to unity.

Be sure of terminology. Amplitude linearity is a measure of the *shape* of the *transfer curve*. It is a function of output luminance levels versus input luminance levels of the system. *Gamma* is the *exponent* of the transfer characteristic. This is the *slope* of the transfer characteristic plotted on a log-log scale. A gamma of unity (dash lines in Fig. 6-19) is a strictly linear transfer slope. If the slope is greater than unity (kinescope), blacks are compressed and whites are stretched. If the slope is less than unity (pickup tube), blacks are stretched and whites are compressed. Overall system gamma is the product of the individual gammas.

For example, the vidicon has a relatively constant gamma of 0.65 over the normal beam-current operating range. The average kinescope gamma is around 2 (color standards assume an exponent of 2.2), which means that the picture-tube high-light brightness increases approximately as the square

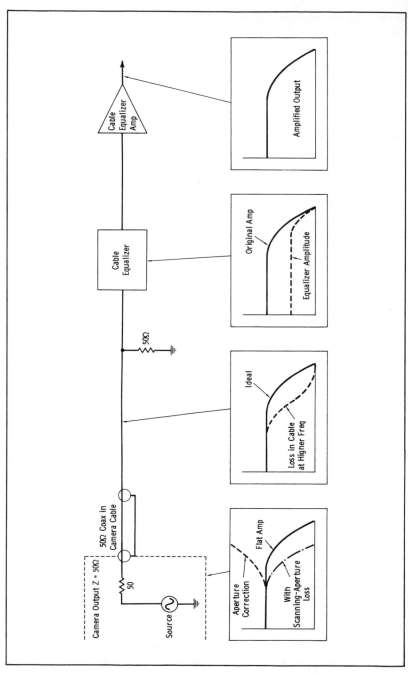

Fig. 6-18. Function of frequency-compensation circuitry.

of the applied video voltage above cutoff. Then assuming all other parts of the system have unity gamma, the overall system gamma is:

$$(0.65) \ (2.2) = 1.43$$

This value is greater than unity. The amount of gamma correction necessary to obtain a unity exponent is:

$$\frac{1}{1.43} = 0.7$$

The product of the system gamma (1.43) and the gamma correction (0.7) is 1, or unity.

(A) Pickup tube.   (B) Picture tube.

Fig. 6-19. Transfer curves of terminal devices of TV system.

A transfer curve is graphically illustrated by the system response to a stair-step signal. Such a curve might be strictly linear except at one extreme end, either black or white. The term "gamma" in photography refers to the *maximum slope* of a curve showing density versus exposure (exposure on a logarithmic scale). The term "gamma" in television usage (and this term is abused by many engineers) should refer to the *overall system* transfer characteristic on a log-log scale. Although an oversimplification, the overall transfer characteristic is given an *exponent* that relates the relative reproduced output luminance to input luminance. Gamma is *not* a constant in a television system, but varies over the contrast range from low lights to high lights.

Although there is considerable variation in gamma-correction circuits, nearly all work on the principle of adjusting the threshold of diode conduction, as shown basically in Fig. 6-20. Some gamma circuits work only in the gray-to-black region; multiple diodes are used so that the transfer curve can be made to "break" at different conduction levels. The circuit of Fig. 6-20 employs both black stretch and white stretch.

When the white-stretch control is rotated clockwise (direction of arrow), a point is reached at which X1 is forward biased. Since white is a positive-going signal at the emitter, this voltage fixes the point of "bend" in the white region. When X1 conducts, the emitter resistance is partially bypassed, reducing degeneration and increasing gain. See the white-stretch curve in Fig. 6-20B.

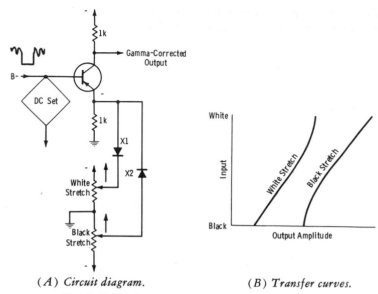

(A) *Circuit diagram.*   (B) *Transfer curves.*

Fig. 6-20. Basic circuit for black and white stretch.

Similarly, as the black-stretch control is adjusted clockwise, the back bias on X2 is reduced and it conducts. Note that a black signal is a negative-going signal at the emitter. Therefore, the gain increases in the gray-to-black region, stretching blacks. See the black-stretch curve in Fig. 6-20B.

Checking Gamma Circuitry for Film Cameras

Regardless of the type of circuitry employed, the amplifier containing gamma correction is most conveniently checked with the aid of a stair-step generator. The following is a complete step-by-step procedure that can be used regardless of the method in which gamma correction is obtained:

1. Observe the linearity-generator output directly with the scope; check for proper linearity of the steps. If the steps from the generator are not perfectly linear, take this condition into account when checking the amplifiers.

2. Remove the camera-signal input from the control chassis, and feed the stairsteps into this point. Increase the amplitude of the stair-step signal until clipping starts to show at the output. This tells you the maximum peak-to-peak signal the control (processing) amplifiers can handle without compression. Keep a record of this amplitude for future reference. Be sure the gamma switch *is on unity* (no gamma correction). Remove any aperture correction employed.

In a vidicon film-camera chain, fixed values of gamma correction normally may be switched into or out of the circuit. Most units have a switch with three positions: unity, 0.7, and 0.5. The 0.5 gamma is often necessary for films originally processed for theater projection in order to fit the wide dynamic range of the film gray scale into the range that can be transmitted without severe compression of low grays and blacks. The vidicon has no knee and will not compress whites (assuming that the beam current is sufficient to discharge the highest high light).

3. For accuracy in checking linearity without danger of erroneous measurements as a result of excessive levels, reduce the input level to one-half the value recorded in Step 2.

4. When observing linearity of steps at the output of the control chassis, be sure the pedestal control is adjusted with sufficient setup so that the bottom (black) step is not compressed. If the black steps are still "pulled down," check any transient-suppressor controls for incorrect adjustment. These circuits affect the black region.

5. Now place the black-stretch (gamma) switch on the position to be checked. Fig. 6-21 shows the values of output steps you should obtain (with linear input) for the two most common values of correction.

As an example of the mathematical relationship involved, assume gamma is 0.5, and figure where the step for a 0.1-volt input step should be:

$$(0.1)^{\frac{1}{2}} = \sqrt{0.1} = 0.316$$

For the 0.2-volt input step:

$$(0.2)^{\frac{1}{2}} = \sqrt{0.2} = 0.447$$

To make computations for the 0.7 gamma correction, it is necessary to use logarithms. For the 0.1-volt step, it is necessary to find the value of $(0.1)^{0.7}$. The method is as follows.

$$\log N = 0.7 \log 0.1$$
$$= (0.7)(-1) = -0.7, \text{ or } -1 + 0.3$$

Since $\log N = -1 + 0.3$, N is 0.1 times the antilogarithm of 0.3:

$$N = (0.1)(2.0) = 0.2 \text{ (approx)}$$

6. If the step-voltage response is distorted or more than a few percent in error with the gamma correction in, the reference diodes (when used) are the most likely source of trouble. A slight departure from theoretical step response is normal in circuits employing diodes that are biased to conduct at various levels of video. This is because the resulting response is in incremental steps from black to white rather than a smooth curve.

NOTE: It was pointed out previously that the camera chain employs high peakers and phase-correction circuits with small trimming capacitors across a cathode (or emitter) resistance. These circuits may cause some tilt on the stairsteps, but normally will not prevent an accurate measurement. If excessive distortion is present, it is a simple matter to bypass such a cathode correction circuit with a capacitance of about 0.47 μF.

7. Now run the gray-scale slide or test loop through the complete film chain. This will give a good indication of the operation of the camera head as it responds to the black-to-white pattern viewed by the vidicon. If any compression (with gamma correction removed) is now present, either the trouble is in the camera head itself, or the peak-to-peak video level from the camera is excessive. Check this level at the input to the camera control chassis. Always keep this input level below that found in Step 2.

Always remember this: The black stretch of gamma-correction circuitry applies to a *linear input signal*. This is the basic function of the amplifier. Then, with the pickup tube looking at a *logarithmic* gray scale (standard), the output *should be linear*.

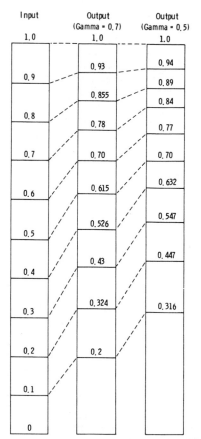

Fig. 6-21. Gamma correction.

Checking Gamma Circuitry for Live Cameras

Gamma circuitry is best checked with the camera looking at the crossed gray scale in the studio. The scope will reveal the range of control, which should be checked against the specification sheet for the camera involved. Obviously, if the gamma correction range of any one channel is low, color balance with the other channels cannot be obtained. In an emergency, you can operate without gamma correction in the channels required to obtain balance.

A troublesome gamma stage is most often the result of faulty transistors or diodes. It is best to make direct substitutions of parts as a check.

Be certain you can distinguish between a gamma fault and some other fault. Fig. 6-22A shows one possible CRO presentation when the crossed gray scale is scanned at full raster. Note that the ascending luminance steps are linear, showing that gamma is proper when a logarithmic chart is observed.

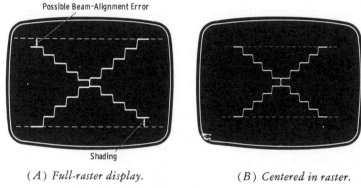

(A) Full-raster display.   (B) Centered in raster.

Fig. 6-22. Waveforms for crossed gray scale.

The mismatch at black level should be corrected with the horizontal shading control. The white levels will still be unequal.

Be certain that the incident-light meter shows equal illumination at the lower left and upper right white chips. Also, be sure the *reflected* light is the same. Sometimes a white chip can be dulled by imperfections or dirt.

The next thing to check is beam alignment. The color camera normally has a "G4-rock" circuit, which is simply a 30-Hz multivibrator; its signal is applied so as to result in a split field if the beam is not properly aligned. Aim the camera at a registration chart, and adjust both the vertical- and horizontal-alignment controls until the images on the viewfinder are superimposed.

You may find that, with the particular pickup tube involved, it is possible to obtain good beam alignment around the central area but not at the corners and edges. This is just one of the factors that can cause unequal white levels at the two extremes of the pickup area. The presence of this effect can be verified by zooming out or moving away from the chip chart so that it is in the center of the raster as in Fig. 6-22B. (The chart should be displayed against a neutral background.) If the levels become even, it is likely that a beam-alignment problem (or possibly black-level shading) is present.

IMPORTANT NOTE: Always check the manufacturer's specifications for the specific pickup tube involved. Some tubes have a photocathode or target sensitivity specification that can vary as much as 20 percent between left and right areas and still meet standards.

6-7. BLANKING AND SYNC INSERTION

Fig. 6-23 illustrates one commercial method of blanking insertion: suppressor-grid injection of the blanking pedestals. It should be noted that negative-black video is combined with negative pedestal voltages. The polarity is important since the video-signal maximum black must occur just under the blanking level. The composite blanking signals from the sync generator are fed to a blanking amplifier, and then to the suppressor grid of the mixing stage. The large negative pulses at this grid result in plate-current cutoff during each pulse. The video signal and the resulting pedestal formed by plate-current cutoff appear as positive excursions (positive black) in the plate circuit, and are impressed on the 1N34 clipper stage. The clipping level (hence pedestal level) of this rectifier is determined by the setting of the pedestal-height control, since any signal amplitude above the dc voltage on the arm of this control is not passed by the clipper diode.

The purpose of *clamping* at this point may now be more clearly understood. It is recalled that clamping occurs during the blanking intervals of the pickup tube. The grid of the blanking-insertion amplifier is the clamp-

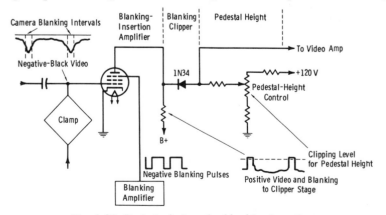

Fig. 6-23. Basic technique for blanking insertion.

ing point, being held to a predetermined grid-voltage level. Therefore, the pedestal control determines the clipping level at a voltage that is fixed in reference to the blanking level of the camera. Thus, any necessary change in the video gain control of the first stage usually does not necessitate another adjustment of the pedestal control, since the *ratio* between black signal level and pedestal height is not affected by the gain control. In this way, the pedestal is fixed with reference to black in the signal rather than with reference to the average signal content.

The synchronizing signal (composite sync from the sync generator) is inserted and controlled by the same technique.

Fig. 6-24 illustrates a common method of blanking insertion in Marconi cameras. The first stage provides a clamp that is dependent on a variable dc potential from the lift control ("lift" is the British terminology for blanking or black-level control). The signal is passed to a three-diode gate, where system blanking is inserted and white clipping occurs.

The gate circuit is followed by a black clipper with a fixed clipping reference. It is interesting to note that in this arrangement the level between the settings of the white clipper and that of the black clipper remains constant regardless of the setting of the black control, and the stability of the two clipping levels is independent of the amplitude of the blanking signal inserted. With the signal at a level of 4 volts, nonlinearity introduced by the diodes in the clippers is kept to less than 3 percent of picture amplitude.

Diode X6 is added to keep the number of forward-biased diodes between point A and the black reference (point B) the same as the number between point A and the lift potential (point C). This together with the use of low-temperature-coefficient components in the derivation of the potentials at points B and C, provides stability in the black level—about a 0.5 percent change for a 20°C change of temperature. Slight transient

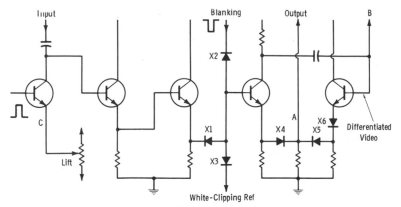

Courtesy Ampex International

Fig. 6-24. Blanking insertion in Marconi camera.

breakthrough at black, caused by the finite switching time of the diodes, is largely offset by adding a small amount of inverted, differentiated signal onto the black reference.

6-8. TESTING AND MAINTENANCE

A detailed description of normal setup and electronic adjustment of the camera chain was presented in a previous volume[2] and will not be repeated here. It remains to examine the special techniques required in testing and maintenance of video processing amplifiers. (The testing of gamma-correction circuitry has been covered already in preceding Section 6-6.)

A typical video path for a camera, or for one channel of a color camera, is shown in Fig. 6-25. The sequence of functions is different for every manufacturer, but the basic function of each block is the same regardless of the sequence. These functions can be outlined briefly as follows:

Preamplifier. This stage normally is used only for vidicons or *Plumbicons.* Its purpose is to convert the video signal current from the pickup tube to an amplitude equal to that from an I.O., or sufficient for satisfactory processing of the signal.

Video Amplifier. Normally this is the first amplifier for an I.O. tube. In the example of Fig. 6-25, the stage includes aperture correction and remotely controlled gain. In a four-channel camera employing an I.O. and three vidicons, aperture correction sometimes is included only in the monochrome channel, since the three chroma channels are narrow-band.

Processing (Proc) Amplifier. This amplifier involves voltage amplification to boost the signal voltage to a level suitable for gamma correction. It includes manual level sets for black and peak white.

Cable Compensation. This section corrects for camera-cable losses at all frequencies across the video passband, for cable lengths of 200 to 1000 or 2000 feet.

Output Amplifier. This amplifier provides multiple 75-ohm outputs for distribution of the video signal to the system, and a 50-ohm output to feed back to the view finder through the camera cable. It also contains circuitry to limit peak black and peak white signal excursions to preselected values under overload conditions. There is provision for insertion of system blanking, and sync insertion is optional at this point.

Buffer Amplifier. This element of the system provides isolation for signal feed to the view finder. It includes cable compensation for the view-finder coaxial line.

View-Finder Amplifier. This section amplifies the video signal to a level suitable to drive the display kinescope. Also, it normally employs a switchable "crispener" circuit as a focusing aid for the cameraman.

---

[2]Harold E. Ennes, *Television Broadcasting: Equipment, Systems, and Operating Fundamentals* (Indianapolis: Howard W. Sams & Co., Inc., 1971).

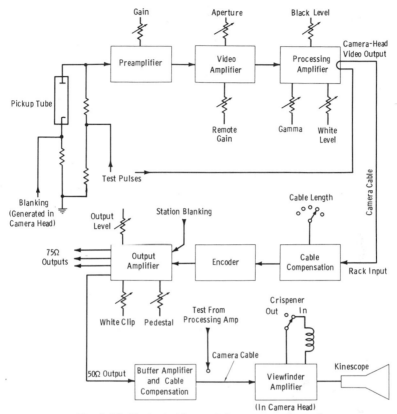

Fig. 6-25. Typical video path for camera channel.

Video Amplifiers

Three typical video-amplifier circuits common to almost all modern cameras are shown in Fig. 6-26. Fig. 6-26A illustrates the well-known method of base biasing in conjunction with emitter-resistance stabilization. This is a popular voltage amplifier when the stage is operated between a low-impedance source and a high-impedance load. The voltage gain in this application normally is limited to less than 20 dB. Resistors R1 and R2 provide the base bias voltage and, in conjunction with R_E, stabilize the transistor operating point. Capacitor C_E (when used) is small to compensate for the limited bandwidth. When amplifiers of this type are cascaded, the $R_E C_E$ time constants normally are staggered to obtain the desired response. (This arrangement is not often used.)

Feedback pairs (Figs. 6-26B and 6-26C) provide improved frequency response, linearity, and ac and dc stability. This circuit eliminates the need for peaking circuits. With the addition of positive feedback (Fig. 6-26D), voltage amplification is increased as a result of feedback loss compensation.

Let us take each circuit in typical applications in a color camera, and do a quick analysis so that you know what to expect in servicing and trouble-shooting.

First consider the single emitter-compensated stage (Fig. 6-26A). Assume an input stage is being fed from a 75-ohm line. The line is terminated at the input, so you would expect a bridging impedance to 75 ohms.

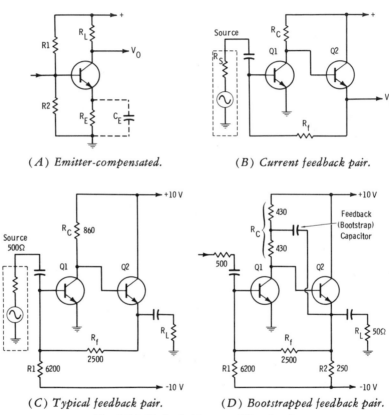

(A) *Emitter-compensated.*   (B) *Current feedback pair.*

(C) *Typical feedback pair.*   (D) *Bootstrapped feedback pair.*

Fig. 6-26. Typical video-amplifier circuits.

Using the values of Fig. 6-27A, perform the dc analysis first. The R1-R2 voltage divider totals 50k, so the current through the divider is $20/50k =$ 0.4 mA. This 0.4 mA through the 5k resistor gives +2 volts at the base. Now redraw the circuit as in Fig. 6-27B so that R_B is the equivalent resistance of R1 and R2 in parallel, and indicate the base voltage of +2 volts (from base to ground).

If the transistor is germanium, a difference of about 0.2 volt exists between base and emitter (if it is silicon, a difference of 0.6 to 0.7 volt exists). So indicate +1.8 volts at the emitter as shown.

The dynamic small-signal emitter resistance (r_e) is approximately $26/I_E$, where I_E is the emitter current in milliamperes. The dynamic base-emitter resistance (R_{EB}) can be assumed to be 4 ohms for all practical purposes in quick analysis. The transresistance (r_{tr}) is the sum of these resistances and any unbypassed external emitter resistance (R_E):

$$r_{tr} = r_e + R_{EB} + R_E$$

When R_E is rather high relative to re_e and R_{EB}, these quantities can be ignored for quick analysis. So in this example, $r_{tr} = 1000$ ohms, the value of resistor R_E.

Since the emitter is at 1.8 volts, the emitter current is 1.8/1000 ampere, or 1.8 mA (Fig. 6-27C). Ignore the transistor alpha (which is about 0.98 or 0.99), and assume the collector current (I_C) is the same as the emitter current. The voltage across the 5k collector load is, therefore, 0.0018 × 5000, or 9, volts, so you should measure 20 − 9, or +11, volts at the collector.

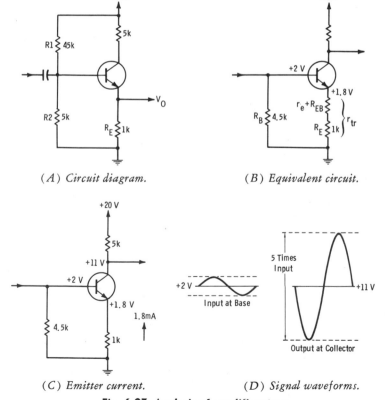

(A) Circuit diagram.

(B) Equivalent circuit.

(C) Emitter current.

(D) Signal waveforms.

Fig. 6-27. Analysis of amplifier stage.

The voltage gain is equal to R_L/r_{tr}. In this case, the gain is $5000/1000$, or 5. Thus, the signal at the collector will have a peak-to-peak value about 5 times that of the signal at the base. The ac axis of the output waveform will be at approximately $+11$ volts (Fig. 6-27D).

It has been assumed that the single stage is feeding a high impedance. In practice, the collector signal load is lowered somewhat by the parallel input impedance of the next stage. Therefore, the gain of the stage under consideration is somewhat less than 5.

When compensating capacitor C_E is used (Fig. 6-26A), the value to provide a nominally flat response curve is:

$$C_E = \frac{1}{2\pi f R_E}$$

where f is the 3-dB frequency for the uncompensated stage. Thus, in the example of Fig. 6-27A, if the transistors used and the circuit impedances are such that the response at the top of the intended passband (assume 8 MHz) is down 3 dB, then:

$$C_E = \frac{1}{(6.28)(8 \times 10^6)(1000)} = 20 \text{ pF (approx)}$$

Next, we will examine the feedback pair with emitter-follower output. Because of the effects of transistor and circuit capacitances, reducing the input and output impedance improves the high-frequency response, but at the expense of gain. When low impedances are obtained by feedback, inverse signal feedback provides another means to improve frequency response and signal linearity. The basic circuit of Fig. 6-26B shows the feedback resistor (R_f). Note that since the output impedance is coupled back to the input, the output and input impedances of the circuit the highly interdependent.

The voltage-gain (A_v) relationship in a feedback pair of this type is:

$$A_v = \frac{R_f}{R_s}$$

where,

R_f is the feedback resistance,

R_s is the source (previous stage, line, or generator) resistance.

This type of circuit is common in modern color cameras, particularly as line drivers for 50- and 75-ohm lines. Fig.6-26C shows a typical circuit.

For the analysis, assume the npn transistors are silicon. Since the emitter of Q1 is grounded, the base of Q1 can be assumed to be at $+0.7$ volt with respect to ground. The difference between the -10-volt supply and the $+0.7$-volt junction is $+10.7$ volts. So the current in R1 is $10.7/6200$ ampere, or 1.72 mA.

The same 1.72 mA is present in the emitter return of Q2 and in the 2500-ohm feedback resistor, R_f. This results in a voltage rise in R_f of 1.72 mA × 2500 ohms, or 4.3 volts. Adding this to the starting-point voltage of 0.7 volt gives a Q2 emitter voltage of +0.7+4.3, or +5 volts.

Since the Q2 emitter voltage is +5 volts, there should be about +5.7 volts at the Q2 base. This is the same as the Q1 collector voltage. With this information, the voltage across R_C and then the current through R_C (5 mA) can be calculated. You now know that the Q1 collector current is about 5 mA, and the Q2 current is about 1.72 mA. (This is incidental information, however, and not pertinent to troubleshooting techniques.)

Assuming the sending (source) impedance is 500 ohms, the expected voltage amplification is 2500/500, or 5. The output signal current is superimposed on the 1.72-mA quiescent operating current of Q2, so the maximum peak-to-peak signal swing without clipping is 2 × 1.72, or 3.44 mA. If R_L is 500 ohms, this current swing results in a signal voltage of 0.00344 × 500, or 1.72 volts peak to peak. If R_L is 50 ohms, the peak-to-peak signal swing is 0.00344 × 50, or 0.172 volt. This is the maximum capability of the circuit in Fig. 6-26C.

Fig. 6-26D shows the previous circuit with minor modifications that drastically affect the operation. The Q1 collector load has been split into two equal resistors that total the original 860 ohms. This provides a tap for the bootstrap capacitor, which provides positive feedback to overcome the signal loss resulting from negative feedback through R_f. Since there is now a greater signal-current swing because of the bootstrapping, an additional current must be supplied to the Q2 emitter. This is done through R2, a 250-ohm resistor. The difference between the −10-volt supply voltage and the +5 volts at the Q2 emitter is 15 volts. Therefore, an additional 15/250 ampere, or 60 mA, is added to the 1.72 mA already in Q2 for a total of 61.72 mA quiescent operating current. The output signal current can now swing ±61.72 mA for a peak-to-peak swing of approximately 120 mA without clipping. This develops a signal of 6 volts peak to peak across a 50-ohm load, or a voltage-gain capability over 30 times that of the circuit in Fig. 6-26C. Note that there is no difference between the voltage readings to ground at the transistor junctions for quiescent operating conditions in the two circuits.

There are two fundamental tests you can make in troubleshooting these circuits; both can be done quickly with the scope. One example will suffice; assume you are scoping the Q2 emitter. First, you know the dc operating point should be about +5 volts. Calibrate the scope for a sensitivity of 1 volt/cm and center the trace at the bottom graticule line with the probe touching a grounded point and the scope set on dc input. Then apply the probe to the Q2 emitter and observe whether the trace moves upward 5 cm. In the case of the circuit of Fig. 6-26C, the signal superimposed on the scope trace will be very small, so to check the signal swing go to ac operation and increase the scope gain to observe the signal. You now have the

complete functional story of conditions at the emitter of Q2. In the case
of the circuit of Fig. 6-26D, you probably could observe the signal swing
without changing the scope settings. This swing should have its ac axis at
the +5-volt level.

Fig. 6-28A shows a typical gain-controlled video stage employing a
feedback pair similar to that of Fig. 6-26C. The difference is that the feedback resistor (R_f) is shunted by a network that includes photoconductive
resistance devices R1 and R2. The filament brightness of the small lamps
determines the resistance of their associated cadmium-sulphide cells, and
therefore the total value of feedback impedance. Voltage is supplied to the
lamps from the master white-level control at the remote-control panel.

If this control should fail, you can, in an emergency, substitute a fixed
resistor of about 1600 ohms for the defective cell. You can shunt the cell

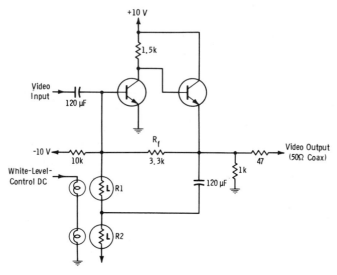

(A) Photoconductive control element.

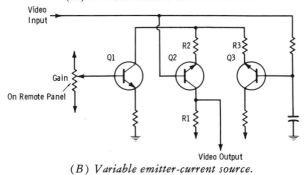

(B) Variable emitter-current source.

Fig. 6-28. Typical remote video-gain controls.

temporarily with the resistor for a quick check to determine whether this is the source of the trouble. If you do not have a direct replacement, it may be necessary to experiment with the value of the substitute resistor to obtain proper control of the output level without overloading a stage prior to the final output-level control in the control room. The temporary resistor should have about one-half the value of the feedback resistor used.

All plug-in modules are susceptible to plug and receptacle contact problems. Intermittent deflection or fluctuations in video black or white level are often the result of dirty or otherwise faulty contacts. If the equipment manufacturer recommends a particular cleaner for these contacts, by all means use it on a regular basis. Otherwise, a good tuner cleaner ordinarily is satisfactory.

Another popular type of remote video-gain control is shown in Fig. 6-28B. This simple and stable circuit depends on the principle that the small-signal emitter resistance (r_e) is approximately equal to $26/I_E$, where I_E is the emitter current in milliamperes. By control of the current in the constant-current supply (Q1), the currents in Q2 and Q3 are controlled; hence, the value of r_e in these transistors (and therefore the stage gain) is varied.

Video Sweeping the Processing Amplifier

Modern solid-state processing amplifiers seldom need the application of video sweep testing except in cases of replacement of critical components. Tube-type processing amplifiers require more frequent sweep alignment.

The basic video-sweep alignment technique was covered for preamplifiers in the previous chapter. The same technique holds for processing amplifiers except for special cases, which we will examine now.

Some amplifier in the system employs circuits in which blanking and sync signals are injected and in which clamping is employed. These stages require a special testing procedure. Fig. 6-29 presents a simplified diagram of a sync-insertion circuit feeding a cathode-follower output stage that has a clamped grid. This arrangement might be in a mixer-amplifier unit following the switcher stage or a line-output amplifier in which sync insertion takes place. Point 1 in the clamper stage is considered first. From the inherent nature of clamping tubes, considerable capacitance is added to the circuit at this point. Therefore, these tubes cannot be removed without upsetting the circuit constants, which would seriously affect the operation of the output stage. Neither can the tubes be left in unless a keyed test signal is employed as described later. This is true because the resultant clamping pulses would give spurious response in the output, since the unkeyed sweep generator contains no blanking pulses and the clamper normally operates on these pulses. For this reason, it is necessary to replace the clamper with a tube of the same type, but with the heater circuit opened by cutting off the heater pins. These "dummy" tubes should be plainly marked in some fashion (such as with paint or fingernail polish) so that they will not be left in place inadvertantly after testing. When the clamp is immobilized in this way, the grid of the output stage is left "floating." It is then necessary to insert temporarily a grid resistor at point 2 in the circuit (Fig. 6-29).

The sweep generator may be connected at point 4 for the purpose of checking this stage and aligning shunt-peaking coil L_1. In this stage, another condition also must be considered. Sync pulses usually are inserted (as shown) by an amplifier that shares a common plate load with the

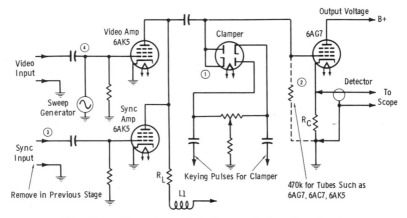

Fig. 6-29. Typical sync-injection and clamping circuit.

video amplifier. Aside from the capacitive effects of the sync amplifier on the video amplifier, the video amplifier steady-state plate voltage is dependent on the load current drawn by the sync-amplifier stage. Sync pulses must not be injected into the video amplifier, however, since the resultant patterns would be meaningless for the purpose of this test. If normal circuit conditions are to be maintained, the sync-amplifier tube obviously cannot be removed. If the amplifier has a plug-in connection for the composite sync signal, this may be removed during the test. If the amplifier is rack-mounted and receives sync from a distribution bus common to a number of amplifiers, the signal should be "killed" in the stage prior to point 3 by use of a dummy tube. Should point 3 obtain its drive directly from a sync-distribution bus, it is necessary to break this connection temporarily.

Often, it is desirable to employ "keyed" test signals phased by the station sync generator to eliminate the test amplitude during horizontal- and vertical-blanking intervals. This permits checking the many types of amplifiers incorporating line-to-line clamps, which otherwise need to be modified if straight test signals are used. Although commercial equipment is available for keyed sine waves, video sweep, stair-step signals, etc., there is an apparent scarcity of available units that process a square-wave signal properly.

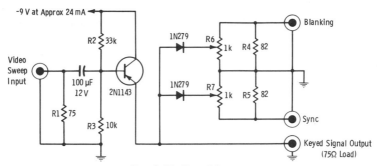

Fig. 6-30. Signal keyer.

Multiburst and window test-signal generators that provide for insertion of an external signal make it possible to insert any test signal over blanking and/or sync signals. For those stations that do not have this facility, the following is useful.

Fig. 6-30 shows a simple transistor circuit devised for this purpose. The 2N1143 transistor is reasonably priced and is effective for video use. Intro-

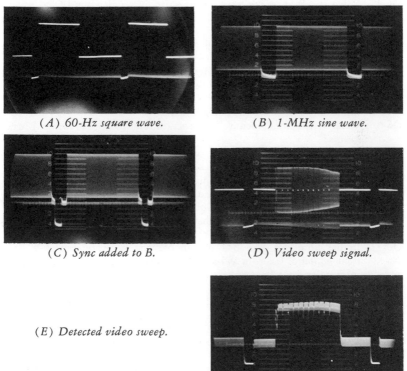

(A) 60-Hz square wave.

(B) 1-MHz sine wave.

(C) Sync added to B.

(D) Video sweep signal.

(E) Detected video sweep.

Fig. 6-31. Waveforms for keyed test signals.

duction of sync and/or blanking pulses of negative polarity drives the transistor to cutoff for the input signal, and the amplitude of the pulse as adjusted by R6 and R7 appears across the output load (input of system to be checked).

Fig. 6-31A illustrates the keyed output when the test signal is a 60-Hz square wave. The setup (blanking) level is adjustable by means of R6 to the desired amount of pedestal. This type of signal results in a clean composite blanking interval, and with the addition of sync no modification is necessary for units employing clamps.

The keyer also may be used for sine waves, as shown by Fig. 6-31B with only blanking inserted. Fig. 6-31C illustrates the waveform after sync is inserted. This unit also enables the engineer to feed keyed video sweep to the system, with the same advantage of being able to leave all clamping circuits in an active condition, just as for any composite picture signal. (See Figs. 6-31D and 6-31E.)

EXERCISES

Q6-1. What is meant by the term "phase shift" in a video amplifier?

Q6-2. What is primarily responsible for phase shift at low frequencies in video amplifiers?

Q6-3. Name five factors that would account for low-frequency losses in a video amplifier.

Q6-4. What detrimental effects can result from trying to employ a coupling capacitor of too large a value between video amplifier stages?

Q6-5. Why should you carefully check the level of a 60-Hz square-wave input to a unit or system being tested?

Q6-6. What is the fundamental frequency of the 2T pulse for a 4-MHz system?

Q6-7. What is the repetition rate of the $\sin^2$ pulse?

Q6-8. In a four-tube camera, is cable compensation used in all four channels?

Q6-9. In a four-tube camera, is aperture correction used in all four channels?

Q6-10. In the circuit of Fig. 6-27A, assume that the collector load is 3.3k, the emitter resistance is 500 ohms, the transistor is germanium, and other values are as shown. What dc voltages would you expect to find at the emitter and the collector? What voltage gain would you expect?

Q6-11. In the circuit of Fig. 6-27A, suppose you measure a peak-to-peak input signal of 0.2 volt. What peak-to-peak signal would you expect to measure at the collector?

Q6-12. In the circuit of Fig. 6-26C, assume that R1 is 10k, R_f is 3k, and other values are as shown. What dc voltages (with respect to ground) would you expect to measure at the base and collector of Q1 and at the base and emitter of Q2? The transistors are silicon.

Q6-13. With the circuit in Q6-12, what would be the maximum peak-to-peak signal voltage developed in a 75-ohm load?

Q6-14. Name all of the factors that can cause the CRO presentation shown in Fig. 6-22A.

CHAPTER 7

Pulse Processing and
Timing Systems

Television camera chains, both monochrome and color, rely heavily on precise pulse-processing subsystems for proper operation. Therefore, it is imperative that the technician understand the principles and circuits involved in these systems.

7-1. THE AUTOMATIC TIMING TECHNIQUE

Older camera chains employed vertical- and horizontal-drive pulses to generate camera deflection and blanking waveforms. These pulses normally were inserted in the camera control unit (in the control room) and relayed to the camera head through the camera cable. The composite blanking signal (the blanking portion of the transmitted signal) also was inserted in the camera control unit and adjusted in level by the pedestal or blanking control for the proper *setup*.

It is important to understand why the driving signals to the camera chains normally are about one-half the width of their respective blanking pulses. This is particularly important at the horizontal frequency, as is shown in Fig. 7-1. Remember that the transmitted composite blanking and the driving signals for the camera are inserted at the camera control unit. However, the camera cable may be as long as 1000 feet in some instances, and allowance must be made for the cable delay, which is roughly 1.5 microseconds per 1000 feet of cable. Since the total path is to and from the camera head (2000 feet), allowance must be made for a total delay of 3 microseconds. It may be observed from Fig. 7-1 that if the horizontal-drive pulses were of the same width as the horizontal-blanking pulses, camera blanking would not be ended at the start of the active line interval in cases where long camera cables are employed (unless the drive is regenerated and narrowed). This effect is not so important at the vertical frequency, since 3 microseconds is negligible compared with a total of about 1250 microseconds.

Fig. 7-2 illustrates an example in which there is a 900-foot difference in the distances between two control units and the system blanking distribution. In this case, the blanking pulse is delayed approximately 1.5 microseconds to control unit 1, and only 0.15 microsecond to control unit 2. If the front-porch-width control in the sync generator is adjusted to obtain a normal front porch in the camera-1 signal after sync insertion, then a switch to the camera-2 signal will result in a lengthened front porch (Fig. 7-2B). Since the receiver retrace is triggered by the leading edge of horizontal sync and the picture is unblanked by the end of horizontal blanking, a lengthened front porch causes the picture area to shift to the left. Similarly, if the front porch is adjusted for normal on the camera-2 signal, a switch to the camera-1 signal will result in a narrowed front porch, and the picture will shift to the right.

Thus, when camera control units are far from each other with respect to the system blanking distribution, it is necessary to add delay lines for the nearest control units to make all delays equal to that of the farthest unit. This is accomplished most conveniently by using the same length of feed line to every control unit; excess cable can be coiled up when necessary.

A similar problem can exist even when centralized control units are used but the lengths of the camera cables are greatly different. Camera blanking normally is formed from horizontal- and vertical-drive pulses. (Vertical-rate pulses are of no consequence in this discussion, since the amount of delay encountered has no bearing on the long vertical-blanking interval of 1250 microseconds.) Camera blanking must "fit under" the

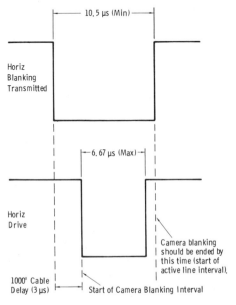

Fig. 7-1. Relationship of pulses for horizontal drive and blanking.

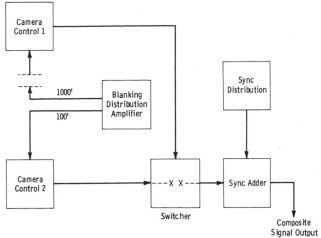

(A) Different cable lengths.

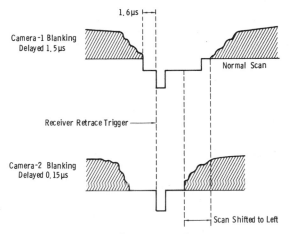

(B) Effect on waveform.

Fig. 7-2. Effect of blanking delay on front-porch width.

composite-signal blanking inserted in the control unit. For this reason, the pickup-tube (camera) blanking normally has a duration of 7 to 9 microseconds compared to the 11-microsecond horizontal-blanking interval transmitted.

If the camera cable should be as much as 1000 feet long, the total camera blanking delay is 3 microseconds, as stated previously. It may be observed that if the duration of camera blanking is greater than 8 microseconds, the interval is not completed by the end of receiver blanking. When this occurs, the front-porch width is not determined by the delay in the sync generator between the leading edge of blanking and the leading edge of sync, but rather is actually determined by the end of camera blanking. Again, if one camera-cable length is only 100 feet, and another camera-cable length is 1000 feet, switching from one camera to the other will cause a picture-area shift on the receiver. In this case, horizontal-drive pulses to the camera units with the shortest cables must be delayed to compensate for the delays in the longer cables.

It should be noted that the example of Fig. 7-1 is for noncomposite switching, in which sync is added following switching. In the case of composite switching (sync as well as blanking inserted in each camera chain), the front-porch width is fixed in each camera chain. However, the sync and blanking must again be coincident at the *input* of the switching system so that mixing and lap-dissolves may be made properly without sync-timing error from two sources.

In color-camera chains (except those of recent design), an additional delay over and above that of monochrome cameras results from the color-processing operations. Thus integration of such color equipment with existing monochrome camera chains requires an extensive timing procedure during initial installation.

Pulse-distribution amplifiers are used to provide isolated runs of the required pulses to various points in the plant. Fig. 7-3 shows the basic pulse distribution for integration of older color systems with monochrome equipment. Note that since monochrome and color must be integrated in this case, all monochrome pulse paths include a delay to match that of the color system. This is necessary to maintain the same front-porch width in the composite transmitted signal for both color and monochrome sources. A shift in front-porch width would cause a shift in the raster (picture) area at the receiver. Since the normal delay through a color system is about 1.2 microseconds, horizontal-drive, -blanking, and-sync pulses must be delayed accordingly for monochrome. It may be observed that with sync pulses inserted after the final switching point in the system, if the front porch were set for the normal 1.6 microseconds for monochrome, only about 0.4 microsecond of front porch would exist for the color system.

Most modern camera chains, both monochrome and color, incorporate an automatic sensing circuit that compensates for system time delay, thus circumventing timing problems for various pulse-cable and processing paths. The sensing circuit detects the time delay in the camera cable, the encoder, and the color-filtering process, and automatically advances an *internally generated* horizontal-drive pulse to avoid introduction of delay in the outgoing video signal.

Fig. 7-4 presents the relationship of some of the pulses discussed in this chapter. Before taking up the technique of how the timing pulse is used to advance the horizontal-scanning waveform in the camera, we must first be sure of an understanding of the "boxcar" circuit.

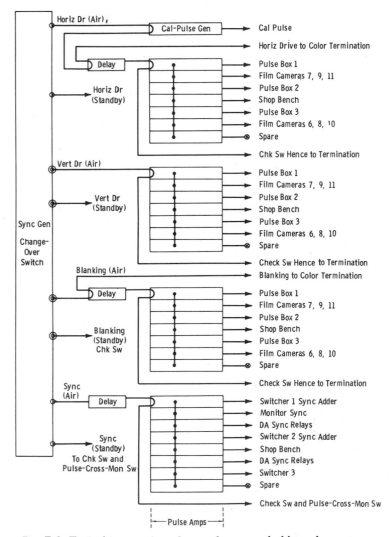

Fig. 7-3. Typical integration of monochrome and older color systems.

The Boxcar Circuit

The boxcar circuit is used extensively in modern color-camera pulse circuitry. In Fig. 7-5A, the input capacitance and resistance form a short time constant and therefore provide differentiator action. In the absence of a pulse, the transistor is held in saturation because the base is returned to -10 volts and the emitter is grounded. The ratio of base resistance to collector resistance (R_L) in a boxcar circuit is about 10 to 1, never more than 20 to 1, so that saturation is assured. In this state, the collector is very nearly at ground potential and is forward biased relative to the base; this is the condition for saturation (both junctions forward biased). Thus, we should expect the dc voltages shown in Fig. 7-5A.

The applied positive pulse (which always has a peak amplitude nearly equal to the boxcar supply voltage) reverse-biases the base-emitter junction and drives the transistor to cutoff. Capacitor C attempts to charge toward −10 volts but is clamped to the base-emitter potential, which again results in transistor saturation.

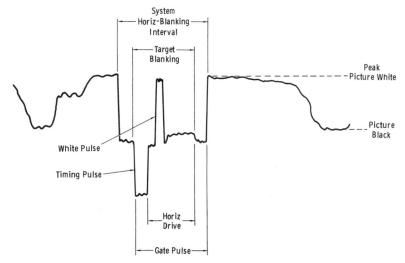

Fig. 7-4. Pulses in horizontal-blanking interval.

For the duration of the time that the transistor is cut off, the collector is practically at the supply voltage of −10 volts, resulting in a rectangular output pulse that is narrower than the input pulse. The output-pulse duration is determined by the time constant of R_B and C. The actual width is 0.7 R_BC. Thus, the output-pulse duration for the values given in Fig. 7-5A is:

$$(0.7) (10,000) (430 \times 10^{-12}) = 3 \, \mu s$$

The actual dc voltages measured between the transistor electrodes and ground depend on the duty cycle of the applied pulse—the length of time the transistor is on relative to the length of time it is off. This is why it is always desirable to use a dc scope for troubleshooting, as described earlier.

A chain of test pulses, and any other application in which a pulse must be triggered from the trailing edge of a preceding pulse, requires boxcars in cascade as in Fig. 7-5B. This is the "delayed" pulse technique. Note that the transistors are now npn. When a positive pulse is applied the base cannot go further positive because of the clamping action of the base-emitter junction. The short time constant causes a large negative signal excursion at the trailing edge of the applied pulse, and the first transistor, Q1, is driven to cutoff. This results in a rectangular, positive pulse at the Q1 collector for the time the base is negative relative to the +0.7-volt

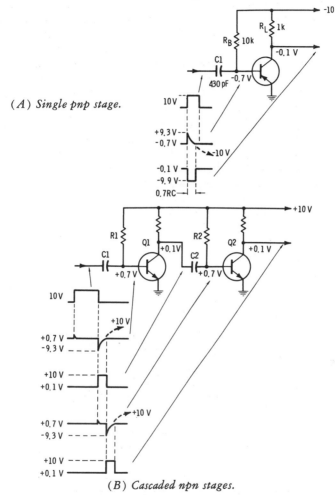

(A) Single pnp stage.

(B) Cascaded npn stages.

Fig. 7-5. Boxcar circuits.

potential (see waveforms in Fig. 7-5B). The same action then occurs at the Q2 base, and the leading edge of the Q2 collector pulse occurs at the trailing edge of the Q1 collector pulse.

You can see that in a circuit using pnp transistors (Fig. 7-5A), if the applied pulse is negative-going the output pulse is delayed. If the transistors are npn and the applied pulse is negative-going, an undelayed output pulse results. By using this knowledge, you should be able to know exactly what to expect, including the pulse duration, in waveform tracing.

Time Advance

See Fig. 7-6. The basic function of this circuit is to sense the delay through the individual camera chain (including the luminance delay in the encoder), compare it to incoming sync, and derive a dc error voltage to start camera drives and blanking to compensate for this total delay.

One output of the horizontal-sync separator in the control-room rack equipment goes down the camera cable to an automatic time-correction (ATC) multivibrator (Fig. 7-6A). This multivibrator turns off after about two-thirds of a line (Fig. 7-6B), depending on a dc control voltage from the comparator. The trailing edge of the multivibrator pulse initiates operation of a delay boxcar. The trailing edge of the delay pulse starts a timing-pulse boxcar, the output of which is inserted into the camera video (during blanking time) and fed back to the encoding equipment in the control room. This pulse is separated from video and fed to the other end of the comparator, where the trailing edge is compared to the leading edge of sync. The resultant dc error signal is the off control for the ATC multivibrator.

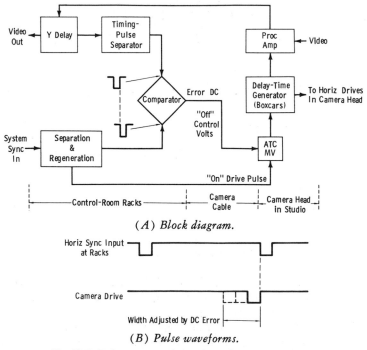

(A) Block diagram.

(B) Pulse waveforms.

Fig. 7-6. Delay compensation for color camera chain.

By using the leading edge of the timing pulse to derive horizontal drive and blanking, the camera signals are advanced relative to incoming sync timing. The amount of advance depends on the corresponding cable length and Y delay. In the camera head, the pulse that corresponds to a given video line is initiated by the sync pulse that corresponds to the preceding line in time.

If a failure occurs in the timing-pulse loop, the picture will not be lost but will shift horizontally, and the front porch will not be correct. Since this timing is a line-to-line function, an intermittent condition can cause erratic shifts of portions of the lines in the raster. Faulty camera cables or connectors can cause this condition. If camera horizontal-drive multivibrators are of the driven type, requiring pulses derived from the timing section, lack of scanning will activate the pickup-tube protection circuits and disable the camera (Section 7-5).

7-2. LEVEL-CONTROL PULSES (MANUAL AND AUTOMATIC)

In modern color-camera chains, pulses are used not only to calibrate individual amplifier gains (Section 7-3) but also for manual and automatic video-level control. Before we examine this circuitry more closely, it is necessary to understand the nonadditive mixing (NAM) technique that is generally employed.

Nonadditive Mixing

The three-channel camera *derives* luminance from the three primary-color channels. In the four-channel camera, the three color tubes operate on color brightness only; the luminance channel provides the "true" luminance information, which *can* be different from chroma luminance. Therefore, to obtain proper color balance in camera setup, proper luminance-to-chrominance ratio, and proper "gain riding" of signal level, the NAM signals are provided.

Review Chapter 2 in *Television Broadcasting: Equipment, Systems, and Operating Fundamentals*. The NAM circuits are fed from a "receiver matrix" based on the principles discussed in that chapter, with this fundamental difference: the brightness levels of B, R, and G are derived as follows:

$$(B - Y') + Y = B$$
$$(R - Y') + Y = R$$
$$(G - Y') + Y = G$$

where,
Y' is the "derived" brightness,
Y is the "true" brightness.

In other words, the color-difference signals are obtained by matrixing from the I and Q circuitry prior to modulation, and then the "true" luminance from the Y (monochrome) channel is added. When the color chain is operated in the test position with color bars, the luminance provided is the "derived" value, since the luminance tube is not contributing to the signal (nor are any of the color tubes). So now the monitoring facilities in NAM positions will be looking at a signal as the color receiver sees it.

Now study the NAM gated detectors of Fig. 7-7. The red, green, and blue signals just described are applied to the detectors as shown in the diagram. Note that the negative detectors are pnp, whereas the positive detectors are npn. Assume for the moment that the gates are saturated and the emitters of all transistors are held at essentially zero (ground) potential. Further note that under this condition positive-going pulses have no effect on the npn transistors.

Now assume that the red signal momentarily has the highest amplitude. The negative excursion will turn Q2 on, and the resultant negative pulse at the emitter (hence across R1) will hold Q1 and Q3 off since their bases are not as negative as the base of Q2. Thus, only the red negative-going signal will appear across R1. This is to point out that the largest negative-going signal at any instant will appear across R1, holding the remaining transistors off.

Similarly, the signal with the largest positive-going swing will appear across R2, holding the other two transistors off. This is the basic principle

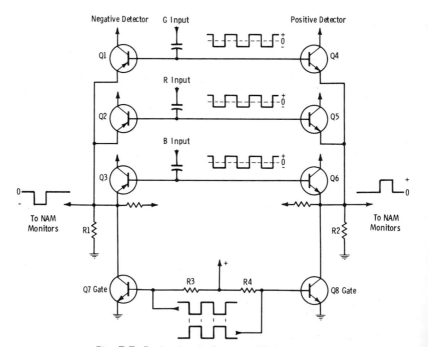

Fig. 7-7. Basic circuit for nonadditive mixing.

of NAM: only that signal with the largest amplitude will be passed, without being added to the other signals (nonadditive mix).

Now look at the gates, Q7 and Q8. They are held in saturation by the base currents through R3 and R4, until negative pulses arrive at the bases to turn them off. Since these pulses are from opposite sides of the NAM flip-flop, one transistor is gated on when the other is gated off. The pulse durations are such that three to four lines of NAM white are on, then three to four lines of NAM black are on, and so on throughout the entire field. You can see that which gate is monitoring white and which gate is monitoring black depends on the video polarity at the point of detection. The important point is that with NAM, the highest-level signal in any of the channels at the time of sampling is monitored. This is one form of electronic switch.

The Nature of Automatic and Manual Control Circuitry

Modern color cameras have one thing in common; either automatic ("semiautomatic" is a more appropriate term) or manual control of white and black levels is available. You will need an orientation in this new type of operation, because it is a different concept in manual control as well as in automatic control.

There are two basic control functions:

1. *Manual control* does not sense any picture information. White level normally is set by a manually adjusted "white pulse" inserted during blanking time.

2. *Automatic control* senses peak white picture information, detects this for a dc control voltage, and uses as a reference a standard set by the operator. (NOTE: Automatic control normally is employed only in film camera chains.)

First, study Fig. 7-8A. Note that the master white-level control sets a reference dc level for all four channels simultaneously. In addition, the master chroma control sets the reference dc voltage for all three color tubes. Note carefully that when the switch is in the "auto" position, the master white control still supplies a reference dc, in this case through the NAM detector circuitry. Remember that the NAM detector supplies only one output at a time, from the channel with the largest instantaneous peak amplitude, and in this instance we are concerned with the peak *white* detector.

NOTE: In some cases, the master controls mentioned above are simply termed "white" and "chroma." The individual balance controls are then designated "mono," "red," "green," and "blue."

Now study Fig. 7-8B for manual control of white level. A white *pilot* pulse is inserted into the video; this pulse is timed from the trailing edge of a master timing pulse that occurs during horizontal-sync time (Fig.

7-4). Notice that the Y-channel control voltage also feeds the white-pulse amplitude modulator stage. This control voltage sets the amplitude of the white pulse at this point.

The gain-controlled amplifier usually contains a pair of transistors with a feedback resistor. Around this feedback resistor a *Raysistor* is used. A *Raysistor* is a light-controlled resistance device consisting of photocells illuminated by lamps. As the lamp current is changed, the resistance changes accordingly. The lamp current is provided by a dc amplifier that receives the output of the peak detector.

Thus far, all we have is a white pulse that occurs during sync intervals and that is controlled in amplitude by the master white control on the remote panel. This combined video and pulse signal is sent through more amplification (not shown) and returned as the sample video shown in

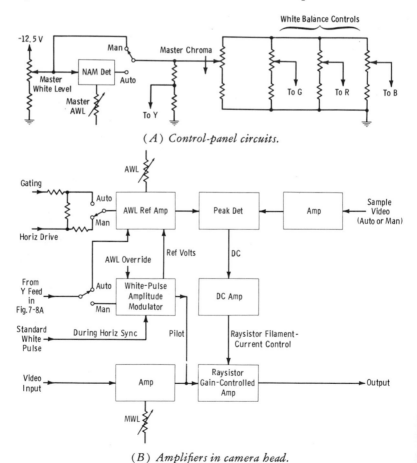

(*A*) *Control-panel circuits.*

(*B*) *Amplifiers in camera head.*

Fig. 7-8. Control of white level.

Fig. 7-8B. Now the closed-loop nature of just the simple manual white-level control begins to become apparent. It is all in the interest of providing automatic level control, as we will see shortly.

We are still concerned with *manual* control at this point. Note that for the manual position of the operating switch the automatic white level (AWL) amplifier is receiving horizontal-drive pulses, which occur during the first part of the horizontal-blanking time (hence sync interval). Therefore, the sampling of the entire video interval being fed back to one side of the peak detector is only during horizontal-blanking time; hence, only the pilot white pulse is sampled. The resultant dc depends on the amplitude determined by the manual adjustment of the white-level control. This direct current is fed to the lamps of the *Raysistor* in the gain-controlled stage, determining the amplitude of the video output.

What we actually have done is to provide a convenient means of going to automatic control, using the same circuitry except for the addition of a reference from peak white video. In the automatic mode, the AWL reference amplifier is gated by a wide pulse that rejects the blanking interval and samples the video interval. If the peak white video being fed back to the peak detector is higher than the reference, the dc error signal that results has a direction to decrease the gain of the controlled stage. The error signal is in the opposite direction when the peak video is lower than the reference. A bias voltage known as AWL override is fed into the white-pulse amplitude modulator; this voltage prevents the gain of the controlled stage from increasing on complete fades to black. Such a gain increase would increase the noise level.

Manual and automatic black-level control works essentially in the same way as white-level controls. Remember the description of NAM circuitry, and the basic function should be clear. For manual control, a black pulse is added to the video during the horizontal-blanking interval. The amplitude of this pulse is governed by the control-panel black-level control.

For automatic black control, the output of the NAM circuit responding to the negative-going signals (signals approaching black) is peak detected. Just as in automatic white control, the black-level control adjusts the ABL detector reference voltage for black. This emphasizes the fact that the white- and black-level controls on the panel are just as functional in the automatic mode as in the manual mode. In the automatic mode, these controls are the source of the reference voltage for the associated automatic level detector.

When optical black is provided, the black-level detector is gated so that it responds to an unilluminated portion of the pickup tubes (this area is provided by an optical mask). Thus, when the automatic black level for each channel is adjusted in initial setup, the black levels of the individual channels are held at a reference regardless of dark-current variation caused by temperature excursions or other factors. This feature normally is used only in nonbroadcast applications such as closed-circuit installations, since the optical-mask black portion must be in the active picture area.

7-3. LEVEL CALIBRATION AND TEST PULSES

Nearly all modern camera chains employ calibration and test pulses for setting amplifier gain and checking amplifier continuity. First, we will consider typical techniques used in film camera chains.

All channel gains, white levels, and black levels must be standardized, or matched, before color-balancing procedures are carried out. Modern color chains (both film and studio) provide for this by inserting reference pulses equivalent to pickup-tube beam currents at the preamplifier inputs.

Fig. 7-9 shows the basic principle of amplifier calibration. Assume a calibration pulse (pulse No. 1) is injected at the output of the camera head. Further assume this calibration pulse is equivalent to the proper peak output signal of the camera, for example 0.7 volt peak to peak. Then a reference pulse (pulse No. 2) is inserted into the pickup-tube target load; this pulse is equivalent to peak high-light target current (normally 0.3 μA for 1-inch vidicons or 0.6 μA for 1½-inch vidicons used in the luminance channel). Now if we adjust the camera amplifier-gain and black-level controls for the channel involved so that this pulse matches the calibration pulse, we have "standardized" the channel gains. If the pickup tube has the electrode voltages applied (but the lens capped), the dark current is present for proper setting of the black level.

Fig. 7-10 shows the usual positioning of pulses in the four-channel film camera. The luminance channel is given the pulses of Fig. 7-9 (the ref-

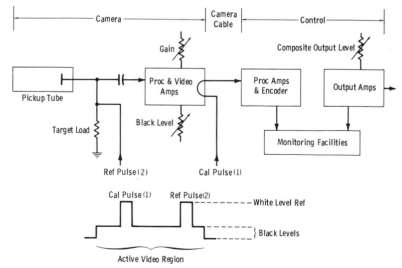

Fig. 7-9. Principles of amplifier calibration.

erence pulse is now pulse No. 3) plus a pulse timed with the green-chan-nel calibration pulse (pulse No. 2). Pulse 2 in the luminance channel is at 0.59 of unity level (the luminance level of green) for proper matrixing (see Fig. 7-11). The blue and red channels receive only the reference pulse (No. 3). (NOTE: this last pulse might be designated No. 5, as in the RCA system, because of spacing.) Such an arrangement makes it possi-ble to carry out balancing procedures with test pulses merely by looking at the picture monitor (on NAM function). When black levels are matched, no "stripes" appear in the No. 1 white pulse on the monitor.

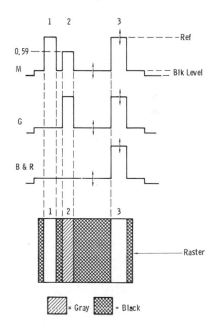

Fig. 7-10. Pulse positions for four-vidicon monitoring setup.

When white levels are matched, no stripes appear in the No. 3 (or No. 5) white pulse on the monitor. A CRO presentation will show all white and black levels identical. White-balance controls are simply amplifier gain controls. Black-balance controls usually set a clipped reference voltage for black-level control.

NOTE: Color-camera setup techniques are covered thoroughly in Chapter 10 of *Television Broadcasting: Equipment, Systems, and Operating Fun-damentals.*

After the channel gains are calibrated, we have a working standard against which to "color balance" the *pickup-tube* operating parameters. The basic requirement (assuming the encoding system has been optimized on color bars) is that zero subcarrier must occur for black, white, and all shades of gray. This, in turn, demands that the pickup-tube outputs be identical in amplitude (at white and all shades of gray to black) when the optical system is looking at a monochrome gray scale.

In this case, a logarithmic transmission-chart slide (for film chains) is used, and the scope is connected to monitor the camera output signal. The slide-projector lamp must be operated at a voltage near the normal line voltage to obtain proper color temperature. Bear in mind also that in this adjustment compensation is made for any film-base color temperature that might exist, and that this can be different for different film.

The basic initial adjustments are:

Target Voltage: Adjust to obtain the required peak white level.

Black-level adjustment: This adjustment is required since any change in target voltage can change the dark current.

NOTE: If the camera chain employs "optical black" circuitry, the vidicon dark-current level is held automatically at a constant pedestal regardless of dark-current variations under target adjustments.

The procedure is simply to adjust the target voltage of one channel at a time to obtain the reference white level for the white chips. Then readjust (if necessary) the black level for each channel to obtain reference black. When these steps are completed, you are ready for the more time-consuming procedure of obtaining tracking at all steps of the gray-scale signal, so that all channel output signals are superimposed on the individual steps from black to white.

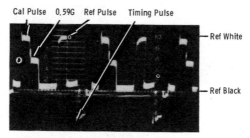

(A) Pulses in camera-head monochrome channel.

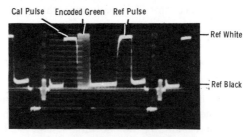

(B) Encoded composite monochrome-green signal.

Fig. 7-11. Test pulses in RCA color camera.

After encoding, the presence of subcarrier in the gray areas but not in the white and black areas indicates mistracking. There are two major causes of lack of tracking: (1) gamma-correction circuitry and (2) spectral sensitivity of the optical paths.

Cameras employing adjustable gamma circuits in the luminance and chrominance channels are readily corrected in case of gamma mismatch. Sometimes only the luminance channel is adjustable, with either fixed gamma correction or no gamma correction in the chrominance channels. In other cases, a fixed gamma correction of 0.7, 0.5, or unity is selectable by means of a switch. Whatever method is used in a particular camera, the best possible tracking should be obtained by experiment before attention is given to the second cause listed above. The second condition must be corrected by using neutral-density discs in front of one or more of the chrominance-tube lenses.

If you have a properly adjusted color monitor (one that produces a truly black and white monochrome picture when color burst is present), you can tell immediately where to start in selecting a proper neutral-density filter to be inserted in front of the pickup-tube lens. For example, if grays are greenish (this is the most common condition), start with a neutral filter of about 0.2 in front of the lens of the green tube. You must, of course, readjust the target voltage of this tube to bring white level back to reference white. Since the target voltage has been readjusted, check for any change in green black level, and adjust if necessary. If grays have now gone "minus green" (purple), the neutral-density filter is too dense, and a 0.1 filter should be tried. Once the optical paths are balanced, it normally is not necessary to change filters when vidicons are changed.

When the film camera has been properly registered and tracked, it is imperative, in practical film operation, to balance whites and blacks of each camera channel for the particular film base and color processing method. For example, a film base can be slightly blue in color temperature. If it is possible to still-frame the color projector, do this on the first frame that has a white and a black area. If the base is blue, the blue channel will have higher setup in the black region than other channels. Balance black and white gains. This is the only way you can keep the flesh-tone concept in telecasting color film. It requires some amount of rehearsal time for previewing the film, but if you are critical in operations, the time is well spent.

A good example of what can happen is illustrated by the following specific example. Assume two studio color cameras have been well balanced to one another. The master color monitor shows good flesh tones for the live pickups. Now stop and think a moment. You have balanced and tracked the camera *on the scene* (essentially) by balancing on a chip-chart under the studio lights for the scenes to be used. You "roll film" for a film clip or commercial; the flesh-tones turn greenish or bluish on the monitor. What has happened? Simply this: Many stations balance on the gray-scale slide, but this is no guarantee that flesh tones will be correct on the film clip, which may have a base with a different color temperature. All encoders can be showing proper phase setup on color bars, but flesh tones may not be correct from all color sources. This is the result of camera color balance. Such a condition can exist between studio cameras not properly color balanced, or between live and film sources not properly color balanced.

Another basic cause of this trouble is lack of proper color balance between slide projectors and film projectors. You should have the gray-scale slide mentioned, and also film loops of the same gray scale for the film projectors. You should balance the film projectors (if necessary) to each other by slight adjustment of projection-lamp voltages. Normally, voltages below about 105 volts cannot be used on projection lamps; less voltage causes the color temperature to go toward red.

You should obtain two Wratten GL filters, one CC10B and one CC20B. Place one of these filters behind the lens of the slide projector and start with a lamp voltage of 100 volts. Adjust the lamp voltage and try the two filters one at a time until the camera stays in balance between the slide-projector and film-projector gray scales. Once you get good balance in this way, a minimum of adjustments should be necessary in going from one film to another.

Some manufacturers now supply specific types of *heat filters* that color-correct slide-projector light sources to specific film projectors. Always investigate current manufacturers' recommendations in color balancing of separate light sources for slide and film projectors.

Fig. 7-12A shows the major difference between the setup of a studio camera and that of a film camera. When a camera has an I.O. and three vidicons, a major difference in signal outputs requires a modification in reference-pulse use. Because of the electron-multiplier output of the I.O., the signal output per lumen is greater for this tube than for the vidicon.

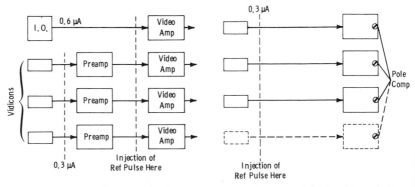

(A) *Camera with I.O. and vidicons.*   (B) *Camera with* Plumbicon *tubes.*

Fig. 7-12. Injection points for test pulses.

The 4½-inch I.O. normally is operated at about 0.6 to 0.7 μA, which is about a 2-μA peak-to-peak current swing on high lights. Therefore, the reference level is at the preamplifier outputs for the vidicons, but at the first amplifier stage for the I.O.

When the camera is placed in the test-pulse mode of operation, continuity of circuit function is complete except for the vidicon preamplifiers. Therefore, variable or switchable gain controls are included in the vidicon preamplifiers to obtain the desired reference *signal* voltage from all channels.

Plumbicons generally are operated at about 0.3 μA peak white current, the same as for vidicons. In a camera employing three or four of these tubes (Fig. 7-12B), all outputs are made equal. The additional control (or fixed circuit) to be found here is the *pole compensation* to accommodate the target capacitance of the *Plumbicon,* which requires compensation around 100 kHz in some camera tubes.

The initial level setups for the live (studio) camera are the same as those described for the film camera. When you set up first on test pulses, you have removed the variables provided by the control panel, and you have standardized white and black levels for all channels. From this point on, the techniques are somewhat different, and these differences will be covered in the next few paragraphs.

The first step in cameras employing the I.O. is to find the knee. It does not matter how much light the camera "sees"; the iris is opened until the steps of the chip chart, as observed on the CRO, start to compress on the top white wedges. This statement assumes the proper neutral-density filter has been installed in front of the I.O. luminance tube (when used with three vidicons in the chrominance channels). For example, the instruction book may call for the knee to occur at exactly $f/8$ with 250 foot-candles of incident illumination on the chip chart. The proper neutral-density filter must be installed for this to occur

Regardless of what you may have heard about color-camera operation employing an I.O. and vidicons, you are going to operate the I.O. at about one-half or even one $f/$ stop over the knee. If you attempt to operate just below the knee, the picture on both monochrome and color receivers will be "washed out." Remember that the vidicon does not have a "natural knee"; the chroma signals keep on increasing as the light is increased or as the iris is opened. "Artificial knees" (a form of signal limiting over a given amplitude) for vidicons have been tried, but, although these help to some extent, you still must use considerable judgement in setup procedures and operations.

You can understand that it is easier to track three or four tubes of the same type than it is to track three vidicons with an I.O. tube. But it *can* be done satisfactorily if you understand what you are doing.

When you have found the I.O. knee, double check the target-voltage setting for 2.3 to 2.5 volts (or whatever voltage is specified) above cutoff. This procedure normally involves a switch that, when thrown to the target-set position, places a negative bias of 2.3 to 2.5 volts on the target. The target voltage is adjusted until the two or three lightest chips can be observed with this bias applied. Then the switch is returned to the normal operating position. Remember that every time you readjust target voltage on any tube (except the *Plumbicon*) you must recheck the black-level setting.

The next step is to measure output signal level of the I.O. channel. This is normally 0.7 volt peak to peak at the camera output, which matches the calibration pulse observable at this point. If the peak white signal is higher or lower than this (operating at ½ stop over the knee), adjust the *dynode gain* for proper peak white output. It is *very important* now to recheck the monochrome black level. If necessary, adjust *both* the automatic black-level (ABL) and manual black-level (MBL) controls (both are involved, even in automatic operation) for proper black levels in automatic and manual operating modes.

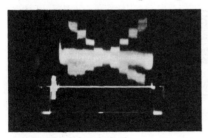

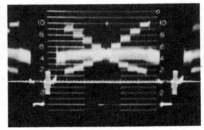

(A) Black-level shading, excessive chroma level.

(B) Minimum subcarrier, even black levels.

Fig. 7-13. Video waveforms for crossed gray scale (wideband scope).

At this time, while you are observing the crossed gray scale of the monochrome channel (only), remember that the scope should show linear steps from the logarithmic gray scale. Adjust the monochrome gamma circuits for the best amplitude linearity from black to white. If the black levels at the two extremes of the scale do not match (Fig. 7-13), adjust the shading slightly for proper balance. (Shading is adjusted initially for flat black levels.) NOTE: If you observe excessive "grass" at black level and even in the blacker-than-black region, either the multiplier-focus voltage is wrong or the vertical sweep of the I.O. tube is not properly centered. The multiplier-focus voltage should be adjusted for maximum output level with the best shading. The I.O. sweep centering should be checked with the size gauge normally provided. This gauge consists of a transparent ring to be inserted temporarily in front of the I.O. tube. The sweeps are centered and sized so that the inner circle just meets the top and bottom of the raster blanking and the outer circle just meets both sides of the raster blanking. This assures proper aspect ratio and size.

IMPORTANT: After resetting the multiplier focus or shading, recheck the black-level adjustments.

Before attempting to track the chroma tubes with the I.O., always be sure the sweeps for each tube are centered and sized properly on the mask of the chroma field lens.

Because of the many variables, the tracking procedure for the I.O.-vidicon combination is a little more involved than that described for the film chain. When vidicon target-current meters are provided, simply adjust the target voltage for 0.3 μA peak high-light current, and then match levels with the I.O. by means of the adjustable vidicon-preamplifier gains. When target-current meters are not provided, the preamplifier gain must be made as low as possible consistent with reasonable target voltage required to match the I.O. output signal level. If the black level for the particular channel involved becomes so high that it is beyond the range of the ABL and MBL controls, excessive preamplifier gain is indicated. Except for this complication, the target-balancing procedures are the same as those described for the film-camera setup.

When the targets in the camera have been balanced, return the test switches to the operate position and adjust the white- and black-balance controls on the remote control panel in the control room. The first step is to select the monochrome channel and adjust for a half stop over the knee to check for proper output signal level. Then select the green-channel signal, and adjust the master chroma control to get the same output level as for monochrome (using the chip chart). Then (normally) go to NAM, and balance the black and white levels for red and blue on the picture monitor as was done for the film chain. Remember that this is proper "color balance"; the subcarrier must be as nearly zero as possible for all shades of gray (Fig. 7-13). The matching of color cameras to one another on an actual color scene will then involve only minor adjustments.

To check "color balance" between two or more cameras, a common, switchable color monitor is used to observe all camera outputs. First of all, what "color" does this monitor have when "looking at" a color camera operating in the monochrome mode (color subcarrier removed)? The picture should be strictly black and white; if it is not, and you do not have time to adjust the monitor, remember the color, and fix it in your mind as your black-and-white reference.

Next, bear this in mind: Even though you have "balanced" all color cameras on a gray scale, there may be a different skin tone from each camera. Why? This results from the difference in gamma correction if the fault is in luminance levels around the skin-tone reflectance area. This is to say that the gray scale must be the same on all cameras as well as being balanced for minimum subcarrier; i.e., all gray-scale steps must be at the same level relative to capped level. To achieve this, it might be necessary to make minor adjustments to gamma and/or target voltages and then rebalance for black and white levels. This problem is common to all types of color cameras, including the three- or four-*Plumbicon* camera. However, only the gamma-correction circuitry is involved in the latter cameras, since target voltage on the *Plumbicon* has a minimum effect on sensitivity or gray-scale tracking.

It should be obvious that all cameras should be color balanced on the *same* gray scale under the *same* light conditions. Then, when you switch between cameras on the common color monitor, slightly different colors of "gray" can be compensated for by slight adjustments of the black-balance controls. The final check is to observe a live model under the lighting to be used (which should be the same as for the gray-scale setup) and select the camera with the most pleasing skin tones as the "standard."

When the studio camera employs amplitude controls in aperture-correction circuitry, the camera picture monitor can be fed from the strobe output of a suitable waveform monitor (such as the Tektronix 529) with the delayed sweep adjusted to the 300-line point of the EIA resolution chart (Fig. 7-14). In most cases, with well designed cameras, this amplitude control can be adjusted to obtain a 100-percent response relative to picture white at 300 lines resolution. Remember that the amount of picture noise introduced is the limiting factor. The vidicon and the *Plumbicon* can stand more aperture correction than can the image orthicon for a given noise level at the output.

NOTE: In Fig. 7-14, the scope time base was adjusted to five TV lines, rather than a single line, so that the brightening pulse would be readily apparent in the photo.

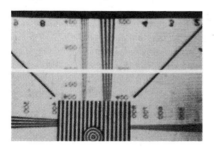

Fig. 7-14. Picture monitor fed from line-strobe oscilloscope.

7-4. CAMERA DEFLECTION CIRCUITRY

Deflection-yoke assemblies were covered in Chapter 4. This section will describe the pulse formation and circuitry necessary to drive sawtooth currents through the yoke assemblies.

Sawtooth generators of the vacuum-tube type generally incorporate one of the two basic circuits, a blocking oscillator (BO) or a multivibrator (MV). The sequence of operation for the blocking oscillator (Fig. 7-15) is as follows:

When the grid of V1A swings positive, plate current *increases,* and plate voltage *decreases* (because of the drop across R_L). The increased voltage drop across the transformer primary is coupled into the secondary winding with such polarity as to reinforce the positive swing of the grid. This is known as a *feedback cycle,* which drives the grid positive to a point at which no further increase in the plate current is obtained. At this time, transformer feedback ceases (no current *change* through the inductance), and the grid voltage falls off rapidly. Therefore, the plate current decreases rapidly, and the plate voltage increases. The new change in plate current causes a voltage of opposite polarity to appear across the primary, initiating a negative voltage on the grid and, therefore, a new feedback cycle of opposite polarity. At this time the tube is biased well beyond cutoff. The large negative charge on grid capacitor C1 (a result of the previous grid current) begins to flow through grid resistor R1 while the tube remains non-conducting. When the capacitor discharges, the grid again reaches a potential at which the tube starts to conduct. The time allowed for the capacitive charge to leak off is determined by the time constant of the RC combination. This time constant is made somewhat longer than the interval between drive pulses to that the pulse triggers exert positive control over the timing of the sawtooth waveform.

The second section of the tube (V1B in Fig. 7-15), is the *discharge* section. The grids of both sections are tied together and therefore receive the same voltages simultaneously. When the oscillator section is conduct-

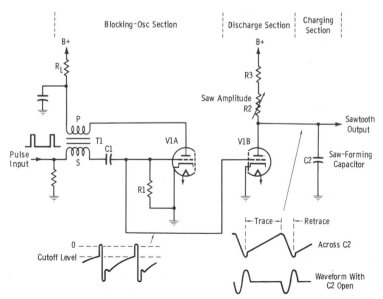

Fig. 7-15. Basic blocking-oscillator circuit.

ing (grid positive), the discharge section also conducts because its grid is also positive. When the grids become negative, capacitor C2 in the plate circuit of the discharge section slowly charges through resistors R2 (adjustable) and R3. This generates the *trace* portion of the sawtooth waveform. When the tube is triggered into conduction by the driving pulse, the capacitor rapidly discharges, generating the *retrace* portion of the waveform.

Variable resistor R2 is used to adjust the voltage toward which C2 can charge; hence it determines the sawtooth amplitude. If the circuit is for vertical deflection, the control is termed "height." If the circuit is for horizontal deflection, the control is a width control.

Fig. 7-16 illustrates a basic multivibrator circuit. Essentially, such a circuit provides feedback action between two tubes (usually a single duo-triode type) so that one tube conducts while the other is nonconducting, then vice-versa on the succeeding alternation. The similarity of circuit action to that of the blocking oscillator described above will become obvious in the following discussion.

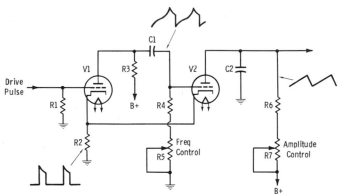

Fig. 7-16. Basic multivibrator circuit.

Assume for the moment that the grid of V2 is swinging in the positive direction. The plate current of V2 will increase, and since this current passes through common cathode resistor R2, V1 receives a negative signal (grid more negative with respect to cathode because of increased voltage drop across R2). Thus the plate current of V1 decreases, and its plate voltage increases. As the V1 plate voltage increases, the grid voltage of V2 increases still further in the positive direction, thus reinforcing the initial increase in this voltage. Here it may be seen that tube V1 is serving as the feedback tube, similar in action to the transformer in the blocking oscillator described above. The grid voltage of V2 increases in the positive direction until no further increase of V2 plate current is able to take place. At this time, since no further change of plate current occurs, feedback ceases, the V2 grid voltage begins to decrease, and the feedback cycle is reversed. Since the negative grid voltage on V1 is now decreased, the V1 plate current increases (tube starts conducting), and the resulting reduced plate voltage on V1 drives the grid of V2 below cutoff. Then V2 remains nonconducting until capacitor C1 discharges sufficiently through R4 and R5. During this interval, capacitor C2 is charging through resistors R6 and R7, and the sweep of the sawtooth is formed as shown in the diagram. When V2 starts conducting, capacitor C2 rapidly discharges through the tube, and the sawtooth waveform returns rapidly to zero as shown. Since R5 is a variable resistance in the grid circuit of V2, it provides a means of determining the rate of discharge of C1 and, hence, the frequency of the sawtooth wave. Resistor R7 in the plate circuit of V2 determines the amount of charge placed on C2 while the tube is not conducting; hence, it provides a means of adjusting the amplitude of the sawtooth waveform.

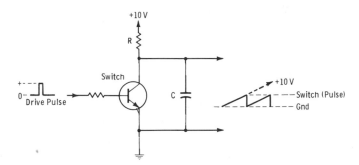

Fig. 7-17. Basic solid-state sawtooth generator.

There are many variations, but Fig. 7-17 illustrates the basic solid-state sawtooth generator. In the interval between drive pulses, capacitor C charges toward the supply voltage, since the transistor is cut off (switch open). This forms the trace portion of the sawtooth. At pulse time, the transistor is saturated (switch closed) to discharge C rapidly. This is the retrace portion of the waveform. The time constant, RC, is made long compared to the interval between pulses so that only a small part of the exponential capacitor charge is used. This assures excellent linearity of the trace portion of the sawtooth.

Deflection-Coil Driving Circuitry

Fig. 7-18 shows a basic tube-type horizontal-deflection output circuit. The sawtooth voltage wave is applied to the grid of V1, termed the *driver* tube. It may be seen from the diagram that the top of deflection transformer T1 is connected to the top of the horizontal-deflection coil (across the damper tube). When the driver tube is conducting, current in the horizontal-deflection coil is increasing, as shown in Fig. 7-19. When the

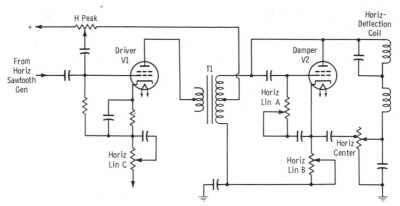

Fig. 7-18. Basic tube-type horizontal-output circuit.

driver tube rapidly decreases conduction, the coil current also rapidly decreases, as shown in Fig. 7-19. This interval is the *flyback time*. It is well known that a rapid change of current through an inductance creates a voltage surge that is dependent on the rate of change of the current and the self-inductance of the coil. The rapid decrease in conduction is sometimes used as the source of voltage for the high-voltage rectifier (flyback type). The yoke inductance and its distributed capacitance are said to *ring* in a damped oscillatory fashion for about a half cycle. The tube which now comes into operation is V2, the *damper* tube. This tube conducts and causes current through the coil in the direction shown by Fig. 7-19. Thus, it may be seen that the scanning (sawtooth) current through the horizontal-deflection coil is supplied alternately by the driver and the damper. In practice, it is found that the driver tube supplies sweep for the right side of the picture, and the damper tube supplies sweep for the left portion of the picture.

The control (labeled "Horiz Lin C" in Fig. 7-18) in the driver-tube cathode, in conjunction with the control labeled "H Peak," adjusts the linearity of the sweep. These controls are largely effective only on the center to right side of the raster (during driver-tube conduction).

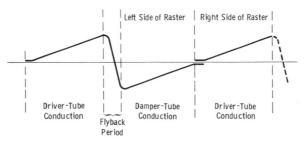

Fig. 7-19. Current waveform for horizontal sweep.

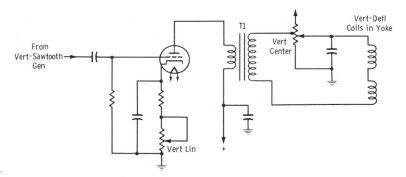

Fig. 7-20. Basic tube-type vertical-deflection circuit.

Two linearity controls are shown in the damper-tube circuit, one in the grid circuit and one in the cathode circuit. These control the phase and extent of conduction of the damper tube, and affect the linearity of the left side of the picture. The horizontal-centering control adjusts the amount of an externally applied direct current through the deflection coil, serving to center the sweep. Another pickup-head control, in the grid circuit of the driver tube, is marked "H Peak." This is a feedback adjustment to prevent the driver tube from conducting too soon, which would upset the proper transition of current from the damper tube to the driver.

A vertical-deflection circuit is shown in the diagram of Fig. 7-20. This circuit operates on the same principle as that described above, but at a much lower frequency (60 Hz as compared to 15,750 Hz); therefore, details of the circuit are somewhat different. The coil of the vertical-deflection yoke is largely resistive at 60 Hz, and no large surge of voltage occurs during retrace. Thus, there is no ring, or oscillation, to be damped, and no damper tube is employed.

Feedback Linearity Correction

Modern deflection circuitry, whether tube type or solid state, employs some form of automatic linearity control of the sawtooth wave for both horizontal and vertical deflection. The basic form is shown in Fig. 7-21. We will analyze this circuit, which, with minor variations, is rather common in all types of cameras.

In this specific circuit, V1 operates between cutoff and conduction. Upon the arrival of the horizontal-drive pulse at the grid, the tube conducts, rapidly charging the saw-forming capacitor to the supply voltage (waveform 2). At the end of the pulse, the tube is cut off, and the capacitor discharges in an exponential manner. During the on period, a portion of the voltage is developed across R1, and this action adds a pulse to the sawtooth waveshape (waveform 2). Linearity control R1 allows variation of the amplitude of this pulse, providing vernier control of sweep linearity.

Because the load presented to the plate of V3 by the horizontal-yoke windings and output transformer T1 is largely reactive at the horizontal frequency, the voltage waveform at the V3 plate resembles a pulse more than a sawtooth (waveform 3). This is necessary to drive a sawtooth *current* waveform through the yoke. A sampling resistance of small value (R2) in series with the deflection coils produces a voltage waveform that is identical to the yoke-current waveform (waveform 4). This provides a source of negative feedback voltage to the comparison amplifier (V2) and to the saw-former capacitor.

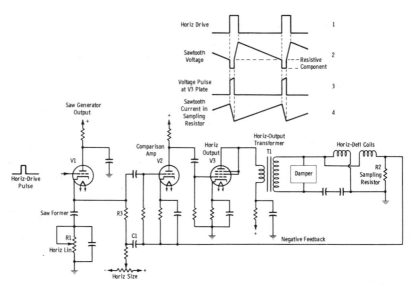

Fig. 7-21. Automatic linearity control.

The comparison amplifier compares the original sawtooth waveform at the grid with the feedback signal at the cathode. Any deviation from a truly sawtooth current through the yoke causes a correction voltage to drive the V2 cathode in such a direction as to compensate for the non-linearity of the original sawtooth applied to the grid. In addition, since the original capacitor charging curve is exponential, capacitor C1 couples the error, or feedback, voltage to the saw-former "charging" capacitor (through resistor R3) with such polarity that it tends to make the exponential curve more linear.

NOTE: The same horizontal- and vertical-drive pulses that energize the deflection circuitry are employed to generate target-blanking pulses. These blanking pulses normally are produced by conventional multivibrator circuits.

7-5. PICKUP-TUBE PROTECTION

Failure of either the horizontal- or vertical-deflection circuitry would limit pickup-tube scanning in the respective dimension to a very narrow path (single horizontal or vertical line) with resultant burning of the overbombarded target area. Thus all cameras employ some form of pickup-tube protection. Regardless of the type of circuit used, the basic idea is common to all: a dc voltage related to the deflection voltage waveforms is used to activate or deactivate the beam current in the pickup tube.

See Fig. 7-22A. Transistors Q1 and Q2 sample the outputs of the vertical- and horizontal-deflection yokes respectively, and control the voltage appearing at the pickup-tube cathode. The biasing arrangement shown causes conduction of both transistors in the absence of a base signal. Under this condition (transistors saturated), the common collector voltage goes to the emitter voltage of +20 volts, cutting off the pickup-tube beam.

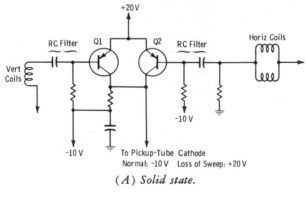

(A) Solid state.

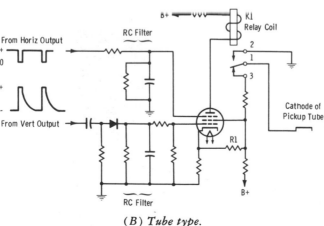

(B) Tube type.

Fig. 7-22. Protection circuits for pickup tube.

When both deflection signals are present, both transistors cut off and the common collector voltage goes to −10 volts. Since this point is connected to the pickup-tube cathode, beam current can exist.

In the absence of either deflection signal, the respective transistor conducts (saturates), and the common collector goes from −10 to +20 volts, again cutting off beam current. This specific circuit treats retrace or flyback duration as absence of deflection signals and triggers the beam to cutoff, thus providing blanking. In this case, blanking and pickup-tube protection are provided by the same circuit.

A common tube-type protection circuit is shown in Fig. 7-22B. Filtered vertical-circuit pulses feed the control grid, and filtered horizontal-circuit pulses feed the suppressor grid. A hold-off resistor (R1) applies a positive

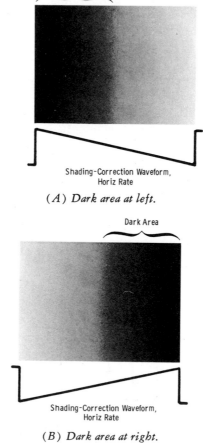

Dark Area

Shading-Correction Waveform, Horiz Rate

(A) Dark area at left.

Dark Area

Shading-Correction Waveform, Horiz Rate

(B) Dark area at right.

Fig. 7-23. Waveforms

Shading-Correction Waveform,
Vert Rate

(C) Dark area at top.

Shading-Correction Waveform,
Vert Rate

(D) Dark area at bottom.

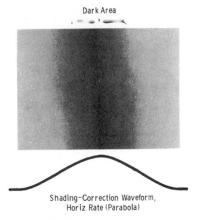

Shading-Correction Waveform,
Horiz Rate (Parabola)

(E) Dark area in center.

for shading correction.

potential to the cathode such that the tube does not conduct in the absence of either the horizontal- or vertical-deflection waveform. Under this condition, the relay coil (K1) in the plate circuit is not energized, and relay contacts 1 and 3 are closed. This applies a positive potential to the pickup-tube cathode, preventing beam current. When both deflection voltages are present, the tube conducts, energizing relay K1. This closes contacts 1 and 2, returning the pickup-tube cathode to ground.

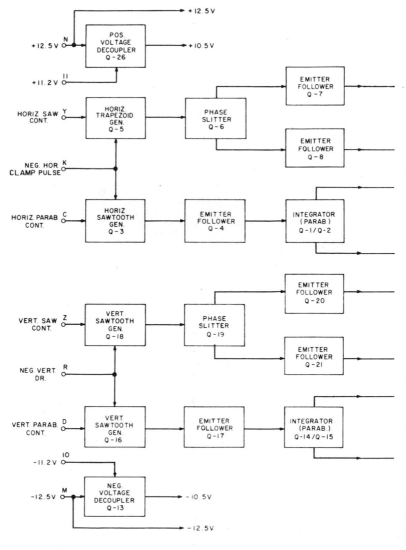

Fig. 7-24. Shading generator

7-6. SHADING-SIGNAL FORMATION

In spite of the coarse and fine adjustments provided for pickup tubes, shading sometimes occurs at various areas of the raster, as illustrated in Fig. 7-23. Correction of a dark area at left (Fig. 7-23A) or right (Fig. 7-23B) of raster requires a horizontal-rate sawtooth of opposite polarity. A dark area at the top (Fig. 7-23C) or bottom (Fig. 7-23D) requires

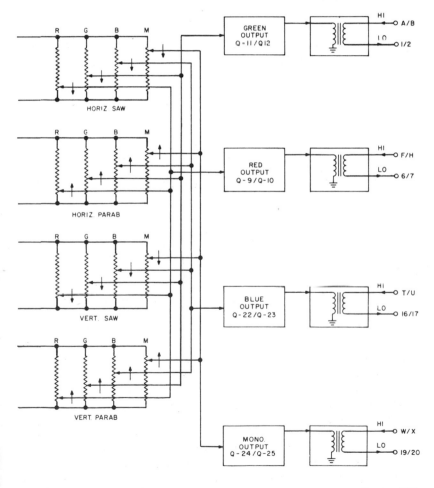

Courtesy RCA

for RCA TK-27 color camera.

a vertical-rate sawtooth of opposite polarity. A vertical dark area in the middle of the raster (Fig. 7-23E) requires a parabolic waveform at the horizontal rate. If the dark areas are at both sides, a parabolic waveform of polarity opposite to that shown would be required. A horizontal dark area in the middle of the raster requires vertical-rate parabolic pulses.

A block diagram of one model of shading generator is illustrated in Fig. 7-24. The functional description of this unit is as follows:

The *horizontal-trapezoid* generator develops a sawtooth waveform, during active scan, that rides on a pulse that occurs during camera blanking. The amplitude and polarity of this waveform are proportial to the incoming horizontal-saw dc control voltage. The pulse is added to the sawtooth to correct for nonlinearities of the output transformer.

The *horizontal-sawtooth* generator produces a sawtooth waveform at a horizontal rate. The amplitude and polarity of the waveform are proportional to the horizontal-parabola dc control voltage.

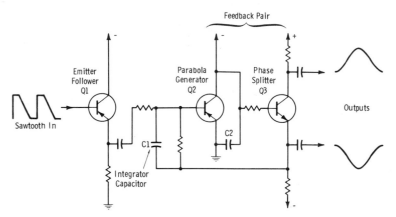

Fig. 7-25. Circuit for generating parabolic waveform.

The *vertical-sawtooth generators* provide sawtooth waveforms at a vertical rate. The amplitude and polarity of the sawtooth are controlled in one case by the vertical-saw dc control voltage, and in the other by the vertical-parabola dc control voltage.

The *phase splitters* provide equal positive-going and negative-going waveforms, which are fed to potentiometers that allow control of the amplitude and polarity.

The *integrators (parabolic)* integrate a sawtooth input to produce a parabolic waveform. They provide equal positive and negative outputs to be fed to potentiometers that allow control of the amplitude and polarity.

The four *output amplifiers* provide a low input impedance for signal mixing. Also, they amplify the mixed signals and provide a low driving impedance to the isolation transformer.

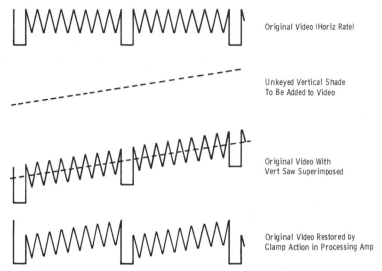

Original Video (Horiz Rate)

Unkeyed Vertical Shade
To Be Added to Video

Original Video With
Vert Saw Superimposed

Original Video Restored by
Clamp Action in Processing Amp

Fig. 7-26. Vertical-shading waveforms without keying signal.

The *voltage decouplers* isolate the dc supply voltages from the rack wiring and other modules.

We have already studied the formation of a sawtooth waveform from pulses. It remains to see how the parabolic waveform is shaped. See Fig. 7-25. A sawtooth signal from emitter follower Q1 is fed into the feedback pair Q2-Q3. This waveform is integrated through the action of capacitor C1 in the feedback path. Then the waveform across C2 is the integral of the sawtooth and is therefore, parabolic in shape. Equal loads are placed in the emitter and collector circuits of output transistor Q3 to ob-

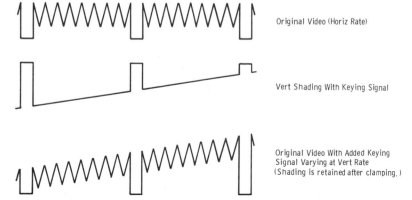

Original Video (Horiz Rate)

Vert Shading With Keying Signal

Original Video With Added Keying
Signal Varying at Vert Rate
(Shading is retained after clamping.)

Fig. 7-27. Vertical-shading waveforms with keying signal.

tain equal and opposite parabolic waveforms for the parabola control potentiometers (Fig. 7-24).

The vertical component superimposed on the video of any signal in the camera will be "clamped out" in the following processing amplifier (see Fig. 7-26). It is necessary therefore to key the vertical-shading signal at a horizontal rate, with the tips of the keying pulses returned to a fixed potential. Since the processing amplifier clamps to the tips of horizontal blanking, the vertical component of the signal will be restored as shown in Fig. 7-27.

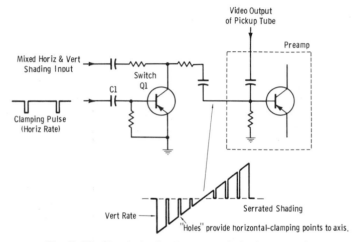

Fig. 7-28. Circuit for keying vertical-shading signal.

Fig. 7-28 shows a typical keying circuit for this purpose. A negative clamp pulse on the base of clamp-reference transistor Q1 charges C1 to the emitter potential (ground in this example) of Q1. With the tips of the pulse clamped at this potential, Q1 cuts off following the negative pulse and is held cut off until the next pulse appears. At cutoff, Q1 is an open circuit and has no effect on the shading signal, but during each pulse Q1 is saturated and switches the signal to ground.

In this circuit, the only collector-to-emitter supply voltage is the shading signal, and the only signal appearing at the collector is the serrated shading signal. When there is a zero shading component, there is also zero output at the Q1 collector. In this circuit, the collector signal is coupled to the preamplifier and added to the pickup-tube video signal.

EXERCISES

Q7-1. Why is the horizontal-drive width made approximately half the horizontal-blanking width?

Q7-2. Do all boxcar circuits narrow the input pulse?

Q7-3. How do you know what output-pulse width to expect in a typical boxcar circuit?

Q7-4. If the input pulse to a pnp boxcar is positive, is the boxcar output delayed or undelayed?

Q7-5. If the input pulse to an npn boxcar is negative, is the boxcar output delayed or undelayed?

Q7-6. Can you change *Plumbicon* sensitivity by target adjustment?

Q7-7. In the automatic mode of operation, are the iris control and black and white level controls on the control panel still operable?

Q7-8. Do automatic level controls affect levels in manual operation?

Q7-9. What are the variables in color-camera balance?

Q7-10. What must be rechecked when the I.O. target voltage is readjusted?

Q7-11. What must be rechecked when the vidicon target voltage is readjusted?

CHAPTER 8

Camera Control and
Setup Circuitry

This chapter covers the remote operating camera-control panel and associated setup controls, which may be physically located at the camera head, on the control panel, or in rack-mounted gear. We also will discuss the usual types of waveform monitors associated with setup and control.

8-1. THE RCA TK-60 MONOCHROME CAMERA CHAIN

Obviously, the simplest form of camera chain is the single-tube monochrome type. We are justified in covering such a camera chain not only because there are many still in use, but also because the color camera is exactly the same, except for the added signal channels and additional associated controls. These will be covered in their proper place later in this chapter.

Fig. 8-1 illustrates the entire TK-60 chain with the exception of the rack-mounted processor unit and the console-mounted picture and waveform monitor. Fig. 8-2 illustrates the normal hookup of the complete installation.

Iris control is achieved by a transistor servo amplifier system. The irises of all four lenses are adjusted simultaneously from either the rear of the camera or the remote-control position. In this manner, the iris is always preset as lenses are changed.

Voltage regulation for the image-orthicon tube is achieved by using corona discharge tubes. Where utmost precision is desirable, these are enclosed in a temperature-controlled oven. Other regulating devices that are utilized include zener diodes and *Victoreens.*

The nuvistor type of triode tube is used exclusively in the video pre-amplifier, and in a number of other circuits associated with blanking and deflection. The signal-to-noise performance and freedom from microphonics of this tube make it particularly useful in preamplifier service.

(A) Rear view of camera.

(B) Interior of camera.

(C) Operating control panel.

Courtesy RCA

Fig. 8-1. RCA TK-60 monochrome camera.

This system includes two separate intercom circuits. These may be inter-connected or operated independently. A transistor amplifier and volume control are provided at each point where a headset is plugged into the system.

A 4½-inch I.O. tube is used in the camera. Circuits in the processor provide aperture response to produce 100-percent response at 400 TV lines, phase corrected. These circuits also provide up to 13 dB of aperture correction peaked at 8.0 MHz, with continuously variable amplitude adjustment to compensate for tolerances in pickup tubes.

Video equalization to compensate for different lengths of camera cable is made available on a tap switch. Positions on the switch correspond to 100-foot increments of cable length, up to a maximum of 1000 feet.

A completely solid-state power supply, Type WP-16B, provides all regulated voltages for the complete TK-60 chain. It occupies 7 inches of rack space.

Either of two modes of operation can be selected. In a *clamp-on-black* mode, exposure is governed by the iris control, and the black-to-white level (gray scale) is dependent on settings of the contrast and brightness controls. This is similar to the operating techniques used with standard 3-inch image-orthicon cameras. In a *clamp-on-white* mode, brightness becomes a setup adjustment to establish the desired peak white level. The iris control is adjusted to maintain white level of the scene at the desired setting over the knee of the I.O. characteristic. The contrast control is used to compensate for reflectance variations in the darkest areas of the scene.

Use of the remote-control panel permits a single video operator to handle as many as six TK-60 cameras. In normal operation, only two controls—IRIS (exposure) and CONTRAST—may require attention. For convenience, a brightness setup control, remote lens-cap switch and tally, on-air tally, intercom phone jack, and headset-level control are included.

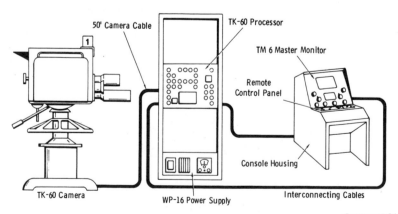

Courtesy RCA

Fig. 8-2. TK-60 monochrome camera chain.

The rear panel of the camera (Fig. 8-1A) contains the main operating controls and jacks for the cameraman, as follows:

Lens cap
Orbitor on-off switch
Viewfinder selector switch: effects preview, camera, or effects (only)
Normal and reverse vertical deflection
Engineering phone and cue volume controls
Engineering phone and cue jacks

The items listed above are all on the left side of the rear panel. The following controls and jacks are on the right side:

Viewfinder contrast, peaking, and brightness controls
Iris remote and local switch
Iris control
Normal and reverse horizontal-deflection switch
Production phone and cue volume controls
Production phone and cue jacks

In the center is the turret-control handle for a four-position lens complement. Immediately above this handle is a six-position filter control to select any of six different neutral-density filters. Thus, if a given depth of field must be retained under varying lighting conditions (dependent on f stop of iris), the exposure for the I.O. can be controlled by this knob.

The setup control panel (lower right corner inside camera in Fig. 9-1B) contains the following controls:

Orth focus
Wall focus
Switch with align, calibrate, set, operate, and B+ off positions
Target voltage
Beam current
Viewfinder focus
Viewfinder contrast
Viewfinder brightness
Beam align No. 1 and No. 2 controls
Video gain
Multiplier focus
Image focus

Once the camera is properly set up, the above controls require no further adjustment under normal operations.

There is only one part of the setup procedure for this particular camera that differs from the steps followed for other cameras: the adjustment of the VIDEO GAIN control on the camera preset panel. The RCA procedure is as follows: Set the selector switch on the preset control panel to the "cal" position. Block off the scene during the white portion of the calibration pulse by inserting the edge of the neutral-density filter disc into the optical path, or, preferably, by placing a black material of very low reflectance (velvet or equivalent) over the proper portion of the scene. Adjust the VIDEO GAIN control for a match of the video white to the white of the calibration pulse. This adjustment will be more accurate if the viewfinder brightness level is set near the kinescope cutoff point. Return the viewfinder brightness to its normal level. Remove the edge of the filter disc or other obstruction from the optical path, and set the selector switch to the "oper" position.

The normal operating procedure for this camera as given by RCA is presented below. Comparable controls in all cameras normally are adjusted by this basic procedure.

Warm-Up

The image orthicon is the determining factor in the camera-chain warm-up time for *on-air operation*. It is recommended that approximately 20 minutes be allowed for the image orthicon to reach optimum operating temperature. The camera circuitry is sufficiently stablized after 2 minutes for *off-air* operation, but the image orthicon will be slightly sticky until it reaches operating temperature. However, the orbiter permits immediate use without serious burn-in.

Lighting

The TK-60 camera is capable of handling wide light-level ranges through the combined action of the IRIS control and neutral-density filters on the filter disc. Except for special effects, the contrast ratio of a specific scene should not be more than 20 to 1. Contrast ratio is best controlled by flat-lighting the scene, which also can reduce undesirable shading effects caused by shadows and unbalanced lighting conditions. It will become more evident from experience that the application of proper lighting techniques contributes substantially to the ease with which camera-chain operation is carried out.

Modes of Operation

A clamp-on-black or clamp-on-white mode of operation is possible for the TK-60 camera chain. Either mode may be selected by a jumper arrangement in the camera and processing amplifier. The following paragraphs further describe the two modes relative to actual camera-chain operation and the specific controls involved.

Clamp-on-Black—In the clamp-on-black mode, changing either the iris setting or the brightness or contrast levels causes a change in all video levels (gray to white) with respect to black. Therefore, depending on the lighting and special effects desired, the proper combination of control adjustments must be made to maintain a constant output level. The following information may be used as a guide for the control procedures required to accommodate various scene conditions.

The IRIS control must be adjusted to maintain scene whites at the proper setting over the knee of the image-orthicon characteristic. This setting is therefore a function of scene white level. During operation, the iris-delegate switch on the camera is in the remote position, since iris control is one of the main control operations at the camera-control position (remote-control panel).

If scene whites are maintained at a constant level, the remote CONTRAST and BRIGHTNESS controls are used to make corrections for contrast-ratio changes. The combined operation of these controls sets the dark areas of a scene to the desired level while maintaining a 0.7-volt output level.

When both the scene white level and the contrast ratio changes, the CONTRAST, BRIGHTNESS, and IRIS controls must be adjusted to maintain the desired black level, output level, and operation of the image orthicon relative to the knee of its characteristics.

Special-Effects Operation—Combinations of the control settings described above may be used to obtain the special effects possible as the result of black clipping or operating the iris below the knee of the image-orthicon characteristic. When the foregoing abnormal settings are to be employed, the normal settings should be recorded to expedite a return to the original operating condition.

NOTE: Under conditions of extreme contrast, such as outdoor sunlight to shade, the gamma-corrector circuit will allow the system to compensate at the sacrifice of true gray-scale rendition. A 0.7-gamma position is provided in the processor for this purpose.

Clamp-on-White—In the clamp-on-white mode of operation, the video white level remains fixed (clamped) relative to black, for scene contrast changes or contrast-control variations, as long as the iris sets the scene whites over the knee of the image-orthicon characteristic. Therefore, a constant peak-to-peak output is maintained without the use of the brightness control. The constant output level facilitates operational simplicity for most operating conditions. This will be evident from the following control-setting procedures given for possible scene conditions.

The IRIS control must be adjusted to keep scene whites at the proper setting over the knee of the image-orthicon characteristic. If scene contrast ratio is maintained at the constant level, the IRIS control is then the only operating control necessary.

The CONTRAST control serves to maintain the effective contrast ratio by compensating for reflectance variations in the darkest areas of a scene. Therefore, if scene white level remains fixed, the CONTRAST control is the only control required.

The use of both the CONTRAST and IRIS controls is necessary when scene white-level and contrast-ratio changes occur simultaneously.

Under normal operating conditions, the BRIGHTNESS control is not used. However, it may be required in some instances as described for special effects. If this control is so employed, the normal setting should be recorded to expedite a return to the original operating condition.

A combination of the foregoing control settings may be employed to obtain special effects such as those resulting from black clipping or operation of the iris below the knee of the image-orthicon characteristic. If the iris is set below the knee, the peak-to-peak output level changes, and this may be compensated for by resetting the brightness level. The note under "Special-Effects Operation" for the clamp-on-black mode of operation also applies to the clamp-on-white mode.

Target Voltage Effect on Operation

With the low target voltage (2.3 volts above cutoff), the knee of the image-orthicon characteristic is rounded. This condition makes operation over the knee less critical and permits the iris to be opened further to lift scene blacks with relatively small loss of white detail. The result of this condition is an improved signal-to-noise ratio. Other advantages of low-target-voltage operation are extended tube life, reduced target flicker, and reduced microphonics.

With a high target voltage (3.0 volts above cutoff), the knee is sharp, and all scene contents are below the knee while the scene whites are at the knee.

Stopping over the knee can cause a loss of white detail, resulting in a chalky appearance. With the proper iris setting, a good gray-scale rendition is possible (operation is more linear).

Camera-Position Operating Controls

The optical-focus and lens-selection controls are the only operational controls required for on-air operation that are located at the camera.

For normal operation, the IRIS control is switched to the remote position. Local operation of this control serves mainly to facilitate camera-setup procedures.

The neutral-density-filter disc normally is kept in the open position so that the lens may be stopped down for greatest depth of field and the light requirements can be kept to a minimum. In bright sunlight, if the iris cannot control the high lights below the knee of the image-orthicon transfer characteristic, neutral-density filters must be inserted in the light path. In cases where the director calls for a given depth of field, the iris must be set, and the high lights must be attenuated to the knee with neutral-density filters.

The orbiter switch normally is placed in the immobilize position. As the image orthicon approaches the end of its useful life, it may develop spots or burns on the dynode. With the switch in the immobilize position, these blemishes will appear to orbit as the picture remains stationary. If this becomes distracting, the orbit position should be used. This causes the blemishes to remain fixed while the picture orbits so slightly as to be unnoticeable. The orbit position cannot be used when absolutely stationary images are required, as in a superimposition or centering a title card.

By use of the deflection-reversal controls (on the camera rear control panels), the image-orthicon horizontal and vertical scans may be reversed for certain special-effects applications.

The LENS CAP switch tends to place the camera in a standby condition by capping the lens electronically and inserting a reduced-amplitude calibration pulse to indicate proper operation.

The viewfinder input is obtained from the output of the processor and is therefore identical to the signal going out on the line. Viewfinder peaking is available as an aid in obtaining a rapid, well-defined focus point without overshooting the correct setting.

The viewfinder picture-selector switch (on the camera rear control panel) permits the cameraman to view his picture as it is inserted or superimposed with other signals coming from the special-effects switcher. When no effects switcher is programmed, the viewfinder picture-selector switch is kept in the camera position, and the cameraman always sees the output of his processor and may set the viewfinder brightness and contrast levels to his individual preference.

Anytime the cameraman desires to see the output of the effects switcher, such as when he is preparing his camera for a superimposing operation with another camera, he may switch to effects preview. Then he can see the output of the effects switcher, regardless of how his camera may be switched at the time.

When the viewfinder is switched to effects, a relay (energized by the effects tally circuit) switches the viewfinder input to the effects switcher only when the camera is punched up on the effects switcher. In this manner, the camera pickup is always on the viewfinder screen together with any other picture superimposed on it by the effects switcher. The camera picture is not disturbed by the effects switcher if it is not one of those being superimposed.

The overscan switch is located on the camera deflection chassis. This switch may be used to prevent raster burn-in when the camera is left on for long periods of time without capping the lens.

The foregoing description has been presented to emphasize the interdependence of controls assigned to the camera and the camera-control operator.

8-2. THE MARCONI MARK VII FOUR-*PLUMBICON* COLOR CAMERA

Fig. 8-3A illustrates the rear control panel of the Mark VII camera. Fig. 8-3B shows the camera-control position as installed at WBBM-TV, and Fig. 8-3C shows the monitor-alignment rack installation at the same station.

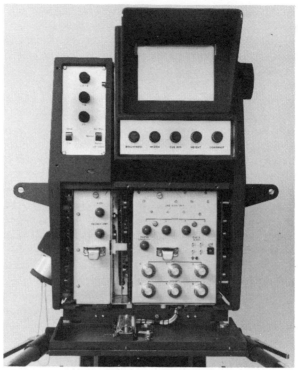

Courtesy Ampex Corporation

(A) Rear view of camera.

Courtesy WBBM-TV

(B) Camera-control console.

Courtesy WBBM-TV

(C) Monitor/alignment rack.

Fig. 8-3. Installation using Marconi Mark VII color camera.

Individual color-channel gain controls normally are not required to be adjusted during a telecast. However, some organizations prefer to have the controls available at the operational position (as in Fig. 8-3B) for correction of color errors in the scene. One example is when light reflected from some colored object in the set falls on the face of an artist.

The block diagrams in Fig. 8-4 show the sequence of the main video signal-processing functions in the camera (Fig. 8-4A) and the camera control unit, or CCU (Fig. 8-4B). The positions in the chain of the major operational and preset video controls, test-signal injection points, and monitoring and bridging points are shown also. The camera has four separate video chains; the block diagram is that of the luminance channel. The differences between the video circuits for the luminance and color channels are shown by dash lines in Fig. 8-4B.

Refer to Fig. 8-4A. The signal current from the pickup tube is amplified by the head amplifier, which is mounted directly on the deflection yoke. The output of the head amplifier is fed to another amplifier, which includes a preset gain control for setting the sensitivity of the channel. Each of the following two stages contains a special remote-gain-control arrangement by means of which the gain of the unit can be controlled from the CCU. The first of these is for the main operational control of gain by means of the master gain control, which is connected to the corresponding unit in all four channels. Operation of the master gain control thus simultaneously varies the gain in the four channels. The second remote gain control is for individual control of the channel gain.

The variation of gain within the remote-gain-control stage is by means of a photosensitive resistor that forms part of a video attenuator. Current for the lamp associated with this resistor is derived from a dc control signal originating from the remotely located gain-control potentiometer. Because the photosensitive resistor is also extremely temperature sensitive, a technique of dc feedback stabilization is employed to ensure stable and linear control of video gains. In the case of the master gain control, this also ensures that the same gain variation is produced in all four channels.

The second remote-gain-control stage is followed by a first clamping stage, which establishes the signal dc level necessary for proper operation of the limiter stage that follows. The clamper also removes spurious low-frequency components of the video signal that might otherwise cause overloading or intermodulation in later stages. The limiter stage also serves to protect later stages from signal components of excessive amplitude. Such a limiter is especially necessary when *Plumbicon* tubes are employed, since this type of tube does not have the inherent self-limiting characteristic of the image orthicon and therefore can produce large-amplitude signals from scene high lights.

The final video stage in the camera is an amplifier that feeds the 75-ohm video coaxial section in the camera cable. Note that the operational control

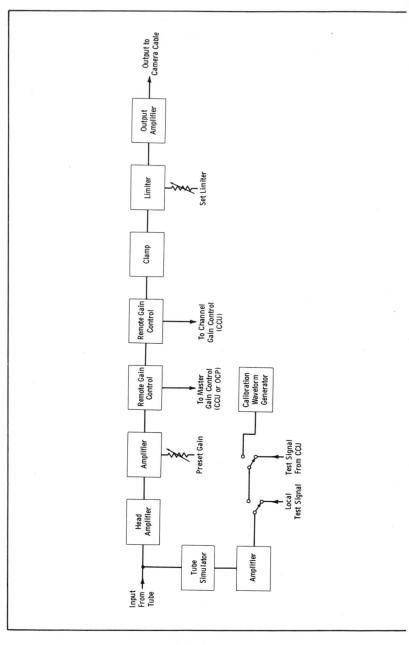

(A) Camera.

Fig. 8-4. Video processing

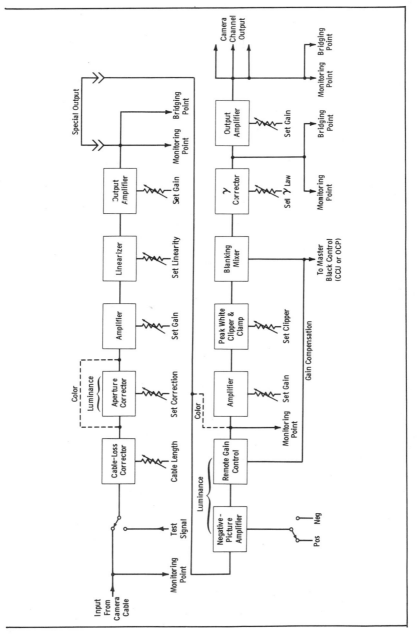

(B) CCU.

Courtesy Ampex Corporation

in Mark VII system.

of gain occurs at an early point in the video chain. This arrangement has the advantage that all the following stages operate at a substantially constant signal level, and no compromise is necessary in the choice between a high level to override spurious signals and a low level to avoid overloading. Thus, the signals fed from the camera to the CCU are at a constant level of 0.7 volt peak to peak.

The first of several waveform monitoring points is at the CCU video input from the camera cable (Fig. 8-4B). Signals from these monitoring points may be fed to the external waveform monitor under control of the waveform-monitor push-button selector switch on the CCU control panel. Following the monitoring point is the changeover arrangement for injection of test signals. This enables the performance of the CCU to be checked independently of that of the camera.

The first CCU stage is the camera-cable corrector, which corrects for the frequency-amplitude characteristic of the cable, and is adjustable to suit the length of cable in use. This is followed by the aperture corrector, which gives an adjustable and substantially phaseless accentuation of the higher video frequencies. (Such correction is necessary to compensate for the finite scanning aperture of the pickup tube.) This stage is not included in the color channels, since the aperture loss within the narrower chrominance band is negligible.

The next block is an amplifier that includes a preset gain control. The purpose of this stage is to provide the correct signal level for the linearizer stage. The linearizer is a nonlinear amplifier that can be adjusted to have an amplitude transfer characteristic complementary to that of the pickup tube; thus, the overall transfer characteristic can be made linear. This facility is included for two reasons. The first is the need to provide the linear camera-channel output that is necessary for certain proposed methods of processing four-tube camera signals for transmission. The linear signal is available at the special output socket, for which an output amplifier of 75-ohm impedance is provided. The second reason for including the linearizer is that the linear signal makes it possible to use a particularly convenient type of gamma corrector (described later).

The next two circuit blocks are included only in the luminance channel. First is a negative-picture amplifier that provides the facility of reversing picture polarity; it is intended for use in black-and-white operation of the camera. The negative-picture amplifier is followed by a remote-gain-control stage of the type already described. The purpose of this stage is to vary the video gain as the master black control is adjusted, so as to maintain the total video signal excursion constant. This is achieved by deriving the dc control signal from the master black control potentiometer. The remote-gain-control stage is required only in the luminance channel because the master black control adjusts the black level of the luminance channel only. Because of the small range of black-level adjustment that is required, and the dominating influence of the black level of the luminance signal in the reproduced picture, it is not necessary to vary the black level of the color channels simultaneously.

The next amplifier stage in the chain has a preset gain control for setting the correct video level for the following stages. This amplifier drives the peak white clipper, which includes the main clamp and serves the usual purpose of preventing the signal excursions from exceeding peak video level. This is followed by the blanking mixer, in which the correct black level is established and the system standard blanking is introduced.

The final processing stage is the gamma corrector, a nonlinear amplifier that gives the outgoing video signal the desired gamma characteristic. The special circuit arrangement employed permits the gamma exponent to be varied continuously by means of a single control while the peak video signal amplitude remains constant. For greater precision, the nonlinear characteristic is approximated by four linear segments. A switch on the CCU control panel provides a choice between two different gamma characteristics. A preset control is provided for each characteristic, one having a gamma range of 0.4 to 0.6 and the other a range of 0.6 to 1.0.

The output amplifiers in each of the four channels provide three outputs at standard level into a 75-ohm impedance.

8-3. THE RCA TK-44A CAMERA CONTROL

Fig. 8-5A illustrates the RCA TK-44A camera-control unit with the normal operating controls exposed and the cover for the setup-panel adjustments closed. Fig. 8-5B shows the setup-panel cover open to expose the setup controls. The TK-44A camera is a three-tube design employing lead-oxide photoconductive tubes and contour-enhancement circuits (Chapter 6).

While there are considerable differences in physical location and (sometimes) nomenclature of controls in various camera systems, all are the same in general function. The following description serves as emphasis and review of adjustments and operation.

Setup Controls

Setup controls are primarily concerned with operating parameters of the pickup tube or tubes. The image orthicon requires the greatest number of adjustments. (The following information is presented through the courtesy of RCA.)

Proper Scanning (Horizontal and Vertical Controls)—Full-size scanning of the target should always be used during operation. Full-size scanning can be assured by first adjusting the deflection circuits to overscan the target sufficiently to cause the edge of the target ring to be visible in the corners of the picture, and then reducing the scanning until the edge of the target ring just disappears. In this way, the maximum signal-to-noise ratio and maximum resolution can be obtained. If the camera employs an orbiter, a

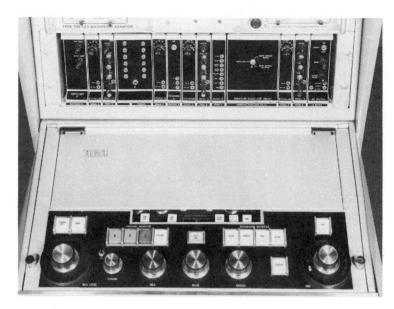

(A) Setup panel covered.

(B) Setup panel exposed.

Courtesy RCA

Fig. 8-5. RCA TK-44A camera control.

scanning-size adjustment jig should be used to assure the proper size of the scanned area.

(This paragraph does not apply to image orthicons employing electronically conducting glass targets [Chapter 4].) Underscanning the target, i.e., scanning less than the proper area of the target, should never be permitted. Underscanning produces a larger-than-normal picture on the monitor. If the target is underscanned for any length of time, a permanent change in target-cutoff voltage of the underscanned area takes place, and the underscanned area then is visible in the picture when full-size scanning is used.

A mask having a diagonal or diameter of 1.8 inches (1.6 inches for 4½-inch types) should be used in front of the photocathode to set limits for the maximum size of scan, and to reduce the amount of light reaching unused parts of the photocathode.

Alignment of Beam—Proper alignment of the beam in an image orthicon is one of the most important steps in obtaining a good picture. Proper alignment for a non-field-mesh image orthicon is obtained when the small white dynode spot does not move when the beam-focus control (grid-4 voltage) is varied, but simply goes in and out of focus. For tubes that have a field mesh, the alignment currents are adjusted so that picture response is maximum and the center of the picture does not move when the beam-focus control is varied, but simply goes in and out of focus. Auto-alignment devices are useful for determining the exact setting of alignment-coil current.

Setup Differences Between Field-Mesh and Non-Field-Mesh I.O.'s— There are two families of image-orthicon camera tubes, one with field mesh and the other without field mesh. Proper setup of any image orthicon is best assured by observing the procedure outlined in the technical bulletin for the individual type. This procedure, however, can be simplified greatly by noting the three principal differences between the setup procedures for non-field-mesh image orthicons and field-mesh image orthicons.

For non-field-mesh types, the technical bulletin explains that a ". . . preliminary alignment adjustment can be easily made by adjusting the alignment current to produce a maximum signal output when the tube is focused on a test pattern. Final adjustment is achieved by regulating the alignment-coil current so that the small white dynode spot appearing on the monitor does not move when the beam-focus control (grid No. 4) is varied, but simply goes in and out of focus."

In the case of field-mesh image orthicons, where no white dynode spot is involved, correct adjustment of alignment-coil current is obtained by regulating this current ". . . so that the center region of the picture does not move when the beam-focus control (grid No. 4) is varied, but simply goes in and out of focus." Note that this generally occurs at the point where the center of the picture is brightest.

The second difference in setup procedures for the two types of image orthicons lies in proper application of the dc operating voltages. In non-field-mesh types, the dc voltages may be applied with the lens capped. If the lens is capped, however, it should be uncapped momentarily while the grid-1 voltage is adjusted to setting that provides a slight amount of beam current.

Insofar as field-mesh orthicons are concerned, *under no circumstances should the lens be capped during application of dc operating voltages.* The lens always must be uncapped and the lens iris opened to allow light to fall on the photocathode before application of the dc voltages.

The third difference to remember is that field-mesh types generally operate properly at only one particular mode of focus of the scanning beam. If a large, coarse-mesh background is evident either with the lens capped or in the low lights of the scene, grid 4 probably is being operated at the wrong mode of focus. For three-inch-diameter field-mesh image orthicons, the proper mode of focus generally occurs within the range of 140 to 180 volts on grid 4.

Beam Current (Beam Control)—During alignment of the beam, and also during operation of the tube, always keep the beam current as low as possible to give the best picture quality and also to prevent excessive noise. Lack of sufficient beam results in improper resolution of high lights.

Target Voltage (Target Control)—Focus the camera on a test pattern. Then adjust the target voltage to the point at which a reproduction of the test pattern is just discernible on the monitor. This value of target voltage is known as the target-cutoff voltage. The target voltage should then be increased to the recommended value above cutoff for standard lime-glass target types and to the selected target voltage (according to operating needs) for electronically conducting glass target types. The beam-current control should then be adjusted to give just sufficient beam current to discharge the high lights. The interrelationship among tube sensitivity, signal-to-noise ratio, and resolution may be used to obtain optimum camera performance for different lighting conditions. The determining parameter is target voltage. At high target voltages, signal-to-noise ratio is enhanced at the expense of resolution. As the target voltage is reduced, this relationship reverses. For a given telecasting session, it is practical and advisable to maintain the target voltage at a single value because such voltage constrains video gain, gamma correction, and other adjustments in the camera chain. Furthermore, it generally is advisable to employ the same target-voltage value for all cameras telecasting a given scene.

The target-voltage control should not be used primarily as an operating control to match pictures from two different cameras. Matching should be accomplished first by individual adjustment of the lens-iris openings. Small changes in target voltage then may be used to produce picture matching. The target-control voltage calibration should be checked periodically to assure that the target-voltage adjustment is correct.

Resolution (Image [Photocathode]-Focus, Orth [Beam]-Focus, and Aperture-Correction Controls)—Adjust the lens to produce best optical focus, and adjust the voltages on the image-focus electrode, grid 6, and grid 4 to produce the sharpest picture.

A loss in resolution can be caused by operation of the tube at too high a temperature. Sometimes a loss in resolution can be traced to a dirty lens or dirty tube faceplate; both the lens and the faceplate should be cleaned periodically. (Be sure not to scratch the optical surfaces.) Make sure the camera and turret are closed to prevent light leakage, which will "wash out" the picture.

The aperture-correction amplitude control generally can be adjusted to obtain 100-percent response (relative to 100 TV lines) at 300 lines resolution for a 3-inch I.O., or 400 lines for a 4½-inch I.O. The limiting factor is excessive noise in the picture (Chapter 6).

Color-Camera Registration Controls—To superimpose exactly the three or four images on the raster, the following controls are involved: horizontal and vertical size, horizontal and vertical linearity, horizontal and vertical centering, and skew. The registration test chart is employed to permit these adjustments. For proper color balance, the gamma controls must be set to match all channels to one another in light-to-signal transfer characteristics.

Wall Focus (4½-Inch I.O. Only)—The wall focus is adjusted so that no mesh pattern is observed in low-light areas of the picture.

Photoconductive Tubes—Setup controls for the vidicon and lead-oxide tubes consist mainly of the target, beam, and alignment controls. The same color-camera registration controls are necessary. These tubes do not have a knee, except as may be provided in video-processing amplifiers.

Operating Controls

Operating adjustments consist largely of maintaining proper pickup-tube exposure, and small adjustments of black-level and gain (paint-pot) controls for color chains.

Proper Exposure (Lighting, Iris, and/or Neutral-Density Controls)— Proper exposure of the image orthicon is required at all times for consistent production of high-quality pictures. The most common error in lighting and exposure control is to overexpose the image orthicon to "bring up" information in the low lights of the scene. A much better picture can be obtained by filling in the low-light areas of the scene with fill light rather than by opening the lens and overexposing the image orthicon.

In general, as the light level incident on the image orthicon is increased and the signal output reaches the knee of the light transfer characteristic, picture quality is improved because of an increase in resolution, signal-to-noise ratio, and contrast range. Signal-to-noise ratio and contrast range are directly proportional to the square root of the illumination on the faceplate of the image orthicon, and they increase until the high lights reach the knee of the light transfer characteristic. Any further increase in light level does not materially improve the signal-to-noise ratio but does increase resolution slightly. Operation of the tube with the high lights substantially above the knee allows it to handle a wider contrast range, because the whites are compressed without loss of detail and the blacks are raised out of the noise.

Remove the lens cap and focus the camera on a neutral (black-and-white) test pattern consisting of progressive tonal steps from black to white. Open the lens iris just to the point at which the high lights (highest step) of the test pattern do not rise as fast as the low lights (lower steps) when viewed on a video-waveform oscilloscope. This operating point is the knee of the light transfer characteristic.

For black-and-white operation, the camera lens then should be opened approximately one to two stops above the knee for each individual scene. This operating point assures maximum signal, good gray scale, freedom from black borders, the sharpest picture, and the most natural appearance of televised subjects or scenes.

The camera lens should be adjusted continuously to maintain this operating point as the illumination in each scene changes. Operation at this point is especially important for studio pickup in order to obtain the best gray scale in the picture and to reduce the possibility of image retention.

For outdoor and other scenes in which a wide range of illumination may be encountered, the camera should be panned across the scene that has the least amount of illumination, and the lens iris should be adjusted so that the high lights in that area are just above the knee. The camera then will be able to handle all scenes having higher illumination without requiring lens-stop adjustments. When the camera is to be shifted rapidly from a scene of low brightness to a scene of high brightness, or vice versa, as may take place during panning, the camera always should be set for the dark scene.

For color cameras, the lens setting usually is adjusted so that operation takes place with the high lights just barely over the knee. The lens setting should be adjusted continuously as the scene changes so that the exposure will not result in operation substantially above the knee, which causes color dilution and contamination.

Table 8-1. Illumination of Outdoor Scenes

Lighting Conditions	Scene Illumination (Lumens/Ft²)
Direct Sunlight	10,000 - 12,000
Full Daylight*	1000 - 2000
Overcast Day	100
Very Dark Day	10
Twilight	1
Deep Twilight	0.1

*Not Direct Sunlight

Scene Illumination—The image-orthicon camera serves as an exposure meter and is the final judge of scene illumination and lens opening. However, before an attempt is made to televise a particular scene, it is good practice to check the incident illumination with a light meter to determine whether the light level is adequate for a picture of good quality. In general, the illumination should be measured with the light meter pointing toward the camera.

Scene illumination on the camera-tube face may be calculated from the following formula:

$$E_s = \frac{4Ef^2}{0.8R}$$

where,

E_s is the scene illumination in lumens/ft^2 (foot-candles),
E is the tube-face illumination in lumens/ft^2 (foot-candles),
f is the f number of the lens,
R is the reflectance of the scene.

For outdoor scenes, Table 8-1 can be used as a guide in determining the approximate scene illumination.

Because of the high sensitivity of the image orthicon, it may not be possible on very bright days to stop the lens down far enough to reduce the high-light illumination on the photocathode to a value near the knee of the signal-output curve. When such a condition is encountered, the use of a Wratten neutral-density filter selected to give the required reduction in illumination is recommended (Table 8-2). Ordinarily, two filters—one having 1-percent transmission and the other 10-percent transmission—will give sufficient choice. Such filters with lens-adapter rings can be obtained at photographic-supply stores.

Under almost all conditions, the use of a lens shade is beneficial.

8-4. THE CAMERA WAVEFORM MONITOR

The camera waveform monitor (CRO) always incorporates a calibration pulse for proper gain adjustment, and a selector switch for wideband or

Table 8-2. Neutral-Density Filters for Exposure Control

Filter Density	Transmission Percentage	Equivalent Number of Lens Stops
0.30	50.0	1
0.60	25.0	2
0.90	13.0	3
1.00	10.0	3.3
2.00	1.0	6.6
3.00	0.10	10
4.00	0.010	13.2

IRE response. Fig. 8-6 illustrates the Tektronix Type 528 CRO often incorporated in camera control consoles. A somewhat more elaborate waveform monitor (Tektronix Type 529) is visible in Fig. 8-3B. The Type 529 is more useful as a line master monitor since it incorporates vertical-interval test (VIT) signal observation.

The Type 528 television waveform monitor provides video waveform displays on a 5-inch CRT and occupies 5¼ inches of height and ½ rack width. All-solid-state circuitry provides low power consumption and long-term reliability.

Either of two video inputs, selectable from the front panel, may be displayed. The displayed video signal also is provided at a video-output jack for viewing on a picture monitor. Calibrated 1-volt and 4-volt full-scale (140 IRE units) sensitivities are provided for displaying common video and sync signal levels. A variable sensitivity control permits uncalibrated displays from 0.25 volt to 4.0 volts full scale. The built-in 1-volt calibration signal may be switched on to check vertical-sensitivity calibration. Flat, IRE, chroma, and differential-gain frequency-response positions permit observation of various signal characteristics.

Horizontal sweep selection provides 2H (two-line), 1 µs/div (expanded two-line), 2V (two-field), and 2V magnified (expanded two-field) sweeps. Displays of RGB and YRGB waveforms from color-processing amplifiers are provided for by means of interconnection through a rear-panel nine-pin receptacle. A dc restorer maintains the back porch at an essentially constant level despite changes in signal amplitude, APL, and color burst. This function may be turned off when not needed.

A basic function of modern camera waveform monitors is in the form of clamping used. It will be recalled from Chapter 6 that there are two general

Fig. 8-6. Tektronix Type 528 waveform monitor.

types of keyed clamping circuitry: the "fast-acting" clamp used in processing amplifiers to remove all low-frequency (sine-wave) disturbances,
and the "slow-acting" clamp useful in waveform monitors.

All camera waveform CRO's have a switch that allows operation with
the dc restorer off or on. However, since the restorer switch normally is
left in the on position, actual hum components in the waveform might be
ignored by the operator if he is unaware of other indicating devices which
would show this defect. Therefore, modern circuitry of this nature incorporates the slow-acting clamp so that, although the waveform is held
constant in position on the CRO with changes in APL, any hum component is visible, although reduced in amplitude. This alerts the operator to
turn the dc restorer off so that actual low-frequency characteristics may be
observed.

The fast-acting keyed clamp was covered in Chapter 6. Fig. 8-7 illustrates
a basic slow-acting clamp typical of modern waveform monitors. Such a
circuit has three fundamental sections:

1. A comparator that measures any difference between a fixed (or variable) reference voltage and the amplifier output voltage (E_0)

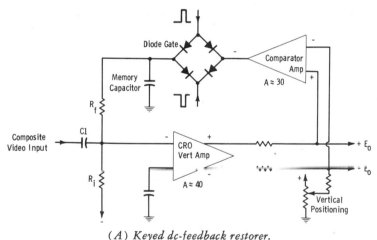

(A) Keyed dc-feedback restorer.

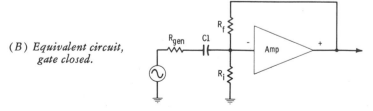

(B) Equivalent circuit,
gate closed.

Fig. 8-7. Slow-acting clamp for waveform monitor.

2. A line-to-line keyed gate to close the diode gate for the duration of the clamping pulses

3. A memory circuit to remember the reference point *between samples,* when the diode gate is open

The line-to-line keyed clamp is a fast-operating type of dc restorer. However, note from Fig. 8-7B what occurs during the time the gate is closed. At this time, C1 is in parallel with R_1, forming a low-pass filter in the feedback loop. Thus, insufficient feedback current is available to charge C1 to the correct value in one sample (one TV line). The equivalent effect on the overall system is that of a "slow" restorer.

The dc-feedback restorer of Fig. 8-7A has two basic functions: to establish dc stability (for drift-free operation) and to establish a selected portion

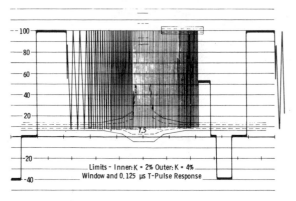

Courtesy Tektronix, Inc.

(A) Flat position.

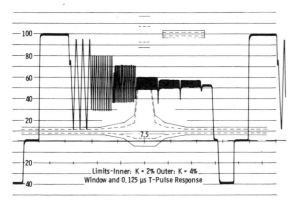

Courtesy Tektronix, Inc.

(B) IEEE position.

Fig. 8-8. Multiburst response

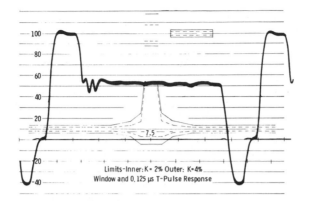

(C) Low-pass position.

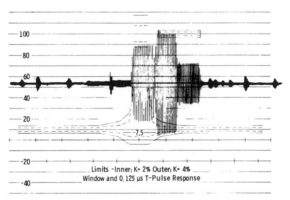

(D) High-pass position.

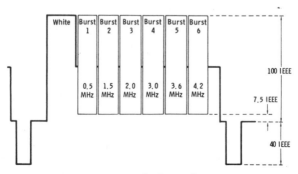

(E) Applied signal.

of waveform monitors.

of the composite video waveform (sync tip or back porch) as a reference point so that the CRO trace does not change vertical position under varying APL.

Under ideal quiescent conditions with pulses applied to the diode bridge, $+E_o$ should equal $-E_o$. If amplifier drift causes $+E_o$ and $-E_o$ to be unequal, the difference voltage applied to the comparator (differential amplifier) is amplified. The amplified error voltage charges the memory capacitor during the next pulse that closes the diode gate. Although the current available through R_f is limited, amplifier drift occurs over a much longer period than the time required to charge C1 through R_f. Thus, the amplifier is made virtually drift-free.

Now consider the composite video waveform. The feedback diode gate is closed during either sync-tip or back-porch time. The absolute voltage level of the sampled part of this waveform is compared to the reference voltage at the center arm of the position control. The difference voltage between the point of sampled video and the positioning-control dc level is amplified and applied to the memory. Because of the low-pass filter formed by C1 and R_f, the error must exist *continuously* for about 1 millisecond (typical time constant of the low-pass filter) before the error will be completely corrected. This time constant prevents complete removal of any 60-Hz hum that may exist in the composite video waveform.

In monitoring the video waveform for the purpose of "riding level," it is important that the operator use the IRE (IEEE) response position of the

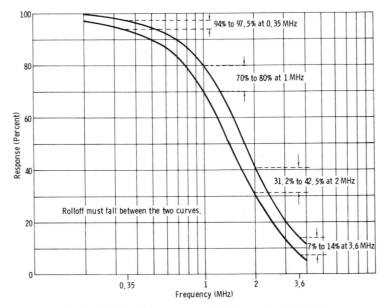

94% to 97.5% at 0.35 MHz

70% to 80% at 1 MHz

31.2% to 42.5% at 2 MHz

Rolloff must fall between the two curves.

7% to 14% at 3.6 MHz

Response (Percent)

Frequency (MHz)

Fig. 8-9. IEEE response curve (1958 standard 23S-1).

selector switch. This removes the possibility of excessive "gain riding" with changes in high-frequency content of a scene. The multiburst response of the Types 528 and 529 waveform monitors for various response-switch positions is shown in Fig. 8-8. Fig. 8-9 illustrates the standard IEEE response curve for all modern waveform monitors when used as level-monitoring devices.

It is very important that the maintenance department check this response with either video-sweep or single-frequency sine-wave response runs at periodic intervals. Fig. 8-10 gives typical response-switch circuitry and representative maintenance adjustments provided.

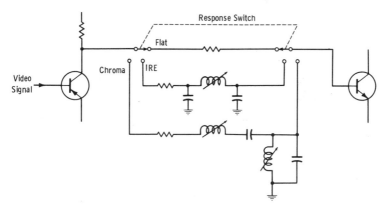

Fig. 8-10. Typical response-switch circuit.

8-5. CAMERA INTERPHONE

A quite important link between the camera-control position and the camera operator is the *interphone* system. This makes possible conversations among the camera and control operators and the director. A schematic diagram of the interphone for the RCA TK-60 camera chain is shown in Fig. 8-11A.

These units are connected internally and may be connected externally to similar units as desired. Each small unit includes an amplifier, a bridge rectifier, and a sidetone-compensation bridge. Sidetone is automatically maintained at a level approximately equal to the received signal level for any number of connected stations.

Each station employs a single-stage transistor amplifier. An unbypassed resistor in series with the emitter of the transistor determines the gain by controlling the amount of inverse feedback. Power to operate the amplifier is derived from the microphone-energizing direct current from the common supply circuit to which the intercom unit is connected.

In this unit, a germanium-diode bridge rectifier is interposed between the line and the amplifier to maintain the correct polarity at the amplifier regardless of the polarity of the intercom supply voltage. Two diodes are biased to a conducting state, and two are biased to a nonconducting state by the direct current on the intercom circuit. Received and transmitted voice frequencies, superimposed as they are on the direct current, pass unimpeded through the conducting diodes to the amplifier input or to the intercom line.

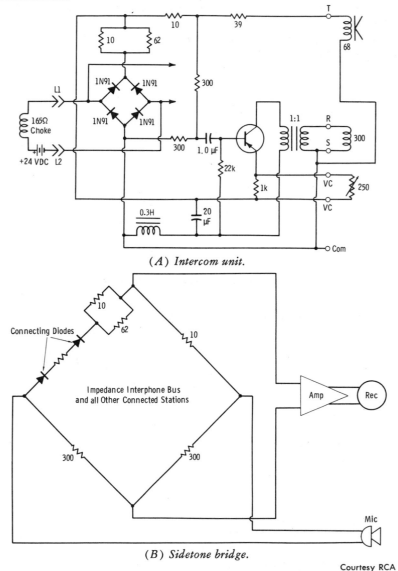

(A) Intercom unit.

(B) Sidetone bridge.

Courtesy RCA

Fig. 8-11. Interphone for camera chain.

In order to maintain the desired relationship between sidetone level and received level, voice frequencies must be kept out of the power supplied to the amplifier. A choke-input filter provides the required decoupling for this purpose.

In the interphone unit, a resistance bridge circuit (see Fig. 8-11B for equivalent circuit) is employed. The intercom line and all other stations connected to it make up a portion of one side of the bridge. A pair of fixed resistors and the forward resistances of the two conducting diodes in series with the line complete this side of the bridge. The remaining three sides of the bridge are provided by three fixed resistors.

The local microphone is connected across one diagonal of the bridge, and the input to the transistor amplifier is connected across the other diagonal. If the bridge were to be perfectly balanced, sidetone would be completely eliminated because the amplifier input would be at the null point of the bridge for signals from the local microphone.

When only two stations are connected to an intercom circuit, the level received at each station is relatively high; however, the sidetone-compensation bridge at each station is substantially unbalanced, permitting a high sidetone level to be applied to the amplifier input. When additional stations are added to the intercom line, the received level decreases, and the impedance of the intercom circuit decreases as well. As the net impedance of the intercom decreases, the bridge approaches, but never reaches, a condition of balance. The sidetone level decreases accordingly. Resistance values in the bridge have been selected to hold the sidetone level to within 2 dB of the received level for any number of conference-connected stations up to 32.

EXERCISES

NOTE: These exercises are largely review of setup and operations techniques covered in previous study. For background information, see Harold E. Ennes, *Television Broadcasting: Equipment, Systems, and Operating Fundamentals* (Indianapolis: Howard W. Sams & Co., Inc., 1971.)

Q8-1. What is the maximum brightness contrast range for optimum control of the camera, and how is this determined?

Q8-2. What is the maximum amount the waveform on an unclamped CRO should shift from scene to scene?

Q8-3. Name the factors that can affect color-picture sharpness.

Q8-4. What is the final check for registration?

Q8-5. When a new color set is used, what is the first thing to check?

Q8-6. Is it possible to have good color balance on a wide shot, then not have good balance on a tight spot in the same scene and with the same camera?

Q8-7. What are the factors pertinent to background effects on skin tone?

Q8-8. How much color-temperature change can occur in lighting before skin tones show error, and what line-voltage change does this represent?

Q8-9. Are light dimmers ever used in color studios?

Q8-10. For color lighting, give the optimum ratio, relative to base light, for (A) back light and (B) key, or modeling, light.

Q8-11. In what part of the visible-light region does the spectral response of the average camera pickup tube peak?

Q8-12. Does incandescent lighting help or hinder the spectral response of the average pickup tube?

Q8-13. Are the deflection yokes for all channels in a color camera connected in series or in parallel?

Q8-14. Are the focus coils in a three-channel image-orthicon camera connected in series or in parallel?

Q8-15. What is skew, and how is it corrected?

Q8-16. What should be the reflectance value of reference white in a color scene?

Q8-17. If there are controls marked "Q" on a color camera, what function do they perform?

CHAPTER 9

The Subcarrier and
Encoding System

Older color systems used an external master color-subcarrier generator, which incorporated count-down circuitry for locking the station sync generator to the frequency (color) standard. More recent systems employ sync generators that include the color-subcarrier generator as an integral part. This also includes the burst-flag generator (burst keyer), which, in older systems, was an additional rack-mounted unit.

NOTE: It is imperative at this point for the reader to review NTSC color fundamentals. This subject is covered extensively in Harold E. Ennes, *Television Broadcasting: Equipment, Systems, and Operating Fundamentals* (Indianapolis: Howard W. Sams & Co., Inc., 1971), Chapter 2. A portion of that chapter deals with the choice of a color-subcarrier frequency that allows minimum interference between luminance and chrominance picture information.

9-1. THE DOT STRUCTURE IN NTSC COLOR

You are probably already aware of the fact that a dot pattern is noticeable on the picture tube (either monochrome or color) when a color signal is viewed. This pattern is particularly noticeable at the edges of vertical transitions, and is more pronounced on a wideband studio monitor than on a home receiver. You may correctly assume that the process of interleaving color information with monochrome information is less than perfect.

From the review mentioned in the note above, we know the reason for the choice of a subcarrier frequency at the 455th harmonic of one-half the line-scanning frequency:

$$455 \times \frac{15{,}734.26}{2} = 3.579545 \text{ MHz}$$

This frequency results in minimum interference, as is emphasized by Fig. 9-1 and the following analysis:

Fig. 9-1A represents the modulated subcarrier signal. The first scan represents scanning of a given line in the picture; this line is not scanned again until the first scan in the next frame (interlaced scanning). The peaks of the subcarrier component cause a "dot pattern" to be laid down, as in Fig. 9-1B. The dots are actually quite small because the subcarrier frequency is high. Also, since the frequency is an odd multiple of one-half the line frequency (as well as being an odd multiple of one-half the frame frequency), the spurious dots on each line in one frame fall midway between those on the lines immediately above and below. This gives the checkerboard pattern shown, rather than a line pattern with rows and columns of dots. The latter would be much more noticeable.

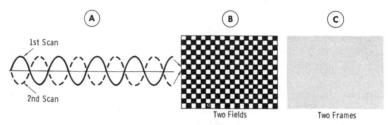

Fig. 9-1. Principle of dot interlace.

Since the subcarrier is an odd multiple of one-half the frame frequency (as well as an odd multiple of one-half the line frequency), the subcarrier component along a given line reverses in polarity between successive scans. Remember that "successive scans" means a given field in the next frame. The subcarrier reverses in polarity between these two scans because it passes through some whole number of cycles plus one-half cycle (180°) during each frame period. This is shown by the dash sine wave of Fig. 9-1A.

The net result is shown by Fig. 9-1C. The peaks of the subcarrier component that caused a dot pattern in one frame are cancelled in the next frame. This process is termed *cancellation interlace,* or simply *dot interlace.*

The above would be entirely true if we had a perfect system. However, in practice, the signals are applied to kinescopes that are inherently non-linear even above cutoff, and certainly below cutoff (negative light). Perfect cancellation is impossible at the present stage of development. Therefore, the cancellation of spurious signal components by dot interlacing is not quite as good as Fig. 9-1C indicates. It is satisfactory, however, at normal viewing distances. When you are very close to a monitor tube (either monochrome or color) that is displaying a color signal you can see the "dot crawl" that occurs from bottom to top of the raster. On a wideband studio picture monitor, this is particularly noticeable at transitions between color bars.

Fig. 9-2 represents a part of a scanning pattern in the presence of a disturbing frequency that is an odd multiple of one-half the line frequency. The letters in the circles have the following significance:

(A) Dots produced by the disturbing signal (subcarrier) in the first field of scanning
(B) Dots produced in the second field
(C) Dots produced in the third field
(D) Dots produced in the fourth field (two frames of picture)

From this we can see that a complete cycle of the dot pattern takes up two full frame scans. Also observe that the dot patterns are opposite each other in successive lines and in successive frames.

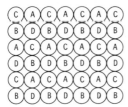

Fig. 9-2. Complete dot-pattern cycle. **Fig. 9-3. Dot crawl.**

Dot crawl, the apparent upward motion of the dot pattern in the picture resulting from the sequence in which dots are laid down, is illustrated by Fig. 9-3. The circles represent points of maximum intensity, and the numbers within them indicate the field to which they correspond. The time sequence (1-2-3-4), which the eye tends to follow, corresponds to a continuous upward motion. In effect, the dot crawls one line per field. There are 60 fields per second, so the dot crawls 60 lines per second. There are about 480 *active* lines displayed on the raster (525 lines minus vertical-blanking lines). Thus, 480/60, or 8, seconds are required for the dot to crawl from the bottom to the top of the raster.

Actually, the field rate for the color standards is not quite 60 per second, but is 59.94 per second. As a result, slightly more than 8 seconds are required for a dot to crawl from the bottom to the top line of the raster. If you concentrate on a dot on the bottom line and follow it to the top line (using a stop watch), and if the elapsed time is between 8 and 8.1 seconds, the subcarrier frequency is correct. In an emergency, you can use this method to set the subcarrier frequency rather accurately. Naturally, you must use an underswept monitor so that all active lines are visible.

9-2. POWER-LINE CRAWL ON COLOR STANDARDS

Since there is a difference (*slip*) between the color field rate and the 60-Hz power-line frequency, a "crawling" hum bar results on receivers or monitors if hum is present either at the transmission or receiving end.

At the receiver, this is no different from the condition that existed for monochrome operation when the receiver was powered from a source not synchronized with the source of transmitter power. Thus, when interference, such as from operation of power tools or rotating machinery in the neighborhood, exists on the power line, a floating "band" of noise flashes can drift from bottom to top of the receiver picture tube.

You can observe this slip by connecting a scope at the composite-sync output of the sync generator (locked to color standards) and adjusting the time base to give one field in 10 centimeters on the graticule. Use a 60-Hz trigger on the scope. Since the difference between 60 and 59.94 is 0.06, there is a shift of 0.06 field per second, and the vertical interval will require about 16 seconds to travel the 10 centimeters on the scope. But do not try to set the actual subcarrier frequency by this method. Because of the large countdown, the error could be considerable.

9-3. ADJUSTING THE SUBCARRIER COUNTDOWN

For color operation, the master oscillator in the sync generator is no longer the frequency standard of the system. To maintain the proper harmonic relationship between the subcarrier frequency and the scanning frequencies, the subcarrier generator becomes the frequency standard for the system, and is so termed by most manufacturers.

The subcarrier-frequency oscillator is normally a crystal-controlled type with the crystal in a precision temperature-controlled oven. A trimmer allows adjustment of the frequency with reference to an external standard.

See Fig. 9-4. The crystal oscillator feeds a conventional buffer amplifier peaked at the subcarrier frequency. Note the low-value resistor (2 ohms) across the secondary of T1. This provides a very low sending-end impedance for conventional 75-ohm distribution lines.

Tube-type subcarrier generators almost universally employ locked oscillators for counters. This type of oscillator operates at the desired output

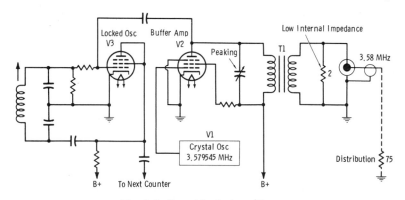

Fig. 9-4. Use of locked oscillator.

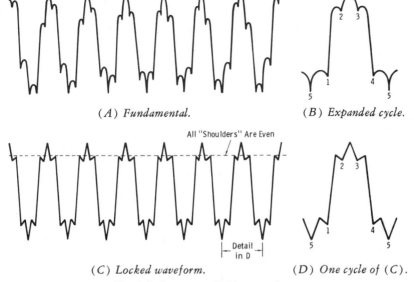

(A) Fundamental.

(B) Expanded cycle.

All "Shoulders" Are Even

|— Detail —|
in D

(C) Locked waveform.

(D) One cycle of (C).

Fig. 9-5. Frequency-divider waveforms.

frequency, with provision for injecting a small voltage into its grid circuit to lock the output at a subharmonic of the injected frequency. A typical circuit is that of V3 of Fig. 9-4. This is a Colpitts circuit, in which the cathode and the midpoint of the tank-circuit capacitors are at ground potential. (The Hartley circuit also may be used in this application. The action is the same.) The tank circuit is adjusted so that, in the absence of the 3.58-MHz signal, V3 oscillates at a frequency very close to the required counter frequency. If this is a 5-to-1 counter, the frequency is 715.909 kHz. The waveform looks something like that in Fig. 9-5A.

Application of the 3.58-MHz signal to the V3 grid should lock the oscillator in precise phase and maintain a stable frequency division of (in this example) 5 to 1. Such will be the case if the peaking of T1 is correct and the tank coil is properly tuned. The trick is knowing how to recognize the proper locked frequency. Fig. 9-5B shows one lower-frequency (715.909 kHz) cycle with the higher-frequency subcarrier "lock." This is on an expanded time base to show details. Count all the *negative*-going peaks to check actual count. Trigger the scope on the negative slope of the signal so that the sweep starts at the end of the previous negative peak. We can see here that the count is five.

Unfortunately, a locked oscillator can produce a nonintegral count, for example 4.5 to 1. You should always use a time base that will display seven or more complete lower-frequency cycles, as illustrated in Fig. 9-5C. A properly locked 5-to-1 (or 7-to-1) waveform has identical symmetry over all cycles. A gradual change over the cycles reveals a nonintegral countdown.

Locking action is a positive function and will occur with a sharp increase of "crispness" as lock-in is achieved. A smeared presentation on the scope indicates that the stage is not synchronized in the presence of the higher-frequency input. (NOTE: Always check the stability of the scope trigger function.)

Fig. 9-5D shows how to interpret the count on the longer time base of Fig. 9-5C. Remember to count the negative peaks, as shown. On this time base, the negative excursion looks more like a slight notch.

The 7-to-1 locked oscillator receives the 5-to-1 countdown and is usually the same type of circuit. Again, it can lock to a nonintegral count such as 6.50 to 1 or even 6.75 to 1. The best way to check this is to observe the waveform of the 5-to-1 counter and adjust the scope time base to obtain exactly seven of the low-frequency cycles in 10 centimeters on the scope graticule. (If you prefer, you can obtain seven complete cycles in 7 centimeters.) Then without changing the scope time base, observe the test

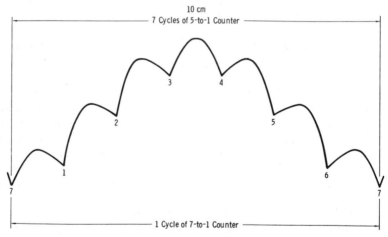

(A) Method of determining count.

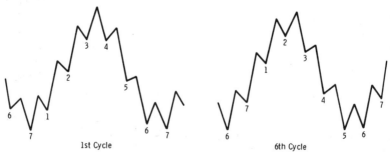

(B) Double-check for exact count.

Fig. 9-6. Waveform checks of divider count.

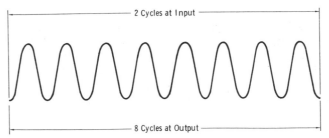

Fig. 9-7. Check of multiplier action.

point at the output of the 7-to-1 counter. There should be exactly one low-frequency cycle in the same space on the scope graticule (Fig. 9-6A). Remember to use the trigger on the negative slope so that the sweep-start "dot" occurs on the last negative peak of the previous cycle.

Fig. 9-6B shows the importance of double checking the frequency by readjusting the scope time base to display seven or more cycles of the lower-frequency component. In the example shown, the oscillator is locking to a 6.75-to-1 count instead of 7 to 1. Always adjust the oscillator tank coil to a point midrange between proper counts on the lower and upper limits. A number of consecutive cycles should appear the same.

The four-times multiplier is usually a conventional circuit with the plate circuit resonated at four times the input frequency. There is really only one suitable method for adjusting this stage, and this is as follows:

1. While observing the 7-to-1 counter, adjust the scope time base to get exactly two cycles of the low-frequency component in exactly 10 centimeters on the scope graticule. Trigger the scope on the negative slope of the signal.

2. Without changing the scope time base, observe the output of the four-times multiplier. Adjust it to obtain exactly eight cycles in the 10 centimeters (Fig. 9-7). Adjust the frequency to the center of the range between the upper and lower limits of the proper count. In most cases, it is only necessary to adjust the tank circuit for maximum output. The stage is usually noncritical, being capable of tuning only to the proper frequency of four times the input signal.

NOTE: The exact sequence in counting can be different. Therefore, the input frequency to a given counter (or the multiplier) may differ from the above examples. The thing to remember is to trigger the sweep with the counter preceding the one to be checked, and adjust the time base for the number of cycles that equals the division of the counter to be checked. You should use the same procedure in checking the operation of the sync-generator counters.

Counters with ratios higher than 7 to 1 (such as 13 to 1) are usually of more straightforward design, such as a conventional cathode-coupled

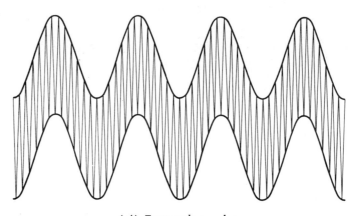

(A) Four-cycle envelope.

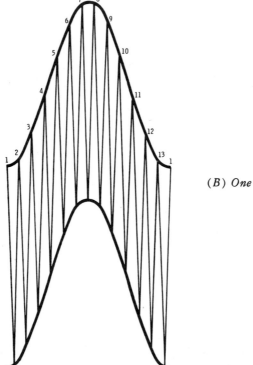

(B) One cycle expanded.

Fig. 9-8. Waveforms of 13-to-1 counter.

oscillator. The higher-frequency component is injected into the grid to obtain the locking action. Fig. 9-8A shows a four-cycle "envelope" of the output of a 13-to-1 counter. Although it is a rather difficult task, you are assured of better accuracy of adjustment if you count the high-frequency peaks occurring over the four low-frequency cycles. There should be 13×4, or 52, peaks. Fig. 9-8B illustrates one cycle on an expanded time base.

9-4. SOLID-STATE COUNTERS

In the interest of completely "automatic" circuitry, modern solid-state subcarrier counter circuits often incorporate no adjustments. It is therefore necessary for the maintenance engineer to be able to analyze such circuitry in the event of miscounting or complete failure.

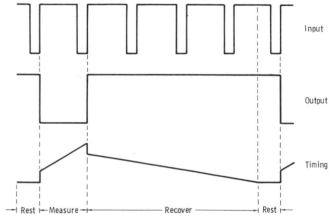

Fig. 9-9. Timing chart of typical solid-state divider.

A counter that is gaining great popularity is a special form of monostable or bistable multivibrator circuit employing noncritical timing elements. It has three basic states of operation—rest, measure, and recover—as shown for a typical 5-to-1 counter in Fig. 9-9. Note that the measure ramp is initiated by one positive-going trigger (trailing edge of one sync pulse) and is terminated by the next positive-going trigger (trailing edge of next sync pulse). The rate (slope) of the recover period is such that it automatically terminates about midway between the last input trigger to be rejected and the next one to count. The remaining time is the rest period.

Fig. 9-10A shows a typical 7-to-1 divider. Transistors Q1 and Q3 form a bistable multivibrator with diode trigger steering (X2 and X5). Transistor Q2 is the ramp generator with a collector-catching network (to prevent saturation) and a bias network controlled by switching transistor Q4.

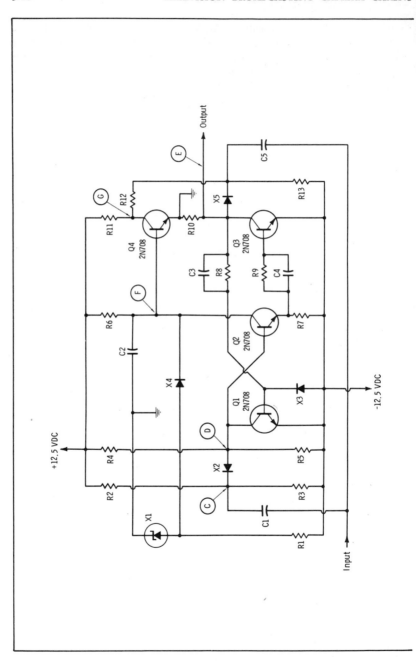

(A) Circuit diagram.

Fig. 9-10. Typical transistor

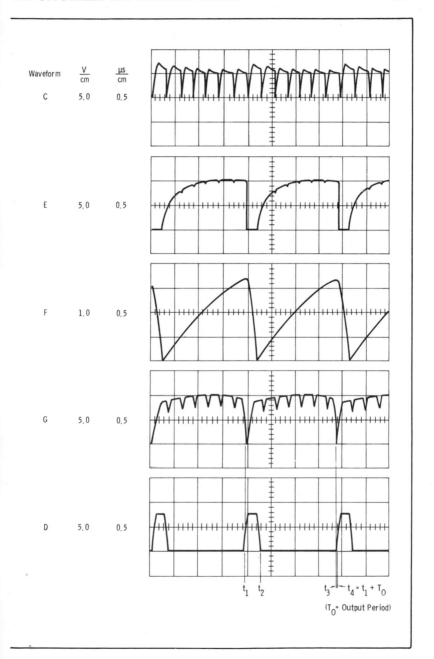

Waveform	$\frac{V}{cm}$	$\frac{\mu s}{cm}$
C	5.0	0.5
E	5.0	0.5
F	1.0	0.5
G	5.0	0.5
D	5.0	0.5

$t_1 \quad t_2 \qquad t_3 \; t_4 = t_1 + T_0$

(T_0 = Output Period)

(B) Waveforms.

frequency divider.

Although you should already have a good basic background in transistor circuitry, the analysis of this circuit can be tricky, so we will study it briefly.

Since a bistable multivibrator has two stable conditions, assume the initial state to be just prior to t_1. Transistors Q1 and Q4 are on, and Q2 and Q3 are off. The negative input trigger at t_1 (waveform C, Fig. 9-10B), has no immediate effect at X2 since the anode of this diode is at a negative potential through the saturated Q1. But the coupling through C5 causes X5 to pass the negative trigger. This drives Q1 and Q4 off, reversing Q2 and Q3 to the on condition. A negative-going ramp develops at the Q2 collector (waveform F) so that at t_2, when the next trigger appears, X5 blocks it but X2 passes it. Diode X5 is not affected because its anode is negative through Q3. But Q1 is off (anode of X2 is positive), so X2 passes the negative trigger. This drives Q2 and Q3 off, reversing the operation of Q1 to the on condition. But note this: Q4, which is controlled by the collector of Q2, remains off, preventing further triggering through X5. A positive-going ramp develops at the Q2 collector (waveform F), and, after the time corresponding to the predetermined number of triggers, this ramp causes Q4 to go on at t_3. Since X5 will now pass the next trigger, the cycle is repeated.

NOTE: Observe through the above anlysis that negative-going triggers are used. These are npn transistors. The example of Fig. 9-9 applies if pnp transistors are employed.

In the collector circuit of Q2, R1, X1, and X4 prevent Q1 and Q4 from being off and Q2 and Q3 from being in saturation upon application of supply voltages. This collector-catching network prevents Q2 from saturating and assures a starting condition with Q1 and Q4 on. If Q2 is saturated, an otherwise normal pulse passed by X2 might not be sufficient to trigger the circuit out of conduction.

9-5. FINAL SUBCARRIER GENERATOR COUNTDOWN CHECK

The following steps constitute a final check on the countdown of the subcarrier generator:

1. Connect the vertical input of the scope to the 3.58-MHz generator output.
2. Use external trigger for the scope horizontal sweep and trigger from the 31.468-kHz generator countdown.
3. The pattern on the scope should lock on a steady trace with no tendency to drift or jump. Temporarily remove the external trigger to be certain the scope is under external trigger control. Critically adjust the trigger-amplitude control on the scope for a "clean" trace.
4. Reversing the above connections to the scope should also produce a steady trace.

NOTE: Some of the more recent generators (solid-state) count down as follows:

Divide by 5 to 715.909 kHz
Divide by 7 to 102.3 kHz
Divide by 13 to 7867 Hz

This sequence gives one-half the line frequency, or a total division of 455. This signal then locks a 31.468-kHz master oscillator in the sync generator by means of an automatic phase-frequency control (APFC) using a voltage-controlled circuit. In this case, Step 2 above simply involves taking the scope external trigger from the master oscillator in the sync generator.

9-6. FINAL CHECK ON SYNC-GENERATOR COLOR LOCK

For a final check on color lock of the sync generator, these steps may be followed:

1. Place the sync generator on internal crystal control, and check all counters in the generator for proper centering.
2. Place the sync generator on free-run operation. Observe any vertical output (composite sync, blanking, or vertical drive) with the scope on 60-Hz trigger. Adjust the master-oscillator frequency for a very slow drift of the trace from left to right on the scope. Check the sync-generator AFC operation with 60-Hz line lock.
3. Place the sync generator under external control with the color-subcarrier countdown (31.468 kHz) feeding the external input of the sync generator.
4. Connect the scope vertical input to the 3.58-MHz output of the color-subcarrier generator. Trigger the scope horizontal sweep with horizontal drive from the sync generator. The resulting pattern should be perfectly stationary with no erratic jumps.
5. If this condition does not exist, recheck the color-subcarrier generator as in Section 9-5. If the subcarrier generator is not at fault, the trouble is in the AFC (or APFC) circuitry of the sync generator.

9-7. SETTING THE COLOR-SUBCARRIER FREQUENCY

There are four quite satisfactory methods for setting the subcarrier frequency. Of these, only Method 1 gives positive assurance that the FCC frequency tolerance is being satisfied by the station.

Method 1: Frequency Standard and Primary Measuring Service

With this method, the station may or may not have a color-subcarrier frequency meter. The procedure is to contact the frequency-measuring service by telephone and transmit a composite color signal for purposes of measuring the subcarrier burst frequency. If the station has a subcarrier-frequency meter, adjust it to agree with the measuring-service reading, and then adjust the subcarrier frequency for zero reading. Or, the service can "talk you in" to the proper frequency adjustment.

Method 2: Vectorscope

If the station is affiliated with a network, or if the output of a good color receiver can be fed to the alternate input of the vectorscope, you can make use of a rapid and convenient way of adjusting the color-subcarrier frequency. In this method you must assume, of course, that the network or the received station is using the proper subcarrier frequency. Feed the external composite color signal to the B input of the vectorscope, and feed the local composite color signal to the A input. Place the vectorscope on external 3.58-MHz lock from the subcarrier generator (normal operation in station use). Place the input switch in the "A shared with B" position. The local burst, locked to the local subcarrier generator, will be stationary, and the signal being used as a standard will have its burst rotating at a rate dependent on the frequency difference. Adjust the local subcarrier frequency until the "standard" burst is as nearly stationary as possible. You cannot obtain a steady lock unless you have color genlock facilities.

This is an extremely quick and convenient procedure if a signal-selector switch is incorporated with the B vectorscope input, so that you can "punch up" the reference frequency at a moment's notice.

Method 3: Oscilloscope

If you do not have a vectorscope, the next best procedure is to feed the reference composite color signal to the external trigger input of an oscilloscope. Trigger the scope at the horizontal frequency. Observe 10 to 12 cycles of the local 3.58-MHz subcarrier on the scope triggered by the reference signal. Adjust the subcarrier frequency to obtain a stable trace.

Method 4: Dot Crawl

This method should be used only in an emergency when you do not have any other means of checking the subcarrier frequency. With a good stop watch, time the travel of a single dot from the bottom to the top of the raster on an *underscanned* monitor. The dots are quite apparent at color-bar transitions. Adjust the subcarrier frequency until this time is between 8 and 8.1 seconds. (Review Section 9-1.)

9-8. THE COLOR-SYNC TIMING SYSTEM

The burst keyer supplies a gating pulse (burst flag) to the encoder (colorplexer) for the purpose of keying on a burst of the subcarrier sine wave. This burst (used as a frequency and phase reference in color monitors and receivers) consists of 8 to 10 cycles of the 3.579545-MHz subcarrier, delayed 0.39 to 0.64 μs from the trailing edge of horizontal sync.

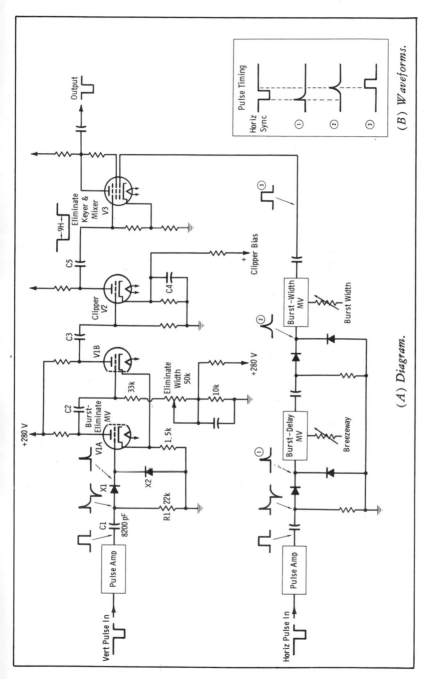

Fig. 9-11. Tube-type burst-key generator.

The burst therefore occurs on the back porch of horizontal blanking. Furthermore, this gating pulse must be defeated during the 9H interval of equalizing and vertical-sync pulses.

Fig. 9-11A is a simplified functional schematic diagram of a typical tube-type burst-key generator. The vertical- and horizontal-pulse inputs may come from composite station sync by way of a sync separator, or they may be the sync-generator drive-pulse outputs. In some cases, horizontal pulses are derived from composite sync, but vertical pulses are supplied from sync-generator vertical drive.

The circuit of V1 is a typical cathode-coupled multivibrator; in this case it is used as the burst-eliminate pulse generator. The network composed of C1 and R1 differentiates the pulse, and diodes X1 and X2 pass positive triggers to the grid of V1A. Note that this grid is returned to ground through the diodes, but the grid of V1B has a positive potential. Therefore, V1B is normally on, while V1A is cut off by the positive potential at its cathode (conventional cathode-coupled arrangement). The positive trigger applied at the grid of V1A drives this section to conduction; the resulting negative pulse at the plate drives V1B to cutoff. The charge on C2 holds this state for a time determined by the grid potential of V1B, which is adjusted by the eliminate-width control. This control normally is adjusted for a pulse width of 9H.

This stage usually is followed by a conventional clipper (V2) to get a good, flat top over the 9-line duration. The resulting negative pulse of 9H duration is fed to the suppressor grid of the mixer tube and must be of sufficient amplitude to hold this tube nonconducting during the 9H interval.

The horizontal circuitry is a duplicate (except for time constants) of that described above. In this case, two multivibrators must be used: One generates a delayed trigger for the actual burst-width multivibrator so that the color-sync burst will be gated on at the appropriate time following the trailing edge of horizontal sync. The timing waveforms are shown by number in Fig. 9-11B. The burst-delay multivibrator is adjusted so that the trailing edge triggers the burst-width multivibrator approximately 0.5 μs following the trailing edge of horizontal sync. The burst-width multivibrator is adjusted to allow 8 or 9 cycles of the subcarrier to appear in the burst. Since one cycle at 3.579545 MHz has a duration of 0.28 μs, then:

$$8 \text{ cycles} = (8)(0.28) = 2.24 \ \mu s \ (\text{minimum})$$
$$9 \text{ cycles} = (9)(0.28) = 2.52 \ \mu s \ \text{nominal}$$

The resultant positive pulse is applied to the control grid of the mixer tube and is passed except during the time the tube is held at cutoff by the burst-eliminate (negative) pulse on the suppressor grid.

Some burst keyers employ a tapped delay line and a bistable multivibrator. The first tap on the delay line goes to an on-trigger tube, and the second tap goes to an off-trigger tube. These triggers control the two sections of the bistable multivibrator so that the duration of the pulse is set by the delay line rather than multivibrator time constants.

Fig. 9-12 presents typical circuitry in transistor burst-key generators. The horizontal pulse is amplified and inverted by Q1. The inverted pulse is integrated by the combination of R1, R2, and C1. Note that the Q2

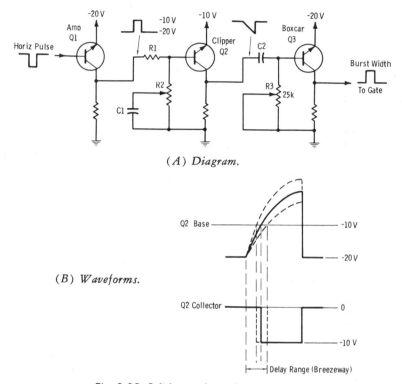

(A) Diagram.

(B) Waveforms.

Fig. 9-12. Solid-state burst-key generator.

emitter is returned to −10 volts, and the voltage at the Q1 collector holds Q2 in the off condition *until* the integrated pulse at the Q2 base becomes more positive than the Q2 emitter voltage. When this occurs, Q2 saturates, and its collector goes essentially to −10 volts. So the Q2 collector is at zero volts prior to the pulse (cutoff) and swings to −10 volts for that portion of the base pulse more positive than −10 volts. The delay control (R2) adjusts the rise time of the integration (Fig. 9-12B). Therefore the output pulse from the Q2 collector is delayed the proper amount for the breezeway interval.

Transistor Q3 is connected in a "boxcar" circuit, in which the clamped base pulse width (and therefore the output pulse width) is determined by the C2R3 product. Width control R3 is adjusted to obtain the proper number of cycles in the burst interval.

A typical transistor gate, which performs the same function as V3 in Fig. 9-11, is shown in Fig. 9-13. Note that when either transistor is saturated, the common collector is at ground (zero) potential. Note also that both transistors must be cut off for the collector to reach the supply voltage. In the absence of input pulses, Q1 is saturated and Q2 is cut off.

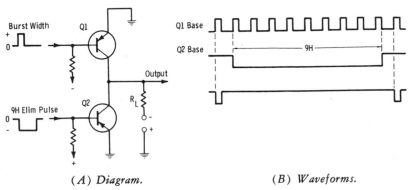

(A) Diagram.   (B) Waveforms.

Fig. 9-13. Burst-key gate.

The collector is at zero (ground) potential. When the positive burst-width (burst-key) pulse arrives at the Q1 base, this transistor is cut off. Since Q2 is already at cutoff, the collector rises to the negative supply voltage. Therefore an inverted key pulse is passed to the output across R_L. This action continues until the 9H eliminate pulse arrives at the Q2 base. For the duration of this pulse, Q2 is saturated and the collector voltage remains at ground, eliminating the burst-key pulses at Q1.

> NOTE: Recall that all synchronizing generators inherently employ a 9H gating pulse in the formation of composite sync. In modern color-sync circuits, you will find the burst flag is generated in the main portion of the sync generator without need of additional equipment. In older equipment, it is formed in a separate unit.

9-9. ADJUSTMENT OF BURST-KEY GENERATOR WITH ENCODED SIGNAL

An encoded signal may be used in the adjustment of the burst-key generator according to the following procedure:

1. Observe the composite color signal (sync and blanking added) at some point after encoding. Trigger the scope externally with horizontal drive to obtain a steady trace.
2. Adjust the burst-key delay control for proper breezeway (Fig. 9-14).

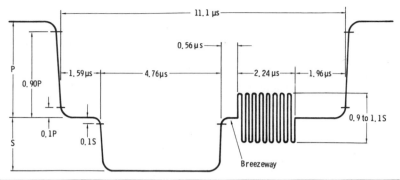

	Nominal Microseconds	Tolerance Microseconds
Blanking	11.1	+0.3 −0.6
Sync	4.76	±0.32
Front Porch	1.59	+0.13 −0.32
Back Porch	4.76	+0.96 −0.61
Sync to Burst	0.56	+0.08 −0.17
Burst	2.24	+0.27 −0
Blanking to Burst[1]	6.91	+0.08 −0.17
Sync & Burst	7.56	+0.38 −0.49
Sync & Back Porch	9.54	±0.32

[1]Blanking-to-burst tolerances apply only to signal before addition of sync.

Fig. 9-14. Time intervals for horizontal sync and burst.

3. Adjust the burst-key width control for 8 or 9 *complete cycles.* Do not count cycles of less than 50 percent of the nominal peak-to-peak value (Fig. 9-15A). If the first cycle starts with a positive-going alternation, count the negative peaks. If the first cycle starts with a negative-going alternation, count the positive peaks (Fig. 9-15B).

NOTE: If you use straightforward internal triggering of the scope instead of an external trigger, you can obtain the "interlaced" burst pattern of Fig. 9-15C. You should count 16 peaks for 8 cycles of noninterlaced burst.

4. Trigger the scope externally with vertical drive, and use the delayed sweep on the scope while observing the vertical interval for one field. The time base should be such as to permit observation of the 9H interval plus a few horizontal-sync pulses following the trailing equalizing pulses. Adjust the burst-eliminate control to eliminate all bursts in the 9H interval. Operate the field-shift key on the scope, and observe the alternate field. If necessary, readjust the burst-eliminate control to obtain the 9H "key out" on this field. Go back and forth between the two fields and adjust until the 9H elimination is correct for both fields (Fig. 9-16).

5. In certain tube-type key generators, the tubes for the delay, burst-width, and burst-eliminate multivibrators must be selected for best stability of adjustment. Initially, check these adjustments daily. Experience with the particular equipment will then indicate whether these checks can be made on a weekly or monthly schedule.

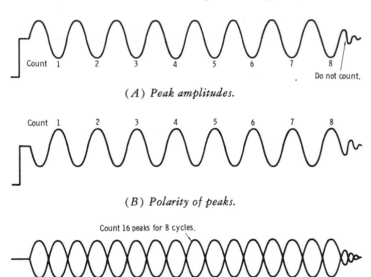

(A) Peak amplitudes.

(B) Polarity of peaks.

(C) Interlaced display.

Fig. 9-15. Counting of cycles in burst.

9-10. ADJUSTMENT OF BURST-KEY GENERATOR BY ITSELF

In some instances, after servicing it might be necessary to adjust the burst-key generator as a unit before placing it in service with the encoder unit. The procedure is as follows:

1. Always check first for proper output level. This is usually 4 volts peak-to-peak across a 75-ohm load.

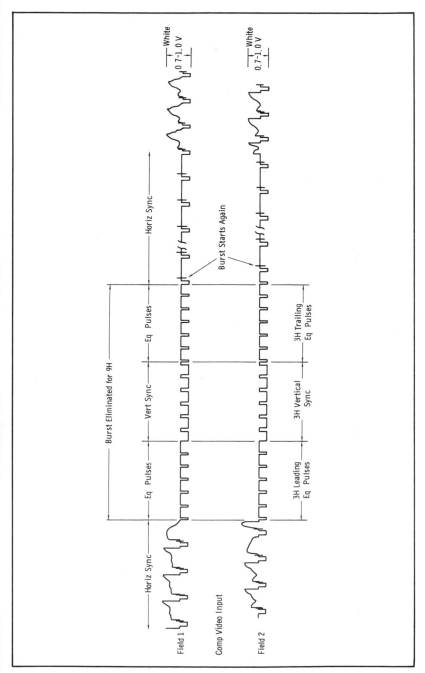

Fig. 9-16. Adjustment of burst-eliminate control.

2. Mix station sync with the output of the burst-key generator. You can do this quite simply by using a coax "T" on the scope; insert sync in one input, and connect the scope probe to the 75-ohm termination on the key-generator output.
3. Trigger the scope externally with horizontal drive. Adjust the delay control for a breezeway (in this case, the interval between the trailing edge of horizontal sync and the leading edge of the key pulse) of about 0.5 μs.
4. Adjust the burst-width control for a key-pulse (flag pulse) width of 2.4 μs.
5. Set the scope time base to vertical rate and trigger with vertical drive. Adjust the burst-eliminate control so that no flag pulses appear during the 9H vertical interval, but pulses start immediately after the first horizontal sync pulse following the last trailing equalizing pulse in both fields.

9-11. THE ENCODING PROCESS

Operations performed by the encoder (colorplexer) are:

(A) Matrixing of R, G, and B video signals from the camera processing amplifiers to produce luminance and chrominance signals. (In four-tube cameras, Y, R, G, and B signals are involved.)
(B) Filtering of the chrominance signals to the required bandwidth.
(C) Delay compensation in the Y and I channels to correct for the delay of the Q signal (narrowest bandwidth, hence most delayed).
(D) Modulation of the 3.58-MHz carrier by the chrominance signals.
(E) Insertion of the color-sync burst.
(F) Aperture compensation of the luminance (Y) signal.
(G) Mixing of the Y, I, and Q signals to form the complete color signal.
(H) Insertion of composite sync (optional).

Fig. 9-17 is a functional block diagram of the encoding system. We will elaborate each of the blocks as we go along. For a review of the mathematics of the encoding system, see Harold E. Ennes, *Television Broadcasting: Equipment, Systems, and Operating Fundamentals* (Indianapolis: Howard W. Sams & Co., Inc., 1971), Chapter 2. It is important to understand at the start that any signal inserted at the red input of the encoder will activate the red gun in the color picture tube. The same is true for the green and blue inputs of the encoder and their respective guns in the color picture-tube.

Fig. 9-18A illustrates the Cohu color video encoder. The operation of the color encoder is discussed in general terms with reference to Fig. 9-18B. This diagram shows in block form the various sections of the encoder, except the power supply, and indicates the signal flow between sections.

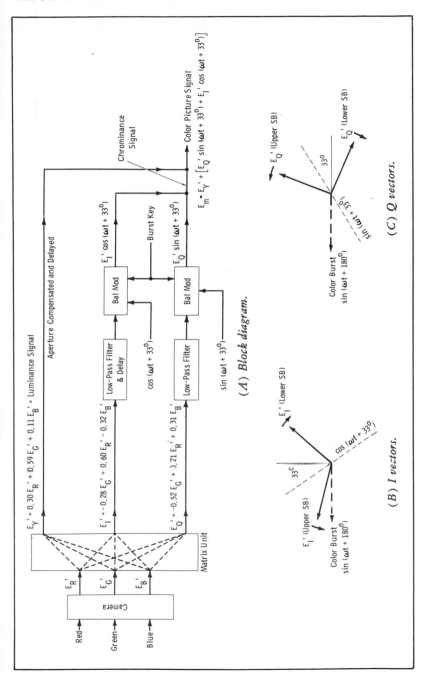

Fig. 9-17. Basic encoding system.

The function of the encoder is to process and combine individual inputs to produce a compatible color signal. In addition, the encoder has integral full-bar and split-bar generators, and various other test and operational features. It also contains circuits for optional relays for the remote control of certain functions.

The signal, which is provided at three separate outputs of the encoder, is suitable for use in color and monochrome systems that include video display devices, recorders, and transmitters. The encoder requires a 3.58-MHz subcarrier, sync, blanking, and RGB or YRGB video inputs.

The encoder pulse and subcarrier inputs pass through isolation circuits on the input board to the matrix, modulator, and processor boards. The video inputs pass through isolation and switching circuits on the input board to the matrix board. When the internal bar generators are being used, the video signals are blocked at the input board. The switching circuits are controlled from the matrix board.

At the matrix board, the red, green, and blue video signals from the input board (or from the bar generators) are matrixed and amplified to provide Y, I, and Q video signals. The blanking-signal input of the matrix board is used in the generation of full-bar or split-bar outputs. The I and Q video signals are filtered and fed to the modulator board. Before reaching the modulator board, the I signal is delayed to compensate for the greater delay of the Q signal, which has a narrower bandwidth.

If RGB inputs or the bar generators are used, the Y signal, which is derived from matrixed RGB video, provides the luminance. The Y signal may or may not pass through a 3.58-MHz notch filter on the matrix board, depending on the connection of a patch cord. If YRGB inputs and luminance correction are used, the luminance signal is derived from a network that uses the Y signal from the matrix and the monochrome input. Without luminance correction, the monochrome input is the luminance signal and is amplified and fed to the Y output of the matrix board.

The luminance signal, which is fed to the processor board, also passes through a delay line to bring it into time coincidence with the Q signal. The luminance line has a termination adjustment (R1), which, together with the two delay lines, is on the interconnection board.

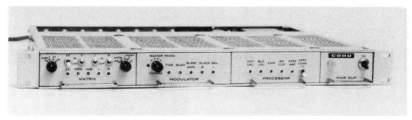

(A) Photograph.

Fig. 9-18. Cohu

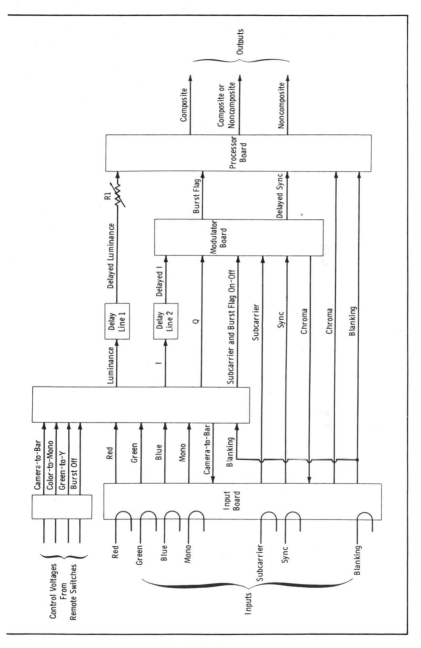

(B) Block diagram.

Courtesy Cohu Electronics, Inc.

color video encoder.

In addition to the matrixing circuits, the bar-generator circuits, the luminance-correction network, and the notch filter, the matrix board contains the input, mode, and various other switching circuits including (when installed) the relays for remote control. Also included are switching circuits for disabling the subcarrier and burst-flag circuits on the modulator board when monochrome-without-burst outputs are required from the encoder.

At the modulator board, the subcarrier signal is amplified and limited to obtain constant amplitude. The sync input of the modulator board passes through a limiting circuit, and is delayed to bring it into coincidence with the I, Q, and luminance signals. The output of a burst-flag generator, which also is fed to the processor board, adds burst-flag pulses to the I and Q signals, which are amplified and clamped at the black reference level (blanking) by the sync signal.

The combined I video and burst-flag signal modulates a subcarrier signal, and the combined Q video and burst-flag signal modulates another subcarrier signal that has been passed through a 90° phase-shifter network. The suppressed-carrier outputs of the modulators are passed through tuned amplifiers, combined, and fed as the chroma signal to the processor board through the chroma-switching circuit on the input board.

The modulator board also has a 360° phase-shifter circuit that includes coarse and fine phase-shifting controls for matching subcarriers to two or more encoders.

At the processor board, the luminance signal is amplified, clamped, and white-peak clipped before passing to a circuit in which aperture correction can be performed. After further amplification, the luminance signal is again clamped, and blanking is then added. The blanking signal, which comes from an amplifier and delay circuit, is also combined with the burst-flag pulses after they have passed through a delay and pulse-shaping circuit.

The Matrix

Fig. 9-19 illustrates one example of a matrix (cross-connected voltage divider) for proportioning the red, green, and blue video inputs to obtain Y, I, and Q signals. The matrix consists of three sets of three resistors. The matrix input signals, R, −R, B, G, and −G, are applied to the resistors as indicated in Fig. 9-19, with each resistor receiving one input. The output of each set of resistors is capacitively coupled to an amplifier (Y, I, or Q). The resistance values in each set of resistors are proportioned so that the currents in the resistors are summed at the amplifier inputs in the following standard proportions:

$$Y = 30\%R + 59\%G + 11\%B$$
$$-I = -60\%R + 28\%G + 32\%B$$
$$Q = 21\%R - 52\%G + 31\%B$$

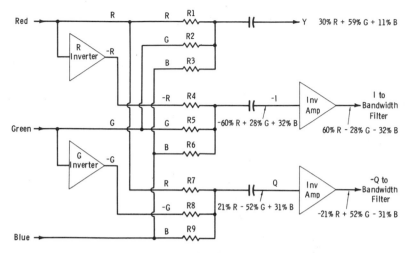

Fig. 9-19. Example of matrix circuitry.

The Y output of the matrix results from the addition of the red, green, and blue signals developed across R1, R2, and R3. The Y signal has component amplitudes, proportioned as indicated in the Y equation, corresponding to the brightness factors of the colors red, green, and blue relative to the 100-percent brightness characteristic of white. Similarly, the I and Q signals have component amplitudes proportioned as indicated in the respective equations. The vector sum of I and Q represents the color. The magnitude of the vector carries the saturation information, and the angle carries the hue information. Note that with equal red, green, and blue inputs (white or gray) the algebraic sum of I and Q is zero.

See Fig. 9-20. These are more detailed drawings of the specific I (Fig. 9-20A) and Q (Fig. 9-20B) networks. The I phase inverter inverts the phase of the red signal injected at the grid; since blue and green are injected at the cathode, no phase inversion of these signals takes place. Hence, the video signal at the plate is −I.

Obviously, the gains for blue and green are different from the gain for red. Remember that a white or gray signal means that all inputs are of identical amplitude. Thus, when R = G = B, the white-balance control (actually a red-signal gain control) is adjusted so that all subcarrier is cancelled in the I output.

Note also that since the I-test pulse is injected at this stage, it becomes a −I-test signal. Since this pulse occurs either on a split field (without RGB present at the same time) or by itself (also without RGB present), and since it is not simultaneously matrixed with the luminance channel, it has zero luminance and is inserted on the blanking pedestal. It is matrixed such that with normal input level, its peak-to-peak amplitude at the encoded output is the same as the amplitude of a properly adjusted sync burst. It is filled with subcarrier just as are all of the actual color bars; only white (equal input amplitudes) contains no subcarrier.

You now should be able to correlate the signal processing just described with that of the Q phase inverter (Fig. 9-20B), and note that the output here is a +Q signal. In this circuit, green is the inverted signal and is adjusted for Q white balance.

Bandwidth Limiting and Delay Compensation

The equivalent bandwidths prior to modulation assigned to the I and Q signals are as follows:

Q-Channel Bandwidth
At 400 kHz less than 2 dB down
At 500 kHz less than 6 dB down
At 600 kHz at least 6 dB down

I-Channel Bandwidth
At 1.3 MHz less than 2 dB down
At 3.6 MHz at least 20 dB down

Corresponding to these band limits are three ranges of pictorial detail. Fine details corresponding to video frequencies above 1.3 MHz are reproduced in monochrome; larger areas corresponding to frequencies between 0.5 MHz and 1.3 MHz are reproduced in a two-color orange-cyan system; and still larger areas, corresponding to frequencies below 0.5 MHz, are reproduced in a three-color red-green-blue (full-color) system. This particular division of color reproduction was found to represent a proper com-

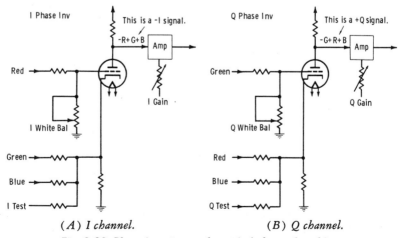

(A) I channel. (B) Q channel.

Fig. 9-20. Phase inverters and matrix balance for white.

promise to assure adequate color fidelity under the bandwidth limitations of the system.

NTSC standards state that the luminance and chrominance signals (the respective carrier envelopes as radiated) shall match each other in time within about half the duration of a picture element, or 0.05 microsecond. Since the luminance signal and chrominance signals pass through circuits of different bandwidth, delay circuits are required in the wideband circuits to bring the respective signals into time coincidence. The time-coincidence standard has the effect of setting a tolerance in these delay circuits.

Fig. 9-21A illustrates the relative bandwidths and necessary delay compensation at the encoder. Since the Q channel has the narrowest bandwidth, it has the most signal delay. Hence the Y and I video channels must be delayed the appropriate amounts with respect to their assigned bandwidths to obtain time coincidence with the Q channel.

Fig. 9-21B shows the bandwidths of the luminance and chrominance signals prior to modulation at the transmitter. Fig. 9-21C illustrates the resultant signal radiated from the transmitting antenna.

Burst Amplitude and Phase

See Fig. 9-22A. The amplitude of the burst-key (flag) pulse is set by the burst-gain control. This pulse is divided into I and Q pulses by the burst-phase control. The I pulse goes to the stage containing +I video, and the Q pulse goes to the stage containing −Q video. The Q modulator receives the color subcarrier delayed 90° from the subcarrier feeding the I modulator. Fig. 9-22B shows the effect on the modulators. If the amplitude of the pulse is increased while the ratio of amplitudes remains the same, the resultant vector amplitude (burst amplitude) increases. If the amplitude remains fixed while the ratio is varied, the phase of the vector sum (burst phase) changes according to the ratio change. When the ratio of burst I pulse to burst Q pulse is −tan 33°, the resultant burst phase is correctly placed along the −x axis.

Since the keying pulse (burst-flag pulse) is timed on the horizontal back porch, no video is present, and only the burst of subcarrier appears at the modulator outputs.

The Modulation Process

The reader should be familiar with the fundamentals of the entire encoding process and the modulation technique from previous study. It remains to examine further the advances in the state of the art with respect to stable video balance and automatic carrier balance. Such modern processing normally includes closed loops between subcarrier, driver (or processing) circuitry, and the modulator.

Fig. 9-23 illustrates the paths to be discussed in the RCA encoding system. The path prior to modulation is shown in Fig. 9-23A. For the color-bar test position, the camera inputs are blocked by loading relays. (In the Cohu

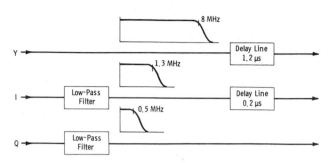

(*A*) *Bandwidths and delays at encoder.*

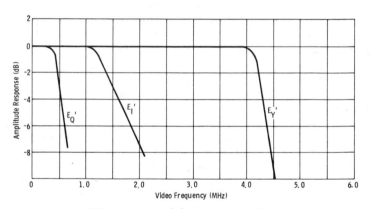

(*B*) *Prior to modulation at transmitter.*

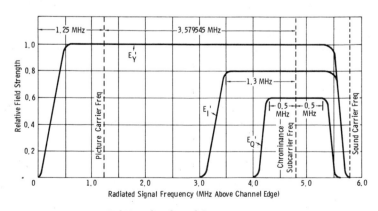

(*C*) *Signal radiated from transmitter.*

Fig. 9-21. Bandwidths of color signals.

encoder, a dc bias blocks the camera video.) Fig. 9-23B illustrates the basic modulation process in the RCA encoder.

The red, green, and blue video signals that are matrixed into I and Q information may originate either from the camera vidicon channels or from the bar generator. The monochrome information from the bar signals is obtained in the matrix module by properly matrixing the red, green, and blue video signals. In normal film-camera operation, the monochrome information is obtained from the 1½-inch vidicon (luminance tube), and the monochrome signal is switched straight through the matrix module. In three-tube camera operation, the matrixing of the red, green, and blue into monochrome is done at the camera in the control module, and the encoder treats this monochrome signal the same as when it originates from the luminance channel.

In the RCA TK-42 studio camera, luminance information is supplied by the 4½-inch image orthicon, and chrominance information by three vidicons. This live camera, unlike the film camera (TK-27) cannot be operated in the three-vidicon mode. The RCA TK-44A studio camera employs only three channels, using lead-oxide vidicons.

The I and Q signals are limited in frequency response in accordance with the FCC transmission specifications. The low-pass filters in the driver module limit the response of the I signal to approximately 1.5 MHz and of the Q signal to approximately 0.5 MHz. A delay line in the driver module is used to match the delay of the I signal to that of the Q filter. The

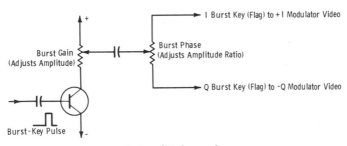

(A) I and Q key pulses.

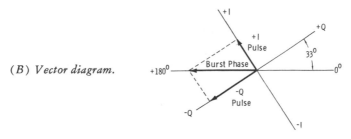

(B) Vector diagram.

Fig. 9-22. Method of generating burst.

monochrome signal is passed through the Y-delay module in order to match the delay of the chroma signals (Fig. 9-23A).

The system always provides a monochrome camera signal from the monochrome-blanker module, even when bars are "punched up" at the color control panel or when the monitor-module test switch is not on "operate." In these cases, the monochrome signal from the camera is switched directly to the monochrome-blanker module, while the monochrome signal matrixed from the color bar goes through the Y-delay module. Since the timing at the camera is adjusted to take care of all delays, a timing shift will occur at the camera when the monochrome signal

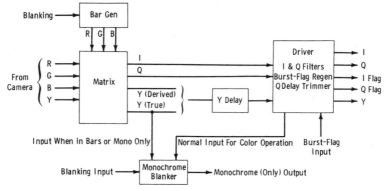

(A) Processing prior to modulation.

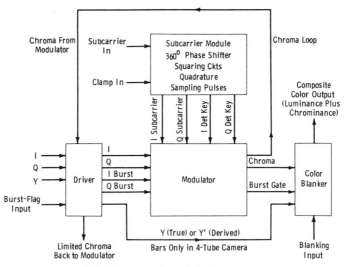

(B) Diagram of modulation process.

Fig. 9-23. Color encoding in RCA four-tube system.

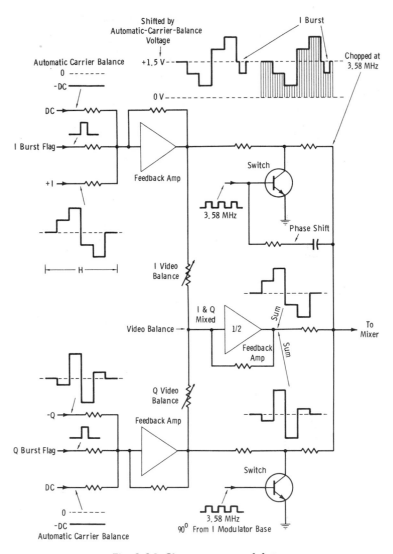

Fig. 9-24. Chopper-type modulator.

is switched directly to the monochrome-blanker module (as in "Bars On"), since the delay through the Y-delay module is eliminated. In this case, the only delay for which the timing circuit must compensate is from the length of cable between the camera and auxiliary (rack unit). This distance is usually short, so if bars are punched up when framing and registration are being performed at the camera, the timing of the scan with respect to system blanking will be the same as at the auxiliary outputs. Otherwise, in order to set horizontal centering properly at the camera, the viewfinder must be switched to observe color in order to see the output timing in proper relationship to system blanking.

The I and Q signals go to the modulator module (Fig. 9-23B), where each of them is chopped to ground at the color-subcarrier rate (Fig. 9-24). The I and Q chopping signals come from the subcarrier module and are phased 90° apart. I burst flag and Q burst flag are adjustable in amplitude with respect to each other by the burst-phase control (Fig. 9-22) and are applied to the chopping transistors mixed with the I and Q video information, respectively. The modulated I and Q are in quadrature and are mixed in the mixer amplifier (Fig. 9-25). The resultant is a vector that varies in amplitude and phase with changes in I and Q amplitude and polarity.

The amplitude of the mixed I and Q modulated signals during the absence of I and Q video information (no color information) depends on the dc output bias of the I and Q amplifiers ahead of the I and Q modulators. This bias (about 1.5 volts) is set by the dc amplifiers that feed the I and Q amplifiers. The effect of this dc component can be removed by feeding the mixer with a subcarrier signal equal but opposite in phase to the resultant of the modulated I and Q dc component. This required subcarrier signal is obtained by shifting the phase of the subcarrier applied to the base of the I modulator transistor through a resistor-capacitor network (Fig. 9-24).

The above phase-shifted subcarrier will balance out the modulator for the absence of I and Q information, but the resultant at the output of the mixer will still contain the video information (represented by a low-frequency component of the modulated subcarrier), unless the following video-balance provisions are included (see Fig. 9-24). The I and Q video signals (ahead of modulation) are mixed, inverted, and fed into the mixer at a level one-half that of the peak-to-peak modulated subcarrier. This removes the video component (or luminance information) to provide a fully balanced modulated signal at the output of the mixer (Fig. 9-26). The video-balance controls provide a limited amount of amplitude variation of the modulated I and Q signal in order to match the video-balance correction signal.

The chroma signal then passes through a low-pass filter to remove the harmonics produced by the square waves of the chopping modulators (Fig. 9-25). One chroma output of the modulator module is mixed with the monochrome video in the color-blanker module, and the other chroma output is amplified in the driver module (Fig. 9-27) to be used in the automatic carrier-balance circuitry described below.

During camera blanking, the chroma output should be balanced out to zero amplitude. If the dc biases of the I and Q amplifiers are not the exact level that results in cancellation of the subcarrier signal inserted directly into the mixer through the series RC phase-shift network, subcarrier will

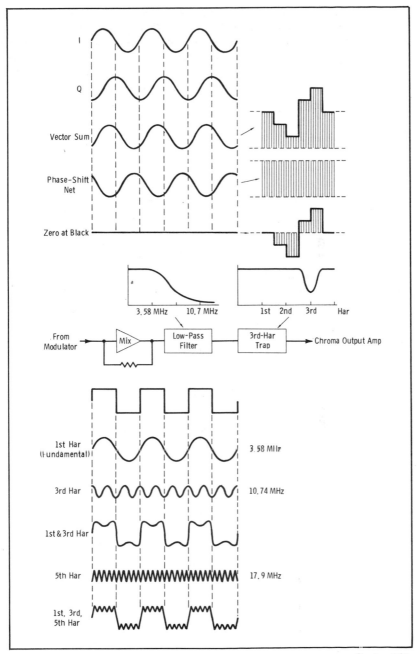

Fig. 9-25. Filtering of modulated signals.

be present during the camera-blanking period. Since I and Q are modulated in quadrature, the dc inputs to the I and Q amplifiers can be adjusted to cancel completely any subcarrier at the output, regardless of amplitude and phase. For this purpose, I and Q black-balance (carrier balance) controls are provided. To maintain this balance, other dc sources to the I and Q amplifiers are obtained (Fig. 9-28). These sources are proportional to the I and Q components of any subcarrier present at the output during

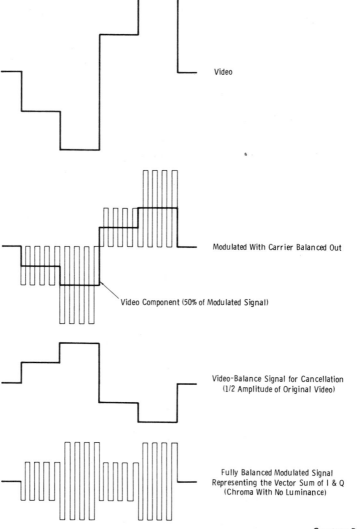

Video

Modulated With Carrier Balanced Out

Video Component (50% of Modulated Signal)

Video-Balance Signal for Cancellation
(1/2 Amplitude of Original Video)

Fully Balanced Modulated Signal
Representing the Vector Sum of I & Q
(Chroma With No Luminance)

Courtesy RCA

Fig. 9-26. Removal of video component.

camera blanking. The loop gain of this feedback system is made high to maintain a high degree of carrier balance.

The chroma amplifier in the driver module has a high gain for low-level inputs, but clips when the output exceeds about 1.4 volts peak-to-peak. Only the low-level information during camera blanking is of interest,

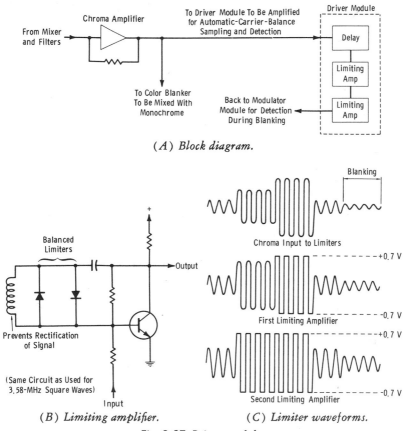

(A) Block diagram.

(B) Limiting amplifier.

(C) Limiter waveforms.

Fig. 9-27. Driver module.

since the I and Q detectors in the modulator module are keyed on only during a portion of the camera-blanking interval. Several cycles of subcarrier occur on the I and Q keying signals, each phased such that the I detector is switched on in phase with the peak of the I component of the amplified chroma signal and the Q detector is switched on in phase with the peak of the Q component (Fig. 9-29). Since the I and Q components are 90° apart, the peak of the I component occurs during the time the Q component is going through zero, so no Q information will be detected by the I detector, and vice-versa. When the detector is keyed on, if the amplified chroma signal is anything except zero or passing through zero, the capacitor is charged to a level either more positive or more negative than the dc component of the chroma signal at the detector. This charge is held throughout the line, amplified, and applied to the respective I and Q amplifier input in proper amplitude to balance out any subcarrier that is present in the output during camera blanking.

The inverted I and Q signal used for video balance is also used in the matrix portion of the modulator module to obtain the color-difference signals, $Y' - R$, $Y' - G$, and $Y' - B$ (Fig. 9-30). When the luminance signal is removed from these color-difference signals in the detector module, by combining with the Y signal from the camera, the proper R, G, and B signals are obtained for NAM monitoring and automatic control (described previously).

Built-in setup facilities are obtained by operation of the test switch on the monitor module. In all positions except position 1 (Operate), the color bars are switched into the matrix module. In position 2, all inputs to the matrix module are tied together. If the bar generator is then pulled, there is no color information, and no subcarrier should be present at the output (except during burst). The black-balance controls on the modulator module can then be adjusted to remove any subcarrier present. If the bar generator is now inserted while the inputs are tied together, all inputs will be of equal amplitude, and there should be no I or Q signal. The adjustments to cause I and Q to cancel out under these conditions are the I and Q white-balance controls in the matrix module. These controls adjust the levels of the inverted red and inverted green signals, respectively (described previously).

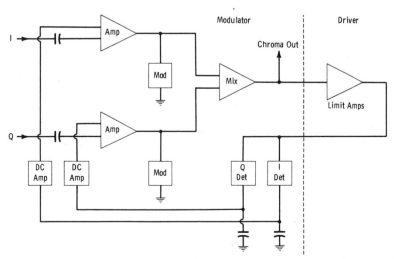

Courtesy RCA

Fig. 9-28. Feedback and detection loop.

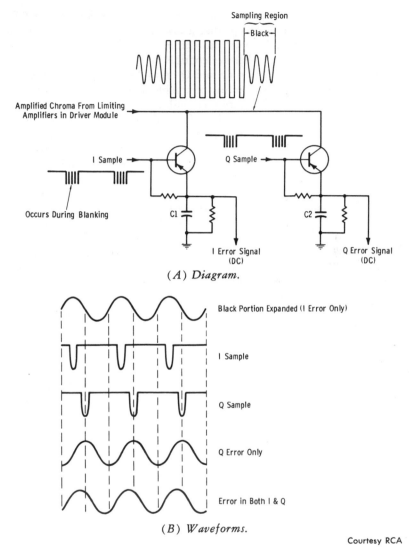

(A) Diagram.

(B) Waveforms.

Courtesy RCA

Fig. 9-29. I and Q detectors.

The above procedure causes the subcarrier to be cancelled out whenever the red, green, and blue inputs are equal in amplitude. If subcarrier appears at the output in test position 3, which disconnects the inputs, it indicates that the three bar amplitudes are not equal. They should be equal during the white bar. Green-gain and red-gain controls are provided in the bar-generator module to set these pulses equal to the blue pulse during the white bar.

Test position 4 produces the same chroma presentation on the CRO that position 3 produces, except that a low-pass filter is used so that any video-frequency information in the chroma signal can be detected and balanced out with the video-balance controls in the modulator module.

Test positions 5 and 6 (I only and Q only) have no adjustments associated with them. They are useful for troubleshooting or double-checking video balance and the quadrature relationship.

Test position 7 sets up a method for making quadrature adjustments with only the CRO display. The Q channel operates in a normal manner with bars applied, but the I channel has applied to it a square wave that goes equally positive and negative around its clamped blanking interval (Fig. 9-31A). The CRO then displays a few lines of the resultant of the Q signal with the positive-I portion of the square wave, and then a few lines of the resultant of the Q signal with the negative-I portion of the square wave. Since the positive-I and negative-I levels are equal, these two resultants will be equal only when I and Q are phased 90° apart (Fig. 9-31B). Since the CRO displays these two resultants superimposed on each other, it is easy to compare the amplitudes and make a quadrature adjustment so that the two CRO presentations become equal.

Test position 8 provides the CRO with the composite color-bar presentation. It is used to set the I and Q signal gains relative to monochrome to obtain the levels shown in Fig. 9-32.

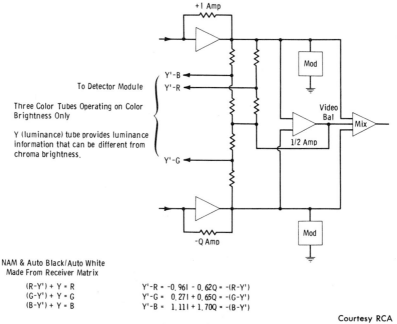

NAM & Auto Black/Auto White
Made From Receiver Matrix

$(R-Y') + Y = R$ $Y'-R = -0.96I - 0.62Q = -(R-Y')$
$(G-Y') + Y = G$ $Y'-G = 0.27I + 0.65Q = -(G-Y')$
$(B-Y') + Y = B$ $Y'-B = 1.11I + 1.70Q = -(B-Y')$

Courtesy RCA

Fig. 9-30. Method of obtaining color-difference signals.

The color-bar patterns of Fig. 9-32 are all those likely to be encountered in various color camera-chain facilities. These are normally different from the standard split-field pattern. Note that in addition to the IEEE units, a voltage scale, such as would be used on an external oscilloscope, is shown. In Fig. 9-32A are 100-percent bars (used only on very special occasions). Fig. 9-32B shows the most usual 75-percent mode, and Fig. 9-32C illustrates the white 100-percent mode. A standard 7.5-percent setup level is used.

Test position 9 is for setting the proper burst phase with the CRO in a manner similar to that used for the quadrature adjustment. To obtain a reference that is phased 90° from burst, a precision voltage divider is used to reduce the amplitude of the blue bar an amount that shifts the cyan bar (third bar) to a phase position 90° from where burst should be (Fig. 9-33A). This reduction of the blue amplitude shifts the phase of all the bars (and unbalances the white bar), but we are concerned only with the third bar in this test. At the same time, the burst is switched from the burst-flag width to full active scan width and is caused to go equally positive and negative, several lines each, by the same blanked square wave

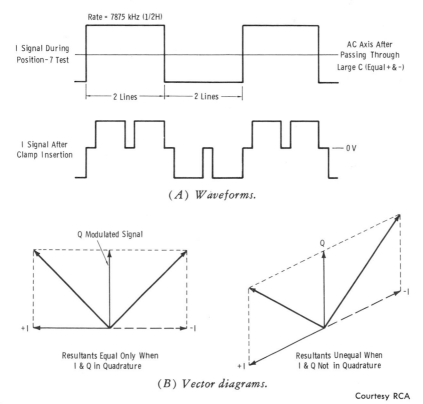

(A) Waveforms.

(B) Vector diagrams.

Fig. 9-31. Method of making quadrature adjustments.

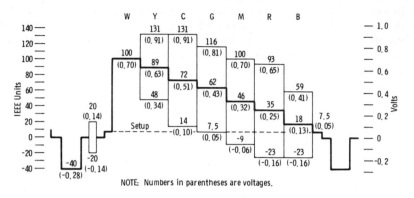

NOTE: Numbers in parentheses are voltages.

(A) 100-percent bars.

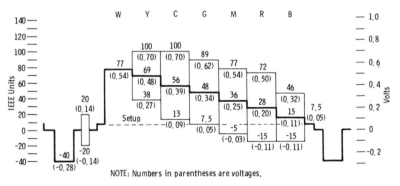

NOTE: Numbers in parentheses are voltages.

(B) 75-percent mode.

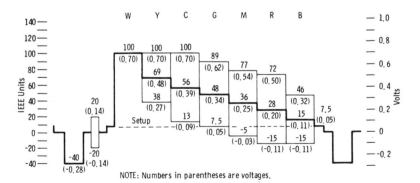

NOTE: Numbers in parentheses are voltages.

(C) White 100-percent mode.

Fig. 9-32. Nominal encoded color-bar signal levels.

used for the quadrature test. The two superimposed signals in the CRO then represent the two resultants of the modified chroma combined with the positive burst for several lines and then with the negative burst for several lines. Only when the burst phase is adjusted so that the two resultants are equal during the third bar (Fig. 9-33B), will the burst be in quadrature with the third bar and properly phased.

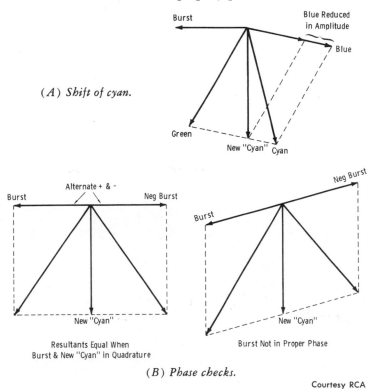

(A) Shift of cyan.

(B) Phase checks.

Courtesy RCA

Fig. 9-33. Method of setting burst phase.

9-12. THE VECTORSCOPE

Fig. 9-34A illustrates the Tektronix Type 520 vectorscope, and Fig. 9-34B illustrates the normal vector display with the selector switch in the vector position. This instrument is designed to measure luminance, hue, and saturation of the NTSC composite color television signal. Self-cancelling push-button switches permit selection of displays for analysis of television-signal characteristics, and to check vectorscope calibration.

Dual inputs provide time-shared displays for comparison of input-output signal phase and gain distortion. A chrominance channel demodulates the chrominance signal to obtain color information from the composite video signal for use in vector, line-sweep, R, G, B, I, Q, differential-gain, and differential-phase displays. A luminance channel separates and displays the luminance (Y) component of the composite color signal. The Y component is combined with the output of the chrominance demodulators for R, G, and B displays at a line rate. A digital line selector permits the display of a single-line vertical-interval test signal from a selected line of either field 1 or field 2.

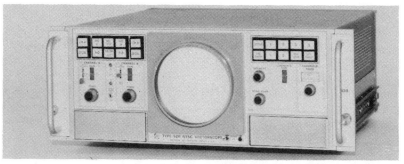

(A) Instrument.

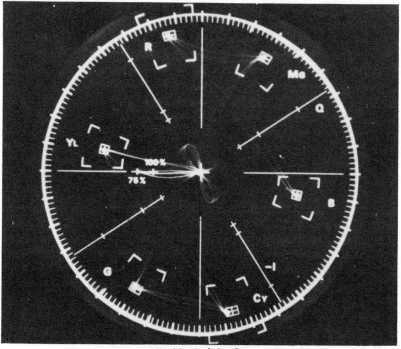

(B) Typical display.

Fig. 9-34. Tektronix Type 520 NTSC vectorscope.

The following basic description, supplied through the courtesy of Tektronix, Inc., serves as an excellent review of the color system for the student:

In color television, the visual sensation of color is described in terms of three quantities: luminance, hue, and saturation. Fig. 9-35 shows a conical representation of these concepts. Black-and-white (monochrome) TV receivers respond to the brightness (or luminance) signal only. Color-TV receivers respond to all three signals, luminance, hue, and saturation. Hue and saturation together are called the *chrominance signal*. With the VECTOR button depressed, the vectorscope displays the hue and saturation quantities. With the Y button depressed, luminance amplitude is displayed.

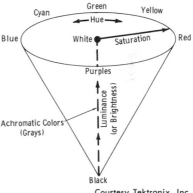

Fig. 9-35. Conical representation of color concepts.

Courtesy Tektronix, Inc.

Luminance is brightness as perceived by the eye. Because the eye is most sensitive to green and least sensitive to blue light of equal energy, green is a bright color, and blue is a dark color as conveyed by the luminance signal to monochrome TV receivers. Color-TV receivers utilize the luminance signal to produce both monochrome and color pictures.

Chrominance consists of two additional quantities: hue and saturation. Hue is the attribute of color perception that determines whether the color is red, blue, green, etc. White, black, and gray are not considered hues. Hue is presented on the vectorscope CRT as a phase angle and not in terms of wavelength. For example, red (which has a wavelength of 610 nanometers) is indicated as 104° on the standard color-phase vector diagram (Fig. 9-36) and the vector graticule (Fig. 9-37).

Saturation is the degree to which a color (or hue) is diluted by white light in order to distinguish between vivid and weak shades of the same hue. For example, vivid red is highly saturated and pastel red has little saturation. On the vectorscope, saturation is represented by the radial distance from the center (where zero saturation exists) to the end of the color vector (where 75-percent or 100-percent saturation exists for a particular color). If the burst-vector amplitude corresponds to the 75-percent marking (Fig. 9-37), the colors are 75-percent saturated. If the burst-vector amplitude corresponds to the 100-percent marking, the colors are 100-percent saturated.

In an NTSC color-television transmission system, the hue information and saturation information are carried on a single color subcarrier, at 3.579545 MHz. These signals, in modulated-subcarrier form, are called chrominance. The hue information is carried by means of amplitude modulation with the subcarrier suppressed. A subcarrier that supplies phase information is required for demodulation. No chrominance signals are present during the horizontal-blanking interval, and a sample of the subcarrier (called *burst*) is provided within this interval .

To recover the hue information, phase demodulators are employed in the vectorscope. The phase reference is a color subcarrier that is regenerated by an oscillator in the instrument. The oscillator is locked in both phase and frequency to the color-burst signal. When the VECTOR button is pressed, the vectorscope displays the relative phase and amplitude of the chrominance signal on polar coordinates. To identify these coordinates, the vector graticule (Fig. 9-37) has points that correspond to the proper phase and amplitude of the three primary colors, red (R), green (G),

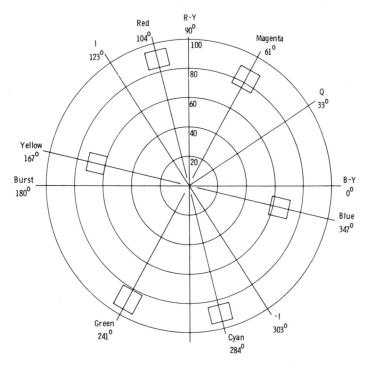

Courtesy Tektronix, Inc.

Fig. 9-36. Vector relationship among color components.

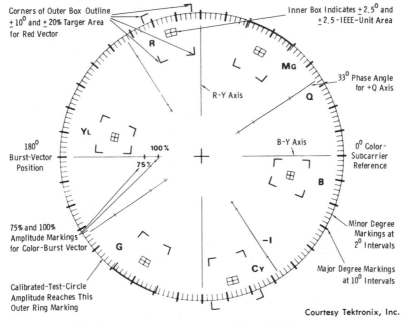

Corners of Outer Box Outline
± 10° and ± 20% Targer Area
for Red Vector

Inner Box Indicates ± 2.5° and
± 2.5 - IEEE–Unit Area

R

MG

33° Phase Angle
for +Q Axis

R-Y Axis

Q

YL

180°
Burst-Vector
Position

100%

75%

B-Y Axis

0° Color-
Subcarrier
Reference

B

75% and 100%
Amplitude Markings
for Color-Burst Vector

G

−I

Minor Degree
Markings at
2° Intervals

CY

Major Degree Markings
at 10° Intervals

Calibrated-Test-Circle
Amplitude Reaches This
Outer Ring Marking

Courtesy Tektronix, Inc.

Fig. 9-37. Internal vector graticule of Type R520 vectorscope.

and blue (B). In addition, the complements of the primary colors are indicated as follows: cyan (CY), magenta (MG), and yellow (YL).

Any errors in the color-encoding, video-tape recording, or transmission processes that change these phase and/or amplitude relationships cause color errors in the received picture. The polar-coordinate type of display has proved to be the best method for portraying these errors.

The polar display permits measurement of hue in terms of relative phase of the chrominance signal with respect to the color burst. Relative amplitude of the chrominance with respect to the burst is expressed in terms of the displacement from the center (radial dimension of amplitude) toward the point that corresponds to 75-percent (or 100-percent) saturation of the particular color being measured.

The outer boxes around the color points correspond to phase and amplitude error limits set by FCC requirements (±10°, ±20 percent). The inner boxes indicate ±2.5° and ±2.5 IEEE units. These limits correspond to phase and amplitude error limits set by EIA specification RS-189, amended for 7.5-percent setup. Fig. 9-38 shows the purpose of the small marks that intersect the I and Q axes.

The two major distortions to which the chrominance signal is subject are *differential gain* and *differential phase*. Differential gain is a change in color-subcarrier amplitude resulting from a change in the luminance signal while the hue and saturation of the original signal are held constant. In the reproduced picture, the saturation will be distorted in the areas between the light and dark portions of the scene.

Differential phase is a phase change of the chrominance signal that results from changing the luminance signal while the original chrominance signal is held constant. In the reproduced picture, the hue will vary with scene brightness. Differential gain and differential phase may occur separately or together. The causes of these distortions are chrominance non-linearities caused by luminance-amplitude variations. To measure differential phase using the Type R520 vectorscope, no graticule is needed. Instead, the trace overlay and slide-back technique using the CALIBRATED PHASE control provides the means for performing the measurement.

The IEEE graticule is used primarily for measuring differential gain and video-signal amplitude. For the measurement of video-signal amplitude, the graticule is marked in IEEE units. In standard TV practice, 140 IEEE units equal 1 volt. Hence, with the aid of the IEEE graticule, the com-

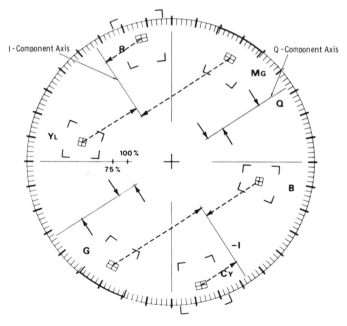

– – –► Dash line with arrow associates the color vectors with the I-axis markings.
If the Q component of the signal is absent, linear color vectors will move to their associated markings on the I axis.

——► Short arrows along the Q axis are intended to show the relationship between the Q-axis markings and the associated color vectors.
If the I component of the signal is absent, linear-color-vector points will appear at the appropriate marks along the Q axis.

Courtesy Tektronix, Inc.

Fig. 9-38. Relationships between standard color-phase vector diagram and Type-R520 vector graticule.

posite video signal will be exactly 1 volt in amplitude when the equipment is adjusted to obtain a display amplitude of exactly 140 IEEE units. Next, the graticule is used as a guide for checking and adjusting the composite video signal for the following typical proportions:

1. The white level should correspond to the graticule marking for +100 IEEE units.
2. The reference black level should correspond to the 7.5-percent setup marking.
3. The blanking level of the video signal should coincide with the graticule line for 0 IEEE units.
4. The sync-pulse amplitude should correspond to the graticule line for −40 IEEE units.

Adjustments to the encoder can be made with the Type R520 vectorscope. By use of the vector mode, the I matrix may be checked for accuracy by turning off the Q channel and observing that all six dots align with the cross marks along the I axis of the vector graticule.

Check that the 100%-75%-MAX GAIN switch is set to 75% and the GAIN control for the channel is set to "cal." Set the I chroma gain on the encoder so that the largest-amplitude (red and cyan) dots align with the I-axis cross marks. All the other dots should align with the appropriate marks, also.

Some typical causes of I-axis dot misalignment are incorrect matrixing from the R, G, and B signals to the I signal; value changes of matrix resistor(s); incorrect gain of an inverting amplifier associated with the matrix function; nonlinear amplification of the matrixed signals; or nonlinear amplification of the doubly balanced I modulators.

The remarks given for the I channel apply equally to the Q channel.

After the correct operation of the I or Q channel has been established independently, with the other channel temporarily disabled, the quadrature phasing between the I and Q channels is facilitated as follows:

1. Turn on both channels in the encoder.
2. Adjust the encoder quadrature phasing and the vectorscope A PHASE (or B PHASE) control so that all color dots lie within their respective inner boxes on the vector graticule.
3. Burst phasing of the encoder then may be adjusted so that the burst vector is at exactly 180° on the vector graticule.
4. The amplitude of the burst vector may be set in the encoder.
5. The luminance levels in the encoder now can be set to their correct amplitudes by using the Y (luminance) mode of operation.

The saturation of the displayed colors can be checked by using the 75-percent and 100-percent burst-amplitude markings on the vector graticule and by noting the position of the 100%-75%-MAX GAIN switch. The general procedure for making this check is as follows:

1. While obtaining a normal vector display, check that the GAIN control for the channel is set to "cal."
2. Set the 100%-75%-MAX GAIN switch to a position that causes the displayed color vectors to appear within the target areas. Note the location of the burst tip along the 180° axis on the vector graticule and the position of the 100%-75%-MAX GAIN switch. If both the burst-tip location and the switch position indicate 75 percent, the colors are 75-percent saturated. If the burst-tip location and the switch position indicate 100 percent, the colors are 100-percent saturated.

Differential Gain

Differential gain is a change in color-subcarrier amplitude as a function of luminance. In general, any differential gain present in the signal can be checked by using the differential-gain mode of operation of the vectorscope and by setting the 100%-75%-MAX GAIN switch to Max Gain. With a standard 10-step linearity staircase signal (Fig. 9-39) applied to the vectorscope (through the equipment to be checked), any differential gain present will cause a variation in the segment levels.

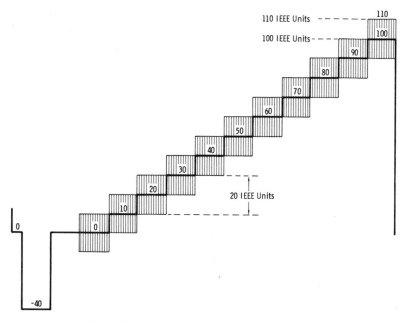

Fig. 9-39. Standard modulated stair-step signal.

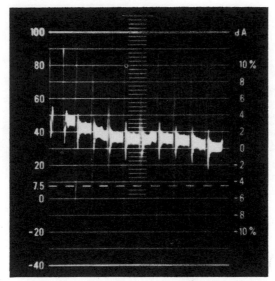

Courtesy Tektronix, Inc.

Fig. 9-40. Differential-gain presentation when modulated stair-step signal is applied.

The major divisions of the IEEE graticule represent percent of signal gain or loss when the displayed 100-unit level coincides with the 0-percent graticule lines. When the right-hand graticule marking for the scale is used in conjunction with the major and minor graticule markings (Fig. 9-40), differential-gain measurements can be made within an accuracy of 1 percent. Fig. 9-40 is an example of a differential-gain presentation using a modulated stair-step signal. The right side of the graticule is marked in percent of gain distortion. From white to black luminance levels the indicated gain distortion is 3 percent.

Differential Phase

Differential phase is a phase modulation of the chrominance signal by the luminance signal. As a result, the hue in the reproduced color picture varies with scene brightness. Differential phase is read from the calibrated phase-shift control. Approximately 1 inch of dial movement represents 1° of phase shift. The vertical deflection of the display is greatly magnified and inverted on alternate lines; this allows the use of a trace-overlay technique and the slide-back method of measuring small phase changes. The CALIBRATED PHASE control provides direct readout of differential phase. When the standard linearity test signal is used, differential phase of 0.2° can be measured.

Fig. 9-41 shows an example of a differential-phase presentation using a modulated stair-step signal. A trace-overlay technique provides excellent

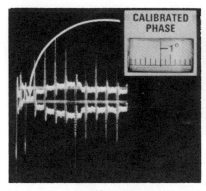 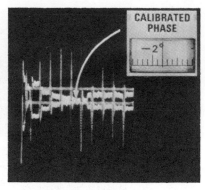

(A) First step overlayed.   *(B) Fifth step overlayed.*

Courtesy Tektronix, Inc.

Fig. 9-41. Differential-phase presentation using modulated stair-step signal.

resolution for measuring small phase changes. The change from the reference point in Fig. 9-41A (first step of stair-step signal overlayed) to the point of measurement in Fig. 9-41B (fifth step overlayed) represents 1.4° of differential-phase distortion.

R, G, B, and Y Observations

The Type 520 vectorscope provides a luminance channel that permits the separation and display of the luminance (Y) component (Fig. 9-42A) of the composite color signal. The Y component also can be combined with the outputs of the chrominance demodulators for R, G, and B displays at a line rate (Figs. 9-42B, 9-42C, and 9-42D). Amplitude measurements of color-signal components can be made with an accuracy of 3 percent.

A great deal of valuable information can be obtained by this line-sweep presentation of Y, R, G, and B. To take full advantage of such measurements, see Fig. 9-43 and Table 9-1. These review the color-bar-generator output pattern.

Note from Fig. 9-42 that the decoded R, G, and B video signals for an encoder that is properly set up have the same maximum amplitude and minimum amplitude per step. The latter falls on the 7.5-percent setup level.

All encoders, whether four- or three-channel cameras are used, convert R, G, and B to luminance for color bars. In the three-channel system, the same is true for the camera signals.

Fig. 9-44 shows errors that can be displayed by line-sweep presentations of decoded R, G, and B video on the vectorscope. Errors of this type occur in the matrix that converts the R, G, and B signals to luminance. There may be a change of actual values of matrix resistors, or there may be an improper setting of an adjustable value.

Fig. 9-44A shows decoded red video. Note that the first four steps are incorrect in value and (from Table 9-1) that green is the common factor.

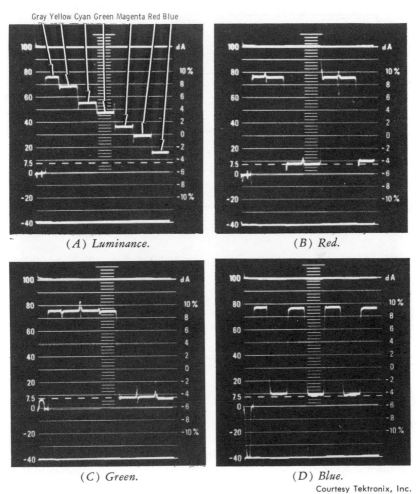

Gray Yellow Cyan Green Magenta Red Blue

(A) Luminance.

(B) Red.

(C) Green.

(D) Blue.

Courtesy Tektronix, Inc.

Fig. 9-42. Line-sweep presentations of Y, R, G, and B decoded video.

Therefore, an error in the green portion of the luminance matrix results in approximately the same error in the first four steps of the *red* or *blue* decoded video display.

NOTE: It is helpful for the technician to memorize the actual sequence of the color-bar pattern: gray, yellow, cyan, green, magenta, red, and blue, and also the primaries used to form yellow, cyan, and magenta.

Fig. 9-44B shows decoded green video. Steps 1, 2, 5, and 6 have an error of approximately the same magnitude. Note from Table 9-1 that red is now the common factor, and an error in the red part of the luminance resistive matrix is indicated.

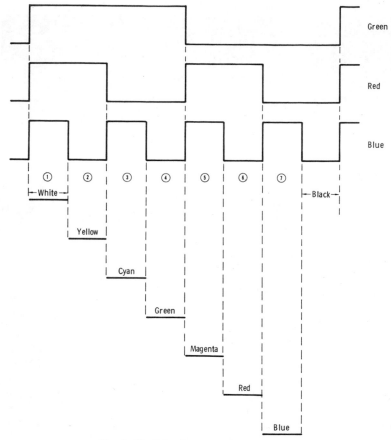

Fig. 9-43. Color-bar generator outputs.

Table 9-1.   Color-Bar Pattern

Steps	Hue	Primaries	Common Factor
1	White	R+G+B	
2	Yellow	R+G	
3	Cyan	G+B	Green
4	Green	G	
5	Magenta	R+B	
6	Red	R	Red
7	Blue	B	Blue
8	Black	0	

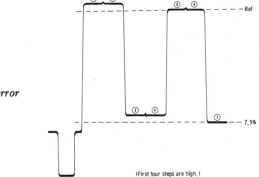

(A) Red decoded output, error
in green channel.

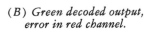

(First four steps are high.)

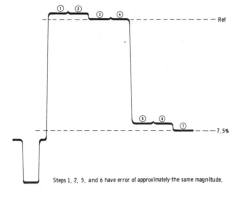

(B) Green decoded output,
error in red channel.

Steps 1, 2, 5, and 6 have error of approximately the same magnitude.

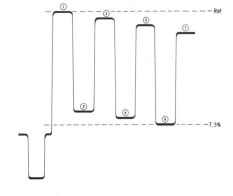

(C) Blue decoded output, error
in luminance-chrominance
amplitude ratio.

Fig. 9-44. Luminance errors in resistive matrix of three-channel encoder.

Fig. 9-44C shows a type of distortion more likely to occur in the video distribution and switching system following the encoder. Remember that the decoded output of the blue channel is made up of only 11 percent luminance. Thus, the majority of the signal contributing to decoded blue is chrominance. Therefore, the decoded blue signal is most sensitive to errors in the luminance-to-chrominance ratio. The downward slope shown in Fig. 9-44C indicates that the chrominance is low in relation to luminance. A downward slope can indicate either that luminance is excessive or that chrominance is deficient. Bear in mind that the first step (gray or white) is all luminance and zero chrominance. Thus, noting the amplitude of the first blue bar on the left indicates whether the luminance signal or the chrominance signal is in error. Then the *slope* of the decoded blue indicates the ratio of luminance to chrominance. An upward slope indicates that chrominance is excessive in relation to luminance.

Phase vs Time Delay

The time delay between two signals can be checked, because the phase difference at any particular frequency can be related to time difference. An example of this is the setup of two color cameras some distance apart. With the outputs of the cameras connected to the inputs of the vectorscope, and the proper push buttons—CH A, CH B, FULL FIELD Aϕ, and VECTOR—depressed, the two signals can be viewed together on a time-shared basis. Any time-delay difference between the two camera links will appear as a phase difference in the vector display.

This time-delay difference can be determined by noting that 360° on the graticule equals 280 nanoseconds of time. The difference can be minimized by adjusting the connecting-cable lengths so that there is no hue or phase difference from one camera to the other.

NOTE: Detailed setup techniques for the encoding system are presented in Harold E. Ennes, *Television Broadcasting: Equipment, Systems, and Operating Fundamentals* (Indianapolis: Howard W. Sams & Co., Inc., 1971). Since the present chapter has covered more advanced details of encoding, it will be helpful for the student to review Chapter 10 of that volume at this time.

EXERCISES

Q9-1. Does a "locked oscillator" require a trigger input to oscillate?

Q9-2. If you notice dots travelling from top to bottom (instead of bottom to top) of a picture monitor when viewing a color signal, what does this indicate?

Q9-3. You are adjusting counter chains before setting the subcarrier frequency. Will this frequency adjustment upset the counter adjustments?

Q9-4. In the answer to question 3, why is it stated that the subcarrier countdown is 113.75?

Q9-5. Does the burst flag have any effect on the phase of the color-subcarrier sync burst?

Q9-6. How would you measure the breezeway?

Q9-7. If you find it impossible to eliminate the flag pulses for the entire 9H interval, and you have replaced tubes with several new tubes to no avail, what are the most likely sources of the trouble? (Assume the circuit of Fig. 9-11.)

Q9-8. If you want to use 9 cycles of color-sync burst, how many peaks should you count on the interlaced color-burst pattern?

Q9-9. If Q2 of Fig. 9-13 should become open, how would this affect the burst flag?

Q9-10. In the encoder, with the Y channel (luminance, or monochrome, channel) turned off and with all inputs tied together on a color-bar signal, what pattern would you expect at the output with I and Q on?

Q9-11. In all instances where we have stated that the Y, I, or Q channel is on or off, exactly what signal is controlled?

Q9-12. What does a white-balance control actually do in the circuit?

Q9-13. Assume you are looking at the encoder output under these conditions: scope in the wideband position, standard (100 percent) bar input signal, IEEE scale calibrated for 1 volt (peak-to-peak) from −40 to plus 100 IEEE units, encoder gains properly adjusted for 0-100 units for white pulse, I and Q gains and gain ratio properly set. What is the maximum peak-to-peak value you should read for the following conditions:

(A) Y and Q channels off, I channel on.

(B) Y and I channels off, Q channel on.

Q9-14. When sync is inserted in the encoder, why must it be inserted after aperture compensation and before Y delay?

Q9-15. When the vectorscope is in the vector display mode, is luminance value indicated?

CHAPTER 10

Color Picture Monitoring
Systems

NOTE: It is imperative for the student to review NTSC color fundamentals at this time. See Harold E. Ennes, *Television Broadcasting: Equipment, Systems, and Operating Fundamentals* (Indianapolis: Howard W. Sams & Co., Inc., 1971), Chapter 2 (in particular, the text associated with Figs. 2-36 through 2-44 concerning color receivers and monitors).

The major divisions of a color monitor are shown in Fig. 10-1. Note that the primary difference between a monitor and a conventional receiver is that the monitor does not have rf and i-f stages. In the monitor, the composite color signal (E_m) is fed (normally from a distribution amplifier) directly to the decoder circuitry. The major difference between types of decoders used in color monitors is that some employ wideband color (I and Q) demodulation, whereas others use narrow-band color-difference (or X-Z) demodulators.

As shown by Fig. 10-1, if the color monitor receives E_m without inserted sync, an external sync drive is required. If the color signal has inserted sync, the monitor normally is operated on the internal sync position, using sync separated from the video itself.

10-1. ANALYSIS OF BASIC I-AND-Q DECODER

The composite color signal (E_m) is connected to J1 or J2 (Fig. 10-2), and the opposite jack is terminated unless loop-through is made to another decoder or terminated line. Signal E_m is amplified by V1. The output of V1 feeds the sync separator (V3) and, through a 1-μs delay line, the Y amplifier (V2). Delay is necessary because the bandwidths of Y, I, and Q all differ, and consequently the channels have different delay characteristics. Since Q has the narrowest bandwidth, the delay is greatest in this channel, and the Y and I signals are delayed accordingly. Time differences greater than $\frac{1}{2}$ picture element (approximately 0.05 μs) in the region close to the subcarrier frequency, and greater than one picture element elsewhere in the spectrum cause luminance and chrominance errors. Therefore, the delays in Y and I are used to cause time coincidence of all three channels for the final matrixing process.

Amplifier V2 is termed a Y amplifier because a burst trap is incorporated to suppress a considerable region around the subcarrier frequency. As previously explained, the subcarrier is an odd multiple of one-half the line frequency as well as an odd multiple of one-half the frame frequency, to minimize beats with luminance video components. However, this holds true only for still pictures. When motion is present in the scene, a small amount of beating takes place. By suppressing the region of strongest chrominance information in the Y channel, such interference is made negligible.

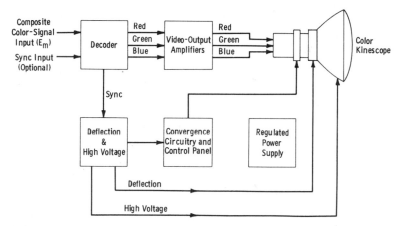

Fig. 10-1. Major divisions of a color monitor.

The encoded signal input also feeds bandpass amplifier V6 through the chroma control. Transformer T2 passes only those frequencies between approximately 2.5 MHz and 4.2 MHz, where the chrominance sidebands are present. These signals are passed to the control grids of synchronous demodulators V7 and V8. The color-sync burst is eliminated from stage V6 (in the particular circuitry under discussion) by a negative keying-out pulse from the monitor flyback transformer.

The sync-separator stage (V3) strips conventional sync from E_m and delivers it to the deflection input channel for the color monitor. A small capacitor in this same stage serves as a color-burst takeoff for delivery of separated burst to amplifier V4. This stage is keyed on at the burst interval following horizontal sync by a positive pulse from the monitor flyback transformer. At other times, the burst amplifier is biased beyond cutoff. A hue capacitor at the grid of V4 adjusts the phase of the burst sine wave to the in-phase condition (I signal) for control of the local oscillator.

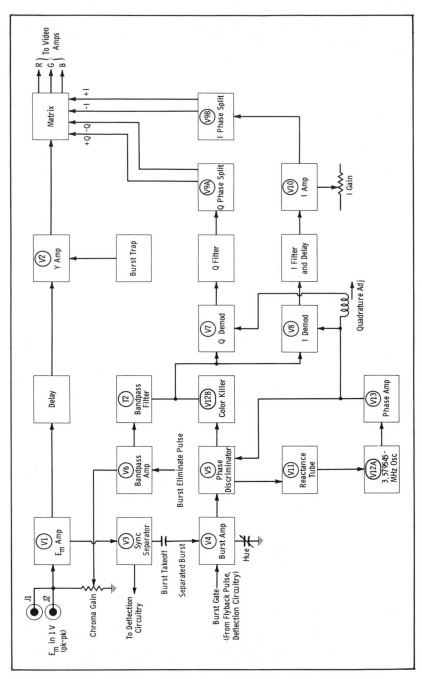

Fig. 10-2. Basic block diagram of I-Q decoder.

Phase discriminator V5 is the heart of the automatic phase-control (APC) loop. This discriminator receives the amplified color burst from V4, and the local-oscillator signal through phase-amplifier stage V13. Any phase difference between the two signals results in a dc voltage change that controls the mutual conductance of reactance tube V11. This tube, in turn, controls the phase of the crystal oscillator (V12A). When no burst is present in the applied signal (monochrome), a control voltage in stage V5 drops below a reference potential and causes the color killer (V12B) to bias the chroma circuit below cutoff.

Synchronous demodulators V7 and V8 are gated by the reinserted subcarrier voltage. Grid 3 of V8 receives the in-phase (I) subcarrier voltage. Grid 3 of V7 receives the quadrature (Q) subcarrier voltage through the quadrature-adjust transformer. Because of the 90° phase relationship, V8 is gated on when V7 is off, and V7 is gated on when V8 is off. The chrominance video is applied to the control grids of both demodulators; the Q demodulator (V7) is sensitive only to Q video components, and the I demodulator (V8) is sensitive only to I video components.

The above statement is true only for double sidebands in quadrature, as exist for the Q band to 0.5 MHz and for the I band to approximately 0.6 MHz. However, the I band has a single sideband extending to 1.5 MHz so that smaller areas of colors along the orange-cyan axis may be defined. To eliminate cross modulation, filters are used following I and Q demodulation. Luminance components also are eliminated by the filters. The chrominance channel carries hue and saturation information, and the Y channel carries the luminance information.

The Q-filter output feeds phase splitter V9A to deliver −Q and +Q signals to the matrix. The I filter and delay feeds amplifier V10 to equalize the wider-band gain with the narrower-band gain of the Q channel. Amplifier V10 then feeds the I phase splitter to deliver −I and +I signals to the matrix.

In the matrix, the minus and plus I and Q signals are added to plus Y signals. The results are shown in Table 10-1.

Automatic Phase Control (APC)

The APC loop in the system of Fig. 10-2 consists of phase discriminator V5, reactance tube V11, crystal oscillator V12A, and phase amplifier V13. The action of the APC discriminator is illustrated by Figs. 10-3 and 10-4. The purpose of this stage is to maintain a balanced condition that exists when the two input sine waves are of the same frequency and phase. When the local oscillator tends to drift, a dc correction voltage adjusts the oscillator phase and returns the circuit to a balanced condition.

The phase detector (V5) receives a signal directly from the burst amplifier (V4), and a second signal from the local oscillator (V12A) through amplifier V13. The burst signal is fed in push-pull by T1 to the cathode of V5A and the plate of V5B. Without local-oscillator voltage, and with the

Table 10-1. Decoded Signal, Fully Saturated Color Bars

Matrix	Color	Relative Amplitudes			
		Y	I	Q	Total
	White	1.0	0	0	1.0
	Yellow	0.89	+0.30	—0.19	1.0
Red	Cyan	0.70	—0.57	—0.13	0
+0.95 I +Y	Green	0.59	—0.26	—0.33	0
+0.63Q	Magenta	0.41	+0.26	+0.33	1.0
	Red	0.30	+0.57	+0.13	1.0
	Blue	0.11	—0.30	+0.19	0
	White	1.0	0	0	1.0
	Yellow	0.89	—0.09	+0.20	1.0
Green	Cyan	0.70	+0.16	+0.14	1.0
—0.28 I +Y	Green	0.59	+0.08	+0.33	1.0
—0.64Q	Magenta	0.41	—0.08	+0.33	0
	Red	0.30	—0.16	—0.14	0
	Blue	0.11	+0.09	—0.20	0
	White	1.0	0	0	1.0
	Yellow	0.89	—0.36	—0.53	0
Blue	Cyan	0.70	+0.66	—0.36	1.0
—1.11 I +Y	Green	0.59	+0.30	—0.89	0
+1.71Q	Magenta	0.41	—0.30	+0.89	1.0
	Red	0.30	—0.66	+0.36	0
	Blue	0.11	+0.36	+0.53	1.0

balance control properly adjusted, the currents (I_A and I_B) through the V5A and V5B branches are equal and 180° out of phase. Therefore, the voltage developed across C3 is zero, and no correction voltage is applied to the grid of reactance tube V11.

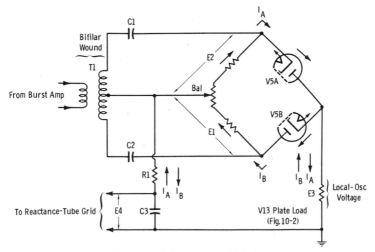

Fig. 10-3. Functional diagram of APC discriminator.

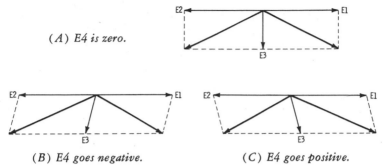

(A) E4 is zero.

(B) E4 goes negative.   (C) E4 goes positive.

Fig. 10-4. Vector diagrams for APC discriminator.

The phase relationship of the voltages in the balanced condition is shown in Fig. 10-4A. As shown by this diagram, the vector sums E1 + E3 and E2 + E3 are equal. Since I_A and I_B are equal, no voltage is developed across C3, and E4 is zero.

If the phase of the local-oscillator signal drifts, E3 will no longer be exactly 90° from the burst signal, as shown in Fig. 10-4B (E3 lags normal phase). In this case, I_A increases relative to I_B, and capacitor C3 charges negatively. The resulting E4 of negative value is applied to the reactance-tube grid to correct the oscillator phase and rebalance the discriminator currents. If E3 leads the normal phase, as indicated in Fig. 10-4C, current I_B increases relative to I_A, and E4 becomes positive.

Bandpass Amplifier and Color Killer

Fig. 10-5 is a partial schematic diagram of a typical monitor and illustrates terminations and circuits associated with the delay line. It also shows a point of burst takeoff that differs from that of Fig. 10-2.

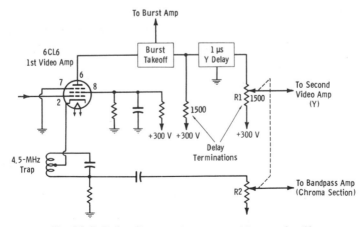

Fig. 10-5. Delay-line terminations and burst takeoff.

Fig. 10-6 shows the first stage of the chroma section. Control R2 of the chroma input is ganged as a second section of R1 of the luminance input (Fig. 10-5). Potentiometers R1 and R2 make up the overall picture-contrast control. Since the proper ratios of Y-channel gain to chroma-channel gain must be maintained throughout the system, these controls operate together in this particular arrangement. Proper balance of respective channel gains then may be adjusted by individual gain controls later in the chroma channels.

The amplified composite signal appears in the plate circuit of the bandpass amplifier, where a filter restricts the bandpass to approximately 2.3 to 4.2 MHz. The color-carrier sidebands and the higher Y-signal frequencies are passed, but the lower frequencies, which include the horizontal- and vertical-sync pulses, are eliminated.

The grid return of this amplifier is through R5 and C5. This network constitutes a bias time constant and is shunted by the color-killer triode through a separate winding on the horizontal-output transformer. The grid of the 6BL7 is returned to plus 300 volts and to a point in the color-sync phase discriminator. During monochrome transmission, no color burst is transmitted. The phase-discriminator circuit is inoperative, and no negative voltage is supplied to the grid of the color killer. The color killer con-

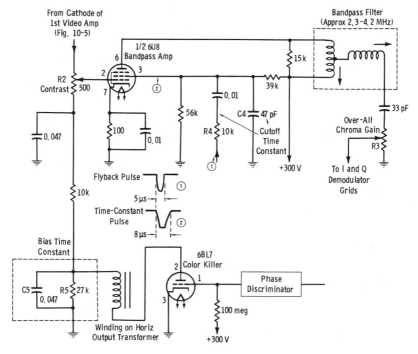

Fig. 10-6. Bandpass amplifier and associated circuits.

ducts because of the positive pulses applied to the plate of the tube. This condition results in a charge on C5. The time constant of C5 and R5 is such that a negative voltage sufficient to cut off the bandpass amplifier is maintained between conduction periods. By cutting off the bandpass amplifier, the possibility of spurious chroma response during monochrome transmission is eliminated.

For color signals, in which the color-sync burst is transmitted, the grid of the color killer is held negative by a potential developed in the phase discriminator. Since the tube then cannot conduct even with a positive plate, C5 is not charged, and the bandpass amplifier functions normally.

It is advantageous to eliminate the color-sync burst from the I and Q demodulation circuits and subsequent chroma amplifiers. In the circuit of Fig. 10-6, this is accomplished by feeding a negative pulse to the screen of the 6U8 at horizontal-flyback time. Since the flyback pulse is only approximately the duration of horizontal sync (5 μs), it is necessary to delay or widen the keying pulse so that the screen is held negative past the occurrence of the color-sync burst. In this circuit, R4 and C4 constitute a time constant sufficiently long that the charge of C4 through R4 holds the screen negative until slightly after the end of the color burst, when active line scanning starts.

Burst Separator and Amplifier

It is necessary to separate the color-sync burst from the composite video signal in order to drive the synchronizing section of the chrominance circuits. One method for doing this is illustrated in Fig. 10-7A. The control grid of the burst amplifier receives the color burst through the tuned takeoff transformer in series with the plate of the first video amplifier (in this particular circuit). Amplification of only the burst is assured by gating the cathode of the 6U8 at the proper time during the blanking interval. During the positive interval of the flyback pulse, the cathode remains too far positive to allow conduction of the tube. During flyback time, the pulse is negative and reaches the conduction potential just before burst time. The cathode is held negative by the gate pulse (hence the tube is conducting) until just after the end of the burst interval. The gated color burst is amplified and fed to the phase discriminator through the tuned plate transformer.

Fig. 10-7B shows some waveforms that illustrate the effect of proper burst-transformer time constant, L/R, on the pulse-passing characteristic. So long as L/R is much greater than the pulse duration, which is 2.5 microseconds, the color-burst envelope is passed without distortion. Should L/R become less than this duration, with a radically increased R or decreased L, the envelope is differentiated, and the color burst is lost.

Another method for extracting the color burst is shown in Fig. 10-8. In this example, a negative gating pulse is fed to the grid of a keyer, is amplified, and is inverted in polarity. The resultant signal is fed to the grid of a burst-amplifier stage. The cathodes of the burst amplifier and the keyer are connected together. The heavy conduction of the keyer tube during scan time produces at the cathode a positive voltage that is sufficient to cut off the burst amplifier. The positive pulse that is fed to the grid of the burst amplifier is sufficient to overcome the bias voltage applied to the cathode, and the tube conducts. Thus, the color burst is extracted from the composite video signal and is passed to the phase-detector circuit.

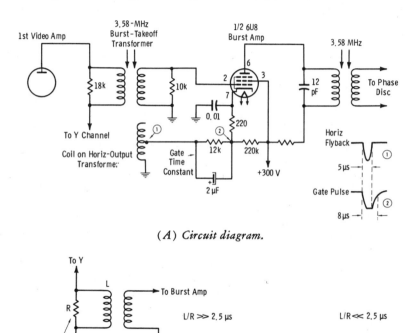

(A) Circuit diagram.

(B) Burst waveforms.

Fig. 10-7. A gated burst amplifier.

Fig. 10-8 includes a circuit that can be used to kill the color-sync burst in the bandpass amplifier. The triode section of the 6U8 is employed as a cathode follower to feed the grid of the pentode section through a 1N34 diode. For the duration of the negative gating pulse, the positive terminal of this diode is held too far negative to allow conduction, and grid 1 of the pentode receives no signal. During intervals between gating pulses, the video, which is of positive polarity at this point, is passed and amplified. Since the gating pulse usually is taken from the horizontal-deflection

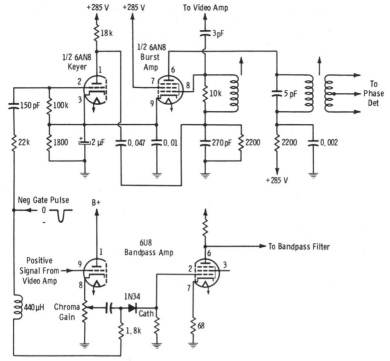

Fig. 10-8. Gating of burst amplifier and bandpass amplifier.

system, it should be realized that the stability of the horizontal AFC is of great importance for color receivers. The phase of the gating pulse must not change radically over the normal range of the horizontal-hold control.

Color Sync and Phase

The output of the burst amplifier feeds a phase-detector circuit that also receives a signal from the local 3.58-MHz oscillator. The phase detector has an output voltage that indicates by changes in polarity and magnitude any difference in phase of the two signals. This correction voltage is fed to a conventional reactance-tube circuit that locks the local oscillator on frequency and at a specific phase relationship with the transmitted color burst.

Before looking at typical circuits in this section of the color monitor, let us review system requirements for the I and Q circuits. Fig. 10-9 aids in this review. Remember that a definite phase relationship is necessary so that I (wideband color) lies along the orange-cyan axis of the color triangle.

The voltage induced in the secondary of the transformer shown in Fig. 10-9A lags the primary reference-burst current by 90°. This is conventional transformer action. Since this secondary is center tapped, the voltages at opposite ends are 180° apart. Thus, E1 and E2 are 180° apart and in quadrature with the reference burst, as shown by the phase diagram in Fig. 10-9A.

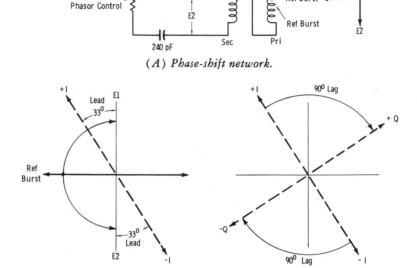

(A) *Phase-shift network.*

A 33° lead at phasor control produces I axis for I demodulator.

A 90° lag from I will produce Q axis for Q demodulator.

(B) *Shift to I axis.*   (C) *I-Q relationship.*

Fig. 10-9. Phasing of cw signals for I-Q demodulation.

We may now see that if voltages E1 and E2 are exactly in quadrature with the color-burst phase, adjustment of the phasor control (R-C combination that produces a leading phase angle) to 33° will cause the output voltage to lie along the I axis (Fig. 10-9B). If E1 and E2 are not exactly in quadrature with the reference burst (as is usually the case because of the leakage reactance), the phasor control covers an adequate range to provide proper phasing. This phase-shifted voltage is fed to the phase discriminator, which compares it with the reference burst. A correction voltage then holds the local oscillator at this phase angle (I axis), as described in the following.

Observe in Fig. 10-9C that a signal lagging the I axis by 90° will supply a carrier along the Q axis. The importance of this relationship lies not so much in practical adjustment of monitor circuits as in the fact that it permits a more accurate visualization of the system function. A phasing control, ordinarily on the front panel, properly phases the local oscillator so that the locally injected carriers fall on the I and Q chrominance-signal axes. Improper adjustment causes inaccurate color reproduction such as reds going blue, blues going green, and, most important, wrong flesh tones.

Fig. 10-10 shows one type of APC loop for color synchronizing. Diodes V1 and V2 constitute the phase-discriminator circuit for comparison of the local-oscillator signal through T1 with the transmitted burst signal through T2. Note that the diodes are connected to conduct on the same half-cycle of the signal. No conduction occurs without voltage from the burst amplifier.

During conduction, the plate of V2 goes negative with respect to ground. This point connects to the color-killer grid (Fig. 10-6) to prevent that tube from conducting during color telecasts.

Transformer T1 and the phase control serve the function described previously. The discriminator then acts to supply a correction voltage for the reactance tube so that any deviation from the I axis is corrected as in conventional AFC circuits.

The in-phase I carrier is taken from the secondary of the transformer in the cathode circuit of the crystal oscillator. The Q amplifier is fed from the cathode of the oscillator.

Actually, APC circuits differ widely. Some monitors use lumped resistance-inductance circuits for the quadrature phasing. The principles, however, remain the same.

A receiver or monitor that demodulates on the R−Y and B−Y axes differs from the one just described in that no 33° phase relationship with the reference burst is necessary. The B − Y signal lies along the sine axis. It is then only necessary to feed the R − Y demodulator through a 90° phasor to demodulate the red color-difference signal. This particular feature does not aid in simplification, since 33° networks are quite simple (Fig. 10-10); and whether this 33° phasor is used or not, a phasing control must be used for accurate placement of the local-carrier phase.

Synchronous Demodulators

Fig. 10-11 illustrates one method of feeding the synchronous demodulators and the matrix. The chrominance signal is fed to the control grids of the demodulators. The suppressor grid of the I demodulator is driven by the in-phase cw, and the suppressor grid of the Q demodulator is driven by the quadrature cw. The chrominance signal, of course, contains both I and Q color information.

In the I demodulator, the output contains the vector sum of the I-signal sidebands from the chrominance channel and the in-phase cw. The Q-signal sidebands in the chrominance signal, since they are in quadrature with the I sidebands, produce zero output in the I demodulator. This is the action of a synchronous demodulator; the output is zero for components 90° in phase from the cw drive.

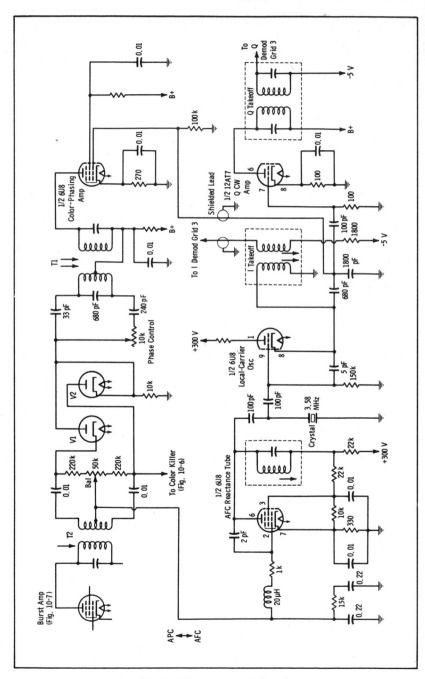

Fig. 10-10. Color-synchronizing circuit.

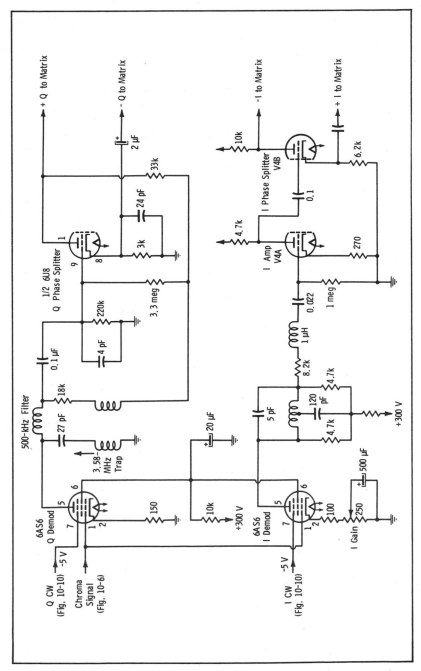

Fig. 10-11. Demodulators and phase inverters for I and Q signals.

The output of the Q demodulator contains the vector sum of the quadrature cw and the Q-chrominance sidebands. The single sideband of the I signal above 500 kHz produces a quadrature component, and therefore introduces cross talk into the Q-demodulator output. This is the reason for the 500-kHz filter in the plate load of the Q demodulator in Fig. 10-11.

Note that the I demodulator has a gain control and an extra amplifying stage. Since the Q sidebands are equal (double sidebands), the Q gain is twice that of the I channel above 500 kHz. The extra I amplifier compensates for this difference, and the gain control allows exact adjustment for proper gain ratio.

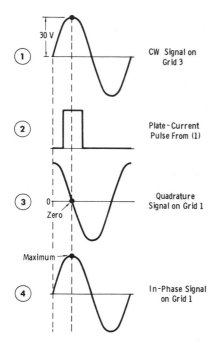

Fig. 10-12. Action of synchronous demodulator.

Fig. 10-12 illustrates the basic operation of a synchronous demodulator. Note from Figs. 10-10 and 10-11 that the suppressor grids (grid 3), which receive the cw signals, are biased at a negative 5 volts. The peak voltage of the cw signals may be approximately 30 volts. Parameters of the circuits are such that the tube conducts heavily on peak regions of the cw signal even though there is no chrominance modulation on grid 1.

Tube conduction with no chrominance signal is shown in waveform 2 of Fig. 10-12. The instantaneous signal voltage on grid 1 determines the amount of that current that reaches the plate. Thus, the instantaneous voltage on grid 1 is multiplied by the instantaneous voltage on grid 3. This action is sometimes referred to as *product demodulation*.

On grid 1, signals that are in phase go positive when grid 3 goes positive. This results in heavy plate current on the positive peaks. The pulses follow the modulation on grid 1.

A quadrature signal on grid 1 (waveform 3 of Fig. 10-12) is zero at the time of the plate-current pulse; therefore no signal is produced in the output. An in-phase signal (waveform 4 of Fig. 10-12) adds vectorially to the cw signal.

If the cw signal is not of the same frequency as the chrominance signal, the usual heterodyne beat frequency results. This undesired beat component is modulated by the amplitude variations on grid 1, and as a result there are spurious products in the output circuit. The fact that these signals pass through the low-pass filters into the chrominance amplifiers emphasizes the importance of proper APC loop-phasing adjustments.

Note that the plate of the Q phase splitter (Fig. 10-11) is dc coupled to the matrix, whereas the cathode is ac coupled. This is necessary to isolate the plate and cathode dc voltages in the matrix. Similarly, the $+I$ signal is ac coupled, and the $-I$ signal is dc coupled to isolate these separate voltages.

Matrixed Signals

The matrix extracts the color-difference signals ($R-Y$, $G-Y$, and $B-Y$) from the filtered output of the I and Q demodulators. These signals, along with the luminance, or Y, signal, excite the color picture tube with instantaneous values such that the overall function matches the corresponding scanned point at the studio camera.

The I-Q matrix at the sending end reduces the $R-Y$ component to:

$$0.877\ (R-Y),\ \text{or}\ \frac{R-Y}{1.14} \qquad (\text{Eq. 10-1.})$$

Similarly, the $B-Y$ component is reduced to:

$$0.493\ (B-Y),\ \text{or}\ \frac{B-Y}{2.03} \qquad (\text{Eq. 10-2.})$$

In this manner, both the I and Q channels contain some of both color-difference components so that only a two-phase color signal is required for three chrominance primaries. It is the purpose of the matrix in a receiver or monitor containing an I-Q demodulator to recover $1.14\ (R-Y)$ and $2.03(B-Y)$. This means that the color-difference components are recovered in their original forms before I and Q matrixing at the transmission end; therefore, the $R-Y$ gain is 1.14 and the $B-Y$ gain is 2.03. In the wideband monitor, this is achieved in the matrix operation.

This action is emphasized in Fig. 10-13, which shows the $R-Y$ and $B-Y$ components multiplied by 1.14 and 2.03, respectively. We may now note the values of I and Q necessary to extract the color-difference components existing before modulation. These are found to be:

$$R - Y = 0.95I + 0.63Q \qquad \text{(Eq. 10-3.)}$$
$$B - Y = -1.11I + 1.71Q \qquad \text{(Eq. 10-4.)}$$

We know that Y is:

$$Y = 0.30R + 0.59G + 0.11B \qquad \text{(Eq. 10-5.)}$$

Rearranging equation 5:

$$G - Y = -0.51\,(R - Y) - 0.19\,(B - Y) \qquad \text{(Eq. 10-6.)}$$

This is the action performed in monitors in which the color-difference signals are demodulated directly in the matrix to extract the $G - Y$ component. In terms of I and Q for the wideband color receiver, substituting Equations 10-3 and 10-4 into Equation 10-6 gives:

$$G - Y = -0.28I - 0.64Q \qquad \text{(Eq. 10-7.)}$$

Equation 10-7 is also shown in Fig. 10-13 for matrix operation necessary to extract the $G - Y$ color-difference signal in terms of I and Q.

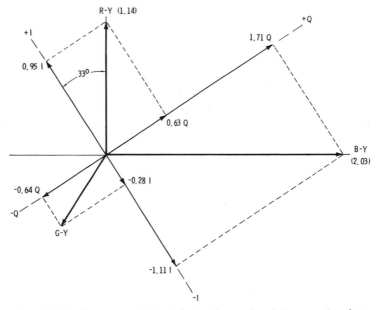

Fig. 10-13. Conversion of I and Q signals to color-difference signals.

The Y signal and both polarities of the I and Q signals are fed to the output matrix to recover the original R, G, and B signals. This matrix must receive the correct *ratios* of signals from the phase-splitter outputs. This depends on the adjustment of the chroma control, the quadrature adjustment of the demodulators, and the setting of the I-gain control.

As previously discussed, the I channel requires an extra amplifier stage to equalize its wideband gain relative to the narrow-band gain of the Q channel. The I-gain control provides control of this amplification.

The result of the combined matrix function is shown by Table 10-1. For example, in order to recover the original red signal, +0.95I and +0.63Q are added to the luminance signal of +0.30 to give unity signal at the red output.

Conditions for proper matrixing are first determined by adjustments as follows:

1. With input representing I and Q signals, the hue control is adjusted for zero Q response in the I channel. Remember that the I and Q sidebands vanish for a white signal, which represents the zero axis. Signals above white are positive, and signals below white are negative. When hue is properly adjusted, the local oscillator is locked to the I-carrier phase, and *no* Q response (either positive or negative) is obtained in the I channel.
2. The quadrature-adjust control must be adjusted for zero I response in the Q channel. This now assures that the demodulators are gated properly with reference to their respective axes, and will not deliver contaminated signals to the matrix.
3. The chroma and I-gain controls must be adjusted so that the blue output has zero I response and maximum Q response. This results in proper luminance-to-chrominance gain ratios so that the fixed matrix network results in correct output signals.

We may now examine the actual dynamics of the matrix function. Fig. 10-14 is a simplified diagram of this section. Relative luminance (Y) and chrominance input levels are shown at the left of the drawing. The Y-channel input is taken as unity, with voltage developed across R1, R2, and R3 in series with their respective output adder resistors, R7, R8, and R10. The chroma and I-gain controls are adjusted so that +Q is 1.71 times Y, and −I is 1.11 times Y.

Now refer to Fig. 10-11, and note that the signal +Q appears at the plate of the Q phase splitter. The plate load is 33k in parallel with R5 (27.2k) in parallel with R4 (10k). This is an effective plate load of about 6k. The −Q signal appears at the cathode. The cathode load is 3k in parallel with R12 (10k), or 2.3k. Thus, the plate provides 6/2.3 or 2.61 times the gain at the cathode. With the chroma gain adjusted so that +Q is 1.71 times Y, −Q is 1.71/2.61, or 0.64, times Y, which appears at one input, R12, of the green matrix. Observation of the required matrix proportions for I and Q in Fig. 10-13 and Table 10-1 shows that the blue matrix requires +1.71Q, and the green matrix requires −0.64Q. These proportions are indicated in Fig. 10-14. Note that the +1.71Q is applied in parallel to blue-matrix resistor R4 and red-matrix resistor R5. Resistor

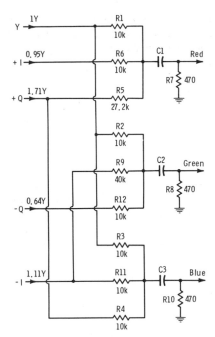

Fig. 10-14. Matrix for color monitor.

R5 provides 2.72 times the resistance of R4. Therefore R5 provides a Q-signal ratio of 1.71/2.72, or 0.63, as required for the red matrix.

For the I channel, −I is available at the plate of V4B (Fig. 10-11), and +I is available at the cathode of V4B. The I gain is adjusted so that the signal at the V4B plate is 1.11Y. As in the Q channel, the total plate and cathode loads are proportioned so that when the I gain is adjusted such that −I is 1.11Y, then +I is 0.95Y for the +I input at the red matrix.

Each output resistor is the adder across which appears the algebraic sum of the inputs to the three preceding resistors. Only those signals containing red appear across R7, since the green and blue components were cancelled by the polarity and amplitude relationship of the Y, I, and Q signals fed to the red channel. The amplitude of the red voltage determines the degree of saturation of the red component. The same principle holds for the green and blue outputs.

It will be noted by serious readers who have followed the actual computations in preceding examples that slight manipulation of values was made to result in required values. The reason is that, in the interest of clarity, absolute NTSC numerical values have not been used, and numbers beyond two decimal places were disregarded. For example, we have used the popular I and Q proportions:

$$I = 0.60R - 0.28G - 0.32B$$
$$Q = 0.21R - 0.52G + 0.31B$$

whereas absolute values are:

$$I = 0.599R - 0.278G - 0.321B$$
$$Q = 0.213R - 0.525G + 0.312B$$

10-2. COLOR PICTURE-TUBE CIRCUITRY

The color picture tube to be discussed is the three-gun, shadow-mask type most common at the time of this writing. The three guns are physically fixed in position relative to mask holes and phosphor dots on a screen. One gun and grid assembly is arranged to strike only phosphor dots that emit red light when struck by the beam; one gun strikes only green dots; and the remaining beam strikes only blue dots. When the beams are properly converged, one picture element is a triangular arrangement of three adjacent dots. When the observer is at normal viewing distance, the dots are not visible as such, and a colored image of satisfactory definition is reproduced. The phosphor screen contains about 7000 dots per square inch. The screen is aluminized, and no ion trap is necessary.

By the nature of this basic functioning of the tube, the physical construction of the gun and electrode assembly must be precisely controlled. Adjustment magnets, coils, and voltages then allow sufficient correction for any slight physical misalignment and characteristics of electrical deflection.

Each gun assembly contains a cathode, a control grid (grid 1), and a screen grid (grid 2). Grid 3 (focus electrode) and grid 4 (convergence electrode) are common to all three assemblies. The ultor is the high-voltage electrode, and this potential is also connected to a wall coating on the large bell of the envelope. External to the picture tube are beam-positioning magnets, the color-purity coil, the deflection yoke, and the field-neutralizing coil (when used).

Fig. 10-15 illustrates the function of one gun assembly of the tri-gun arrangement. Numbers along the top designate the various forces acting on the beam. The voltages shown are typical values for a 17-inch tube.

Obviously, the ultimate goal of the electron beam is the high voltage of the ultor, or phosphor screen. Force 1 is one of varying repulsion depending on the potential of grid 1 relative to that of the cathode. Force 2 is an accelerating force with strength dependent on the grid-2 (screen) potential, which is adjustable.

There is one beam-positioning magnet for each gun. The magnetic field adjusts the individual beam relative to the other beams transversely to the direction of beam travel (force 3). An adjustable current through the color-purity coil creates an electromagnetic field with the same action (force 4), except that all three beams are influenced by this current. Its purpose is to assure that all beams travel straight down the neck of the tube toward the mask and phosphor screen. Construction of' the tube is

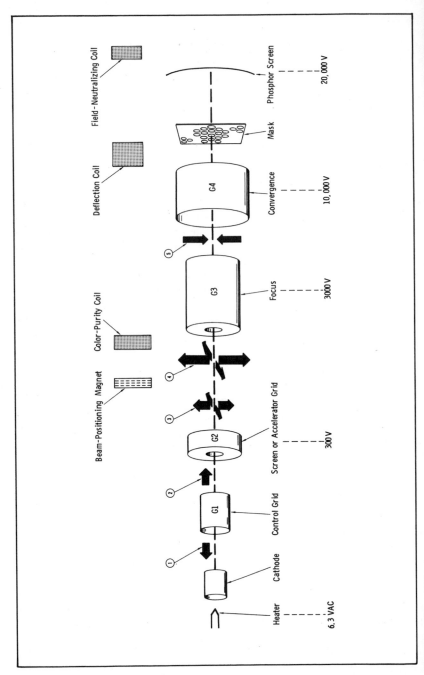

Fig. 10-15. Functional diagram of one gun in three-gun CRT.

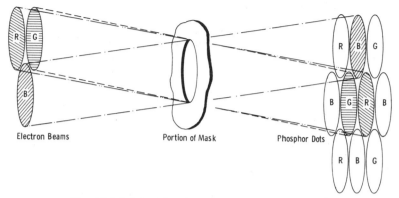

Fig. 10-16. Principle of mask in color picture tube.

such that when this occurs, the beam for red strikes only red phosphors, the beam for green strikes only green phosphors, and the beam for blue strikes only blue phosphors. This is maximum color purity.

Force 5 compresses (focuses) the beam; this force depends on the ratio of the grid-3 and grid-4 potentials. It will be noted that variation of the voltage on grid 3 (focus grid) causes a variation of beam convergence, and a change in the voltage on grid 4 causes a variation of focus.

The mask contains a large number of holes (Fig. 10-16), which allow the beams (when properly converged) to strike only the proper phosphor elements. Proper convergence exists when all three beams converge at the same hole, then diverge to their respective phosphors. The mask and phosphor screen of the tube are flat (in 70° deflection tubes). Fig. 10-17 illustrates the fact that under beam deflection, the point of convergence traces an arc. To maintain both convergence and focus at the portions of the raster away from the central area, horizontal and vertical dynamic waveforms are applied to grids 3 and 4.

The deflection coil performs the same function as it does for monochrome picture tubes. The yoke is huskier because of the greater deflection current required for color tubes.

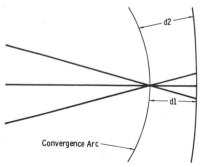

Fig. 10-17. Variation of convergence with deflection.

Current in the field-neutralizing coil around the face of the tube minimizes the effects of stray magnetic fields. Control of the beams must be precise; even the magnetic field of the earth must be nullified.

Fig. 10-18 shows the circuits associated with a three-gun color picture tube. Each grid receives excitation from its specific video channel, is dc controlled, and is returned through individual background controls. The cathodes are driven positive during vertical-blanking intervals by pulses from the vertical-output transformer. This assures complete retrace blanking. The dc voltage of each screen can be adjusted individually to aid in balancing phosphor efficiencies.

The convergence-amplifier grid receives waveforms from the cathodes of the horizontal- and vertical-deflection amplifiers. The shapes of the waveforms at these points are very nearly parabolic. The controls that are shown provide any correction necessary to shape the applied voltages properly. The amplified output modulates the dc applied to the convergence electrode. Since the distances from the guns to the top, bottom, left, and right regions of the screen are greater than the distance to the center of the raster, the beams from the three guns would not register properly without correction. Dynamic convergence corrects the applied voltage in step with the scanning position of the beams.

The dc voltage applied to the focusing electrode is adjustable for optimum control of the scanning-spot size from each gun. The focusing electrode also receives a portion of the parabolic waves to maintain focus during scanning.

The picture-tube high voltage is regulated by a shunt-regulator tube across a flyback voltage-doubler power supply. During a black picture and blanking intervals, when no beam current is drawn, the regulator becomes the power-supply load. For a maximum white picture, the kinescope becomes the load, and the regulator tube absorbs very little power. This method maintains a constant load on the high-voltage power supply and allows good regulation for the picture-tube anode.

In addition to the contrast (video gain) control, controls that affect picture quality are the brightness and background controls. Usually, dc restoration is provided by a triple-diode tube. Note from Fig. 10-18 that the anodes of this tube are connected to the arms of variable potentiometers in the background control circuitry. The anode of the red-channel dc restorer is returned to the arm of a brightness control, to which the blue and green background controls are returned through the picture-tube cathode circuitry. Adjustment of the brightness control makes the picture-tube grids more or less negative relative to the cathode; hence this control determines the beam current and raster brightness.

Assume the blue-background and green-background controls are at the extreme counterclockwise position. This places the three grids at the same bias. Because of differences in the efficiencies of the three phosphor dots of each element, the background of a picture that should be white would

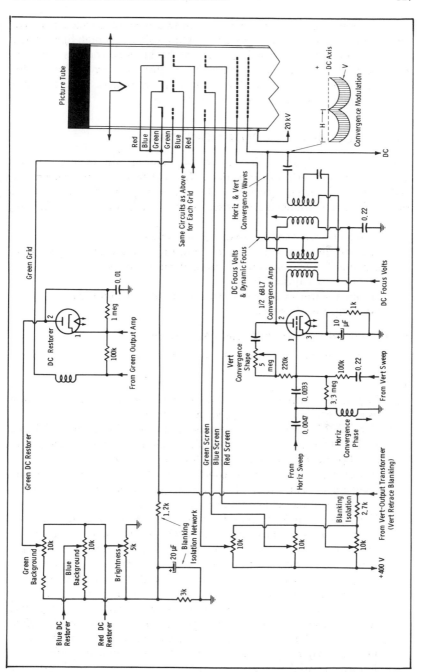

Fig. 10-18. Dynamic-convergence circuits and screen and background controls.

be tinged with a color component. The green-screen, blue-screen, and red-screen controls normally are adjusted for white balance on high lights. Since restorer action refers to blanking level, the background controls normally are adjusted for white balance with the brightness control adjusted for low raster brightness. This method assures proper balance over the range from low lights to high lights of the video signal.

10-3. ADJUSTMENT OF COLOR MONITORS

Following is a basic outline of typical monitor adjustments that are applicable to color monitors in general. We will then consider refinements and additions in modern circuitry. The proper sequence for check-out of the monitor, or adjustments after installation of the kinescope, is as follows:

1. Horizontal-drive and high-voltage adjustment
2. Size and linearity adjustments (adjustment for correct aspect ratio and linearity by conventional bar-generator methods)
3. Color-purity adjustment
4. Screen adjustment for white, and color-balance adjustment
5. Convergence adjustment

Horizontal-Drive and High-Voltage Adjustment

To adjust the horizontal drive and the high voltage, perform the following steps:

1. Plug a 0-1.5 mA or 0-2 mA meter into the regulator-current jack usually provided on the high-voltage chassis.
2. Turn the brightness and contrast controls fully counterclockwise (off). The regulator current should be close to 1 milliampere, or as specified in the instruction book for the monitor.
3. Adjust the horizontal-drive control (rear of chassis) for maximum regulator current. From 0.8 to 1 mA (or slightly more) is normal (with brightness and contrast controls off).
4. If at least 0.8 mA of regulator current is not available with the adjustment in step 3, measure (with a VTVM and high-voltage probe) the potential at the ring on the flyback rectifier. *Lower* the voltage with the high-voltage-adjust control (rear of chassis) until the regulator current reaches at least 0.8 mA. Note that as the ultor voltage is lowered, the regulator current *increases,* and vice versa. *It is more important to obtain good raster brightness without blooming than to have an absolute 20 kV (or other specified voltage) on the ultor.* Because of differences in color kinescopes, a variation of several kilovolts may be observed between tubes for optimum performance.
5. Return the brightness and contrast controls to the normal operating positions.

6. By conventional bar-generator methods, adjust the aspect ratio of the raster and the linearity of the sweeps.

Screen-Purity Adjustment

To avoid shading of the color screen, it is necessary for the individual gun emissions to be properly aligned down the center of the tube structure. If color impurity is noted in the raster, proceed as follows:

1. Loosen the yoke-assembly bracket screws and slide the yoke as far to the rear as possible. Remove the alignment magnets from the neck assembly. Turn the blue-screen and green-screen controls fully counterclockwise. Turn the red-screen control fully clockwise. Since the yoke has been slid back from its normal position, deflection is not complete, and the red beam should be coloring the screen except at the extreme sides. Disregard coloring around the edges at this time. Adjust the color-purity control (or magnet) until approximately the same area of impurity exists around all sides of the raster. A slight rotation of the color-purity coil (or magnet) may be necessary in conjunction with coil-current adjustment. When a color-impurity magnet is used, tabs normally are provided for positioning.

2. Next, slide the yoke forward until the entire screen is red with maximum purity. The yoke may *not* need to go entirely to the front of the slot. If the purity is good except at one extreme side or corner, adjust the field-neutralizing control (when provided) for maximum purity. This control determines the current through the coil at the outer front of the kinescope.

3. Check the blue and green screen purities by turning the other screen controls off and the one to be checked on. A slight adjustment of color purity and/or field-neutralization may be needed to effect the best compromise in screen purities. Tighten the yoke-assembly bracket screws and replace the alignment magnets.

NOTE: It may be necessary to demagnetize the picture tube with a degaussing coil to obtain best color purity.

Color Balancing

Turn the contrast control to minimum (fully counterclockwise). Set the brightness control for maximum raster brightness. Adjust the red-screen, blue-screen, and green-screen controls to obtain a low-brightness white raster.

Feed a black-and-white test-pattern signal to the monitor, and be sure the three channel gains at the camera control are equal, as indicated on the CRO's. Bring the monitor contrast control up slightly over midway, and adjust the brightness for a reasonable presentation. Adjust the monitor green-video gain and blue-video gain until the reference white high light of the test pattern prevails on the monitor screen.

Leave the contrast control at its present setting, and turn the brightness down to low level. Adjust the blue-background and green-background controls (dc-restorer controls) for proper low-brightness white.

Convergence

Check convergence by placing the normal/convergence switch in the convergence position. This ties all three video channels together. With a test-pattern signal (crosshatch or dots), all portions of the image should be black or white. Any color fringing is an indication of misconvergence. Remember that camera misregistration would not contribute to color fringing in this case, since a common video signal is being amplified by the monitor. If slight misconvergence is noted, the dc-convergence control should bring the beams into convergence. When serious misconvergence or bad fringing at the sides and corners requires adjustment of the dynamic-convergence controls, tube or circuit malfunctioning may be indicated. After maintenance, it is desirable to adjust the convergence controls with a dot generator as described below.

Place the normal/convergence switch in the normal position. Feed a dot-generator signal to the monitor. (NOTE: Use a pattern in which the dot size is no more than two raster lines, not the large dot size of ten raster lines.) Adjust the height, width, and linearity controls for proper aspect ratio and linearity. The dots may be used for this purpose, or the more conventional bar generator may be used.

Fig. 10-19 illustrates the basics of convergence. The kinescope is installed with the blue gun on top, the green gun to the left, and the red gun to the right as viewed from the rear (Fig. 10-19A). Fig. 10-19B shows the necessary conditions for proper convergence. With the correct dc-convergence voltage, the beams converge at the proper hole in the mask, then diverge to adjacent dots forming the triangular picture element, as shown. The dc-convergence control has the greatest effect in the central picture area.

Fig. 10-19C represents a front view of one picture element properly converged. The phosphors are so closely spaced that from a normal viewing distance the color dots blend into one white dot. If the dc-convergence voltage is too high, the blue dot will be high, the red dot will be separated several elements to the left, and the green dot will be several elements to the right (from a front view of the screen—Fig. 10-19D). Note that misconvergence does not cause loss of any primary color; the beams always strike their intended phosphors, but they may be as much as ¼ inch or more from the proper triangular area defining adjacent dots. With low dc-convergence voltage, the blue dot is lower than the other two dots.

The first rule is to obtain proper convergence on a horizontal line and a vertical line through the center of the raster. Place the following controls in their approximate midrange position: horizontal dynamic-convergence amplitude, vertical dynamic-convergence amplitude, vertical dynamic-con-

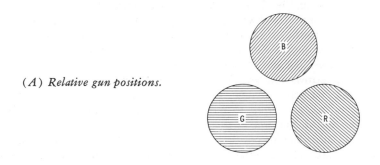

(A) Relative gun positions.

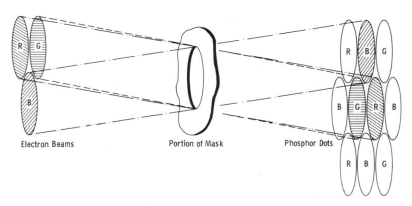

Electron Beams   Portion of Mask   Phosphor Dots

(B) Paths of electron beams.

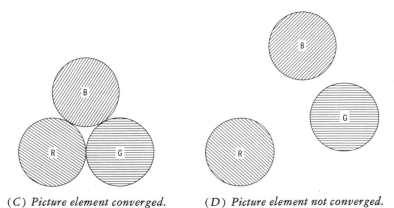

(C) Picture element converged.   (D) Picture element not converged.

Fig. 10-19. Principles of convergence.

vergence phase, and horizontal dynamic-convergence phase. Adjust the dc-convergence control clockwise to a point at which the blue dot is high. Then by adjustment of the individual red, green, and blue beam-positioning magnets, bring the dots into the approximate relative positions shown in Fig. 10-20A. The dots should describe an equilateral triangle. In Fig. 10-20B, the arrows indicate direction of dot movement under influence of the dc-convergence voltage. Raising the voltage causes blue to move upward and red and green to move farther apart. Lowering the voltage causes the dots to converge as shown by the arrows. With a little practice, the technician is able to position the dots accurately so that lowering the dc-convergence voltage perfectly converges the dots through the central area of the picture-tube screen.

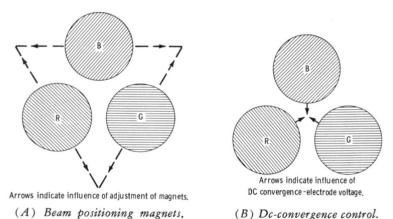

Arrows indicate influence of adjustment of magnets.

(A) Beam positioning magnets.

Arrows indicate influence of
DC convergence-electrode voltage.

(B) Dc-convergence control.

Fig. 10-20. Basic convergence adjustments.

It may now be noticed that convergence occurs in the center of the tube, but not at the top, bottom, and sides of the raster. This requires adjustment of the dynamic-convergence amplitude and phase controls.

A peaking adjustment that affects the total amplitude of the dynamic voltage normally is provided in the convergence transformer. This control should be rotated to find the arc through which the most radical change in spots can be observed, with the horizontal dynamic-convergence amplitude control in the maximum clockwise position. (That is, find the arc of rotation in which the spots shift apart, converge, and go through the opposite phase most rapidly.) Leave the control at the midpoint of the arc. Return the horizontal dynamic-convergence control to midrange.

Observe the extreme right and left dots on the center horizontal line. If the blue dot is high, adjust (counterclockwise) the horizontal dynamic-convergence amplitude control. If the blue dot is low, increase this dynamic voltage. Adjust the control until equal displacement is noted for all dots. Then adjust the horizontal dynamic-convergence phase for best dot coincidence. Readjustment of the dc-convergence control will then converge the dots horizontally. Considerable practice is necessary to be able to accomplish convergence with only a few trials. Ordinarily, a back-and-forth adjustment must be made all through this procedure.

Next, observe the extreme top and bottom dots, and adjust the vertical dynamic-convergence amplitude control for equal displacement error. Another adjustment of the dc-convergence control should converge the dots down the center vertical line. Remember that adjustment of dynamic voltage waveforms affects the total convergence-electrode voltages. A change in one control requires a change in other controls. Note also that convergence depends on the ratio of grid-4 voltage to neck-coating voltage (high voltage), and that focus is established by the ratio of grid-4 voltage to grid-3 (focus grid) voltage. Therefore, it is necessary to readjust the focus control often.

If the raster has become converged at the center and one side of the screen before the other side converges, adjust the horizontal dynamic-convergence phase control for equal displacement, and then readjust the dc-convergence control. If the raster was converged in the center and either the top or bottom (with the other portion misconverged), adjust the vertical dynamic-convergence phase for equal displacement, and then readjust the dc convergence.

There is a slight limitation in some present-day color kinescopes that affects the degree of convergence in one corner of the raster. If convergence is good in all portions except a corner or small edge, the kinscope probably is at fault.

IMPORTANT NOTE: Modern studio-type color monitors employ many additional convergence controls, as will be outlined in the following section.

10-4. THE RCA TM-21 COLOR MONITOR

One of the most important applications for a color monitor is in control rooms where operators face the problems of setting up and matching color cameras. A properly designed and operating color monitor offers the following benefits:

1. It provides a better check of registration during actual programming than the black-and-white master monitor does.
2. If the monitor has good deflection linearity (within 1 percent in both directions), a good check of camera deflection linearity is possible.
3. Provision for underscanning, to show the corners of the picture, permits better checking of camera framing, camera-lens aberrations, and camera deflection transients. Underscanning also makes cue marks in the picture corners readily visible.

4. A highly stabilized method of black-level setting permits better evaluation of camera shading characteristics and clearly indicates the effects of camera pedestal adjustments.

5. Precision decoder circuits and linear output amplifiers produce a picture of improved color fidelity, so that camera color fidelity can evaluate more accurately.

6. Good picture sharpness facilitates checking of camera focus.

Fig. 10-21 is a basic block diagram showing the five major chassis of the RCA TM-21 color monitor. This monitor serves as a studio color standard in many stations, and its description is warranted to point up the details important to a standard color monitor. It demodulates on the I and Q (wideband color) axes.

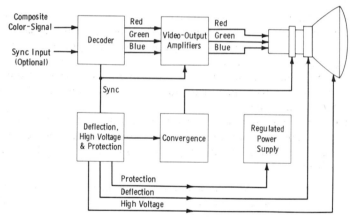

Courtesy RCA

Fig. 10-21. Block diagram of TM-21 monitor.

The heart of the decoder design (Fig. 10-22) is a stabilized video driver stage that drives the monochrome channel and the burst-controlled oscillator from its plate circuit, and the two chrominance demodulators from its cathode. The dc component is restored at this stage by means of a feedback-stabilized clamp. One of the gating stages involved in this feedback clamp has been made to serve as a burst separator as well, thus eliminating a separate tube for this function. In the video driver, the plate signal current is inherently equal to the cathode signal current, so there is no possibility of gain variations in the plate circuit relative to the cathode circuit.

Prior to the video driver stage, the input signal is raised to a relatively high level (about 12 volts peak-to-peak) by an amplifier equipped with a nonselective, or wideband, gain control. Through use of this high level at the driver stage, virtually all of the voltage gain required in the entire decoder is supplied by an amplifier that handles all signal components simul-

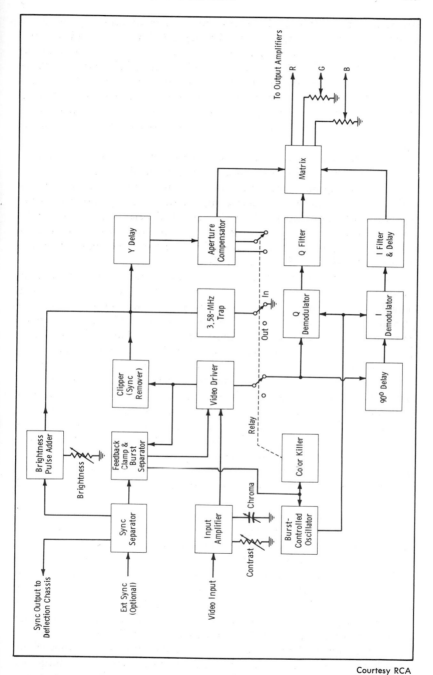

Fig. 10-22. Block diagram of TM-21 decoder.

taneously. This technique eliminates the problem of matching the gains of several individual amplifiers. In the stages following the video driver (which must necessarily be split into separate channels), it is possible to sacrifice voltage gain for the sake of stability and still deliver signals at about a 1-volt level at the output of the decoder. The amount of degeneration (or feedback) that it has been possible to incorporate by following this approach has made practical the elimination of several conventional gain controls (normally provided in decoders to compensate for circuit variations).

One of the channels following the video driver stage is the burst-controlled oscillator, which consists of a crystal-controlled 3.58-MHz oscillator shunted by a reactance tube. The control voltage for the reactance tube is derived from a phase detector that compares the oscillator output with the separated burst provided by the video driver. Special attention has been given to drift problems in this oscillator, so that the phase of its output remains stable relative to the phase of the chrominance signal delivered from the cathode side of the video driver.

In conventional decoder designs, the burst-controlled oscillator normally delivers two subcarrier outputs (90° apart in phase) to the chrominance demodulators. A popular method of deriving the two outputs is to use a pair of tuned circuits, one tuned above resonance and the other below resonance, to achieve the required phase shift. In the TM-21 decoder, however, a potential phase stability problem has been avoided by providing only a single output from the oscillator. This output is tied directly to both demodulators so that there can be no relative phase drift between them. The required 90° phase difference is provided in the *video* channel by passing the input signal to the I demodulator through a precision delay line equivalent to 90° at 3.58 MHz. The delay line is manufactured with a tolerance of ±1°; hence it is possible to eliminate the conventional quadrature phase control. The presence of the delay line in the I video channel poses no problem, because it is very simple to take it into account when adjusting the total delay of the I channel relative to the narrow-band Q channel.

The demodulators themselves are a stabilized diode type, as shown by the simplified schematic in Fig. 10-23. In essence, the circuit is a fast-acting clamp. The diodes are closed periodically at a 3.58-MHz rate, and their effect on the signal is to connect the output side of the 120-pF capacitor to ground through the center tap of the 3.58-MHz transformer. The charge stored in the 0.01-μF capacitor through the rectifying action of the diodes serves to make the diodes conductive only during the extreme peaks of the subcarrier cycle. Because the clamp is closed only momentarily, the output side of the 120-pF capacitor is normally free to follow the variations in the input signal. The average output level, however, is a function of the input level at the instants when the diode conduction occurs, as illustrated by the waveforms in Fig. 10-24. This average level is affected by both the amplitude and the phase of the incoming chrominance signal, and represents the desired demodulated signal.

The major advantages of this demodulator circuit are: (1) it has no video-gain drift problem, since it behaves in principle like a fast-acting switch, and (2) it is insensitive to the level of the cw subcarrier signal, provided the cw signal is always of higher amplitude than the modulated rf signal.

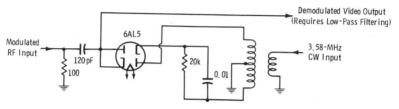

Fig. 10-23. Clamp-type demodulator.

The convergence chassis contains purely passive circuits for modifying certain waveforms derived from the deflection chassis before applying them to the convergence yoke surrounding the kinescope gun structures. The word "convergence," as applied to color-display devices, refers to the process of adjusting the positions of the red, green, and blue beams so that the respective images are registered in all parts of the screen. Because the effective distance between the guns and the screen assembly varies with the deflection angle, it is necessary to control the convergence with dynamic waveforms containing both horizontal-frequency and vertical-frequency components. The basic waveforms consist of a parabola and a sawtooth at each frequency, but these must be mixed in different proportions for each gun.

As indicated in Fig. 10-25, both the output tubes and the output transformers of the TM-21 serve as signal sources for the convergence circuits. Two features of the convergence-circuit design that contribute to a straightforward setup procedure are: (1) The controls are arranged so that the red and green rasters may be adjusted as a pair, relative to each other, after which the blue raster may be brought into registration relative to the redgreen pair. (2) Every control has been made to direct some type of movement in either the horizontal or vertical direction instead of along the 120° axes.

The large number of convergence controls needed for a tricolor tube (16 in the case of the TM-21) need not seem too formidable if each one performs some readily understood function. Those used are so designated and arranged on the control panel that it is easy to visualize them as trimming adjustments for the deflection circuits. There are five basic types of controls: The *position* controls are trim adjustments for the centering function, while *size* and *linearity* carry the same connotation as in conventional deflection systems. The *tilt* and *bow* controls produce these effects on the lines of the grating pattern commonly used to facilitate convergence adjustments; the *bow* control affects the curvature of the lines, whereas the *tilt* controls are used to make the lines parallel.

The controls are grouped in two ways. The vertical, static, and horizontal adjustments are located in separate columns. The upper controls in each column adjust the red and green rasters relative to each other, and the lower controls adjust the blue raster relative to the red-green pair. A screen-selector switch just to the left of the convergence control panel makes it possible to view any of the rasters separately, or to view only the red-green pair.

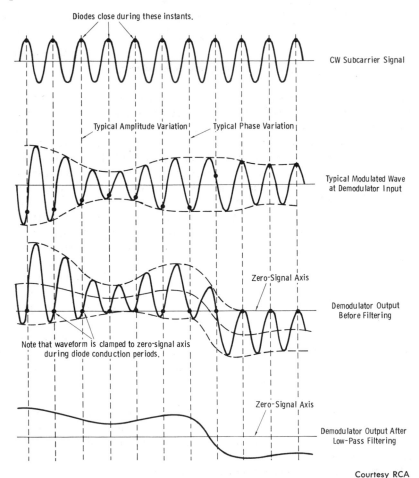

Courtesy RCA

Fig. 10-24. Waveforms in clamp-type demodulator.

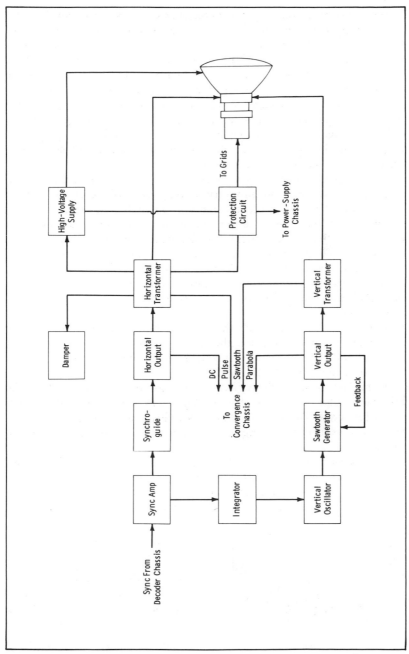

Fig. 10-25. Deflection, high-voltage, and protection circuits in TM-21 monitor.

The BRIGHTNESS control produces the same effect as conventional brightness controls, even though it operates in an unusual manner. Instead of varying the bias on the kinescope, it varies the level of a special pulse added to the signal in place of the normal sync and the burst signals (Fig. 10-26). This technique is made feasible by the use of keyed clamps in the output amplifiers; these clamps operate during the time interval of the added pulses. The advantage of this brightness-control technique is that it eliminates the need for individual red, green, and blue background controls. The single BRIGHTNESS knob automatically exercises the proper degree of control over the three color channels because the added pulse is passed through the standard decoder matrix.

The CONTRAST control varies the gain of the input stage in the decoder. It varies the luminance and chrominance components of the picture equally.

The SYNC SELECTOR enables the operator to switch between external sync and the internal sync pulses separated from the composite video signal. In the remote position, this switch brings the sync interlock into operation so that the use of internal or external sync is controlled from a remote point (such as a switcher) through a dc control lead.

The APERTURE COMPENSATOR adjusts the degree of high-peaking in the luminance channel for optimum picture sharpness without objectionable overshoots.

By moving the TEST switch through its several positions and making specific adjustments at each step, the monitor is brought into proper operating condition. The following paragraphs go through each of these steps with reference to Figs. 10-26 and 10-27.

In the first position, the signal is automatically disconnected, but the brightness pulse remains. In this position, the RED SCREEN control is adjusted for cutoff. The SCREEN SELECTOR switch can be set at R to facilitate this adjustment by cutting off the green and blue beams to avoid confusion.

In the second (screen balance) position, both the signal and the brightness pulse are disconnected, and the green and blue screen controls may be adjusted relative to the previously set red screen to produce a gray screen of approximately 20 percent brightness. The SCREEN SELECTOR

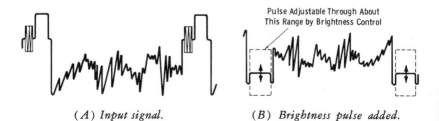

(A) *Input signal.*   (B) *Brightness pulse added.*

Courtesy RCA

Fig. 10-26. Brightness-control waveforms.

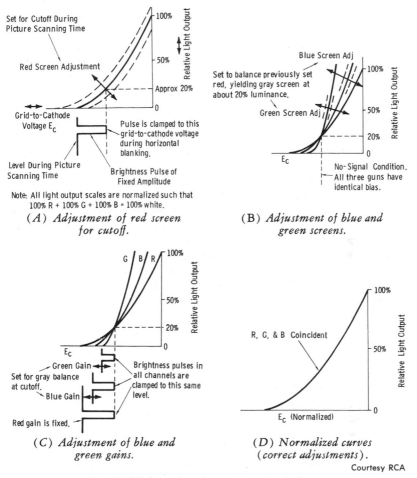

(A) *Adjustment of red screen for cutoff.*

(B) *Adjustment of blue and green screens.*

(C) *Adjustment of blue and green gains.*

(D) *Normalized curves (correct adjustments).*

Courtesy RCA

Fig. 10-27. Procedure for setting color balance.

switch must, of course, be in the RGB position for this adjustment. As shown in Figs. 10-27A and 10-27B, this step brings the kinescope transfer characteristics into coincidence at the point corresponding to about 20 percent of the maximum signal level.

In the next position (monochrome), both the brightness pulse and the signal are applied, but the chrominance circuits are disabled. In this position, the green and blue gain controls may be set to provide proper color balance in all parts of the gray scale. This adjustment is facilitated by the use of a signal containing a gray-scale pattern.

As shown in Fig. 10-27C, the absolute signal amplitudes required for the three guns are different because of different phosphor efficiencies. When the proper adjustments are made, however, and the signal scales are normalized, the effective transfer characteristics are essentially coincident (Fig. 10-27D).

The next position, unity chroma, is the normal operating position, in which the signal is applied to both the luminance and chrominance channels. The CHROMA control is inoperative in this position, and the saturation of the colors in the picture yields a good indication of the quality of the incoming signal. The PHASE control may be set conveniently while the TEST switch is in the unity chroma position by examining the blue component of a standard color-bar signal (use the B position of the SCREEN SELECTOR switch). When the phase adjustment is correct, the standard color-bar signal produces four blue bars of equal brightness. If the phase adjustment is incorrect, the blue bars are of unequal brightness. This test is very sensitive, particularly if the brightness is temporarily reduced to place the blue bars near cutoff on the kinescope characteristic.

In the final position of the TEST switch, variable chroma, the conditions are the same as for the unity-chroma position, except that the CHROMA control is made operative. This position is intended for operation in applications where the monitor is used to make the most pleasing pictures, even though the signals available are slightly substandard. The CHROMA control is simply set for the most pleasing overall effect.

The TEST switch can be used to make a rapid test of the convergence adjustments in the monitor. If there is any uncertainty in the viewing of a color picture as to whether observed misregistration is a fault of the signal or of the monitor, it is only necessary to place the TEST switch in the monochrome position. If the color fringes disappear, they are clearly a fault of the signal, but if they remain, it is necessary to touch up the convergence adjustments of the monitor itself.

10-5. THE X AND Z DEMODULATOR

It will be recalled that the I and Q signal components contain some of each of the $R - Y$ and $B - Y$ components, from which $G - Y$ can be extracted conveniently. The reader should already be familiar with demodulation with respect to the I and Q axes and the $R - Y$ and $B - Y$ axes.[1] The latest and simplest form of "narrow-band color" demodulation is termed X *and Z demodulation.*

See Fig. 10-28. Through transformer T1, the local 3.58-MHz oscillator provides signal voltages to the suppressor grids of demodulators V1 and V2. The signal applied to the suppressor of V1 is in phase with the burst signal and is arbitrarily termed the X signal. The voltage at the suppressor of V2 is shifted approximately 90° by L1, C1, and R1. This signal is

---

[1] For example, see Harold E. Ennes, *Television Broadcasting: Equipment, Systems, and Operating Fundamentals* (Indianapolis: Howard W. Sams & Co., Inc., 1971).

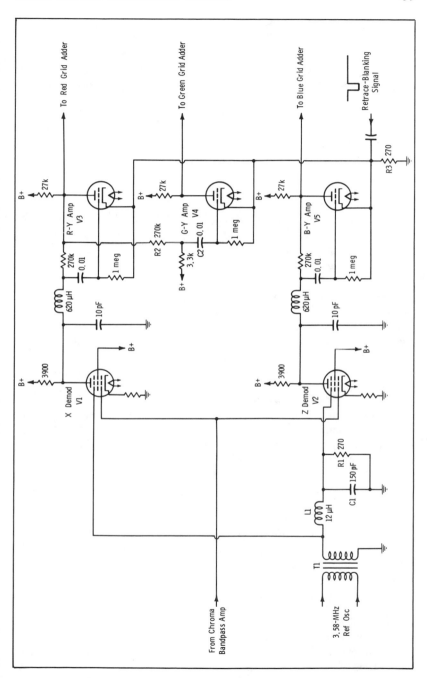

Fig. 10-28. Basic circuit for X and Z demodulation.

termed the Z signal. The cw signals thus supplied provide the gating signals for the demodulators. The chrominance signal from the chroma bandpass amplifier is applied to the control grids of V1 and V2 in parallel.

The demodulated outputs at the plates of V1 and V2, after filtering by the 620-μH coils and 10-pF capacitors, are applied to chroma-difference amplifiers V3 and V5. The signal currents in the R − Y and B − Y amplifiers combine in the common cathode resistor, R3, to develop the signal input voltage at the cathode of G − Y amplifier V4. Note here that R2 and C2 at the grid of V4 serve only to couple a small amount of signal from the plate of V3; this provides slight degenerative feedback to assure the correct phase and amplitude for the G − Y output signal. The picture-tube retrace-blanking signal normally is coupled into the common cathode of the chroma-difference amplifiers as shown.

As a result of the phase relationship of the X and Z demodulators, the R − Y and B − Y currents of V3 and V5 develop across R3 a voltage that is a G − Y signal. Remember that:

$$G - Y = -0.51(R - Y) - 0.19(B - Y)$$

Thus, we note that a positive G − Y value is composed of negative values of both R − Y and B − Y. The cathode currents of V3 and V5 are in phase with the input signal, and the cathode injection at V4 results in a plate signal 180° out of phase with that at the V3 and V5 plates. This meets the polarity requirements.

This type of demodulation occurs at a high level, and the output amplitude at the color-difference amplifiers is sufficient to drive the grids of the picture tube. This fact, coupled with the relatively simple "matrix" circuitry just described, results in a relatively inexpensive but effective color monitor as compared to monitors using the more elaborate I-Q demodulator. Also, the separate filtering and additional gain stages required in I-Q demodulation are unnecessary with the X and Z demodulator.

10-6. USE OF THE COLOR MONITOR IN MATCHING TECHNIQUES

The final check for color balance in multiple-camera setups is accomplished most readily by using a switch bus for the color monitor and vectorscope, as illustrated in Fig. 10-29. With this technique, the effect of slight differences in color monitors themselves is eliminated, since a common color monitor is used for all signal sources.

It is assumed at this point that all cameras have been individually adjusted as correctly as possible. Very briefly, the results of these adjustments should be as follows:

1. Chroma and white level are the same from all cameras, and system phase as well as burst phase are the same for all cameras at the switcher output (on color bars).

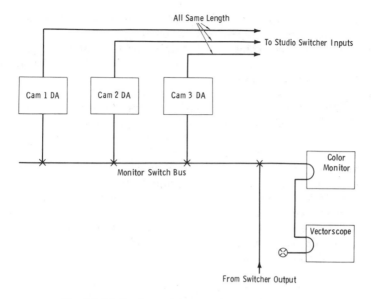

Fig. 10-29. Technique for matching color cameras.

2. The same video output level is present from all cameras when the pickup tubes are operating at the proper potentials and looking at the same reference white chip on the gray scale. When image-orthicon tubes are involved, this will be at the knee or slightly over the knee.

3. The gray-scale steps (looking at the same chip chart under the same lighting conditions) are at the same levels relative to capped level for all cameras. Slight adjustments in target voltage and/or gamma circuitry are required to correct any differences.

4. Luminance-to-chrominance ratio and luminance-to-chrominance tracking are correct and the same for all cameras.

We now are ready to color balance all cameras by using the common color monitor as the final check. All cameras should be color balanced on the *same gray scale* under the *same light conditions*. Use the following procedure:

1. Switch between all cameras on the color monitor. If different "colors" of the gray scale exist, go to the following steps.

2. Turn the camera chroma off. With a properly adjusted color monitor, the background should be gray and show no color. Turn the chroma on. Note any color in the background on the color monitor; balance this output with a slight adjustment of the appropriate black-level control on the camera. For example, if the background is slightly green, adjust the green black-level control for cancellation of color. It

is much easier to see this on a good color monitor than on the wave-form oscilloscope. Repeat this procedure for each camera.

3. If the high-light chips are slightly colored, adjust the appropriate "paint pot" (video vernier gain control) to cancel the color. Repeat this step for all cameras.

4. The final check of the overall system (through the studio switcher) is best made by using a split-screen presentation for two cameras at a time. First check the system by feeding a signal from the *same* camera (on color bars) to both banks of the special-effects bus with vertical-wipe operation. There should be zero hue and saturation shift as the vertical wipe is made on the same signal source. If a hue shift occurs, the switcher paths are not properly phase delayed. If a saturation (color intensity) shift occurs, the two paths of the system have different response at 3.58 MHz.

Then perform the same operation with two cameras looking at the same chip chart. The color monitor should show no color shift as the vertical-wipe transition is made.

EXERCISES

Q10-1. In an R—Y, B—Y receiver or monitor, why is only the luminance-channel signal delayed?

Q10-2. Give the color-system primaries and their respective complementary colors.

Q10-3. Describe the chrominance signal for the corresponding complementary color.

Q10-4. What is the purpose of the color burst?

Q10-5. What is the purpose of the bandpass amplifier in a monitor?

Q10-6. What is the purpose of the color-killer circuit?

Q10-7. Give the basic operation of a synchronous demodulator.

Q10-8. Describe the difference between the cw reference signals used for I-Q demodulation and those used for color-difference (R—Y, B—Y) demodulation.

Q10-9. In a color-encoding system, what would be the effect of carrier unbalance on white and gray areas displayed on a properly adjusted color monitor?

Q10-10. In case of carrier unbalance in the encoder, will dominant hues be shifted on the color monitor?

CHAPTER **11**

Preventive Maintenance

Maintenance procedures for specific parts of a camera chain (such as video amplifiers and preamplifiers, power supplies, processing circuitry, yokes, etc.) have been covered in applicable sections of the preceding chapters. It remains to cover the general system checks of a camera chain, and to suggest pertinent preventive-maintenance scheduling.

11-1. THE VIDICON FILM CHAIN

When performance of a film camera chain deteriorates, there may be some uncertainty concerning where to start a search for the problem. The experienced maintenance technician is able to analyze the symptoms and know approximately which area to explore. The following subsections should orient the reader in the general techniques to be used in checking performance specifications.

Steps in Obtaining Maximum Film-Camera Resolution

NOTE: Before proceding, review "Measuring Detail Contrast" in Section 6-4, and "Checking Gamma Circuitry for Film Cameras" in Section 6-6 of this book.

The problem here is to obtain maximum detail contrast (by means of aperture correction in a properly adjusted camera chain) while maintaining the lowest possible noise level. You should perform the following steps before undertaking the longer procedure of a complete video-sweep alignment; such alignment may not be necessary.

(A) Use open gate on the projector at normal lamp voltage and lens settings. Adjust the target voltage to obtain a beam current of 0.3 μA.

(B) See Fig. 11-1. Turn any manual video-gain control to maximum gain. Connect the oscilloscope at TP 1. You should know what peak-to-peak level the camera head should deliver at 0.3 μA of beam current (this usually is indicated on the schematic or in the instruction book). Usually

437

this level is around 0.4 to 0.5 volt peak to peak. If this signal level is low under the above conditions, replace the camera-head tubes one at a time until proper gain is restored.

If you do not know what this level should be, and/or if the camera chain does not employ either a beam-current meter or a calibration pulse, proceed as follows:

Assume the target load resistance is 47k. For a beam current of 0.3 μA, the voltage swing is: $0.3(10^{-6})(47,000) = 0.014$ volt (white to black). Now see Fig. 11-2. Temporarily disconnect the vidicon target lead (with target voltage off), and substitute the signal as shown. If you adjust the signal-generator output (which can be a 15.75-kHz square wave, pulse, or

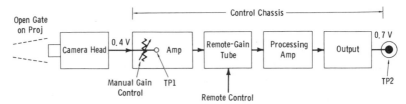

Fig. 11-1. Camera-chain level check.

stairstep) to 1.4 volt peak to peak and use a 40-dB pad, the input level to the preamplifier will be the required 0.014 volt to simulate 0.3 μA of target current. (It may be necessary to bypass temporarily the camera head high-peaker cathode circuitry with a 0.47-μF capacitor.) Now measure the peak-to-peak level at the input to the control chassis. Restore the camera head to normal operation, and (with open projector gate) adjust the target voltage to get the same peak-to-peak level as measured with the signal generator. This target voltage is that which is required to obtain 0.3 μA of target current. Even if the equipment employs a metering circuit, if you doubt its accuracy, check it by the above procedure.

(C) With a test-pattern signal, readjust any manual gain control to obtain 0.1 to 0.15 volt (pk-pk) video (or the safe maximum value of video found by the previous linearity test—Section 6-6).

(D) With a test-pattern slide or film loop, adjust all high-peaker controls for minimum smear or streaking. Aperture compensation should be removed for setting these controls to observe only medium- and low-frequency response.

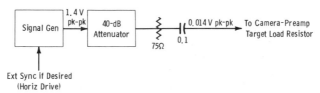

Fig. 11-2. Connection of signal generator to target load resistor.

(E) Restore the aperture compensation, and use the maximum boost required to obtain 600-line response with good detail contrast as observed on a master monitor. If noise is apparent when the aperture compensation is adjusted to obtain just 600 lines of horizontal resolution, do the following:

1. Check for excessive beam current. Reduce the beam until high lights are just resolved.
2. Be certain of camera physical and electrical focusing.
3. Determine that the camera output is normal (Step B above).
4. Replace video amplifier tubes in the control chassis one at a time for reduction of noise without loss of resolution.
5. If there still is no improvement, the camera chain needs a complete video alignment.

NOTE: As a general rule, a "bad" vidicon will show excessive *target lag* before it deteriorates in resolution. Target lag is image retention under vertical or horizontal movement of the images. Such a vidicon can be retired to a slide (only) camera, since no movement (except slide change) is involved. After considerable extended use in the slide camera, the vidicon will go "soft" and lose resolving power.

Excessive overpeaking (by either video alignment or use of high-peak controls to get white-following-black for apparent sharpness) will result in video bounce (excessive on drastic scenic changes) and white or black compression. In case of excessive bounce on scenic changes, check *both* the high-peak and low-frequency compensation controls. Adjust the low-frequency compensation control (where used) for a flat vertical interval as observed on an external CRO at the output of the control chassis. Remember that most master monitors incorporate their own tilt controls (a separate adjustment) for the master monitor CRO circuitry.

Parameters of Target-Voltage Set

A surprisingly large amount of "maintenance" time is spent needlessly when optimum adjustment procedures are not followed in the beginning. Be sure you understand such procedures (normally considered under "operations" rather than "maintenance") so that you know when actual maintenance is required.

Thus far, our discussion of the vidicon film camera serves only as a starting point in obtaining good film telecasts. It has been assumed that the projector-lamp voltage and lens $f/$ stop were set properly. Also, we still have to consider focus current, beam alignment, and possible shading problems.

Here are some "fine points" that help in "squeezing everything possible" from the vidicon film setup:

1. Block all light from the vidicon faceplate, set the target voltage at 20 volts as a start, and turn the beam control off (maximum bias).
2. With the target volts-current meter in the current position, rotate the zero-adjust control so that the meter indicates upward on the scale. (You do not want a zero reading because you are going to take a difference measurement.)
3. While watching the CRO or the monitor kinescope, bring up the beam current until the information at black level is "wiped-in" (dark-current information is discharged). This will be quite evident if you watch closely.
4. Measure the difference in meter readings between 2 and 3 above. For the best overall picture quality, this difference (actually a measurement of dark current) should be no more than 0.01 μA. Use the maximum target voltage possible while maintaining dark current no more than 0.01 μA. This is for film application *only*, not studio use for live pickups.
5. Now zero the target meter with the beam control off, and remove the light block from the vidicon faceplate. Project a resolution slide into the vidicon. The camera lens should always be wide open. Adjust the projector lens wide open (minimum depth of field) to obtain sharpest mechanical focus with the beam adjusted to just discharge the highest high light. Then stop the projector lens down two stops from wide open. This will provide adequate depth of field for warped or buckled film.
6. Now adjust the projector-lamp voltage to obtain 0.3 μA of beam current with the gate open and with the target voltage arrived at in Step 4 above.

You have now arrived at the optimum settings of target voltage, lens aperture, and lamp voltage. NOTE: If the film chain incorporates a color camera, you must maintain the projector-lamp voltage at its normal operating value to preserve color temperature. We are currently discussing monochrome operation only. If a monochrome camera is multiplexed with a color camera, you should use an additional neutral-density filter on the monochrome-camera axis to obtain proper operation and light balance.

If the vidicon camera chain does not employ a target-current meter, or if the metering circuit is not dependable in reading very small values of current as is required in dark-current readings, keep the following in mind: As the target voltage is increased, there is a point at which *edge flare* (excessive dark current) is reached. When the target voltage is reduced below a certain value (depending on the individual vidicon), a point is reached at which *portholing* (darker around edges than at center) occurs. It is possible to adjust the target voltage above the point at which portholing just occurs, and an entirely "flat" raster (with respect to shading) results. This value then can be varied slightly if an unrealistic projector-lamp voltage is required to obtain 0.3 μA of target current at the target voltage being used.

> NOTE: If you attempt to use too low a target voltage, a condition known as *target bounce* can occur on drastic scenic changes. A small amount of dark current (up to 0.01 μA for a film camera) stabilizes voltage excursions caused by target-current variations through the load resistor in accordance with the charge on the photocathode. As a broad general rule, optimum results in a monochrome film camera are obtained with a target voltage somewhere between 18 and 25 volts.

Focus Current vs Focus Voltage

For a vidicon, the magnetic field strength at the center of the focus coil normally should be 40 gausses (produced by 40 mA of focus current). For an image orthicon, the value is 75 gausses (produced by 75 mA of focus current). Therefore, the manufacturer usually recommends a fixed value of focus current. However, with some tubes (either vidicons or I.O.'s) it is possible to vary the ratio of focus current to focus-electrode voltage slightly and obtain sharper delineation of the high-frequency wedges of the test pattern. The focus-electrode voltage is, of course, what is varied by adjustment of the beam-focus control (or orth-focus control in RCA image-orthicon terminology). This assumes that the beam current has been aligned properly for optimum conditions, and that the proper mode of focus exists for an image orthicon.

Remember that when the focus *current* is varied, the scanning size varies. More current (greater focus field) stiffens the beam, and the photocathode or target area is underscanned (kinescope image increases in size). As the focus field is reduced, the scanning beam travels farther for a given deflection voltage, and overscanning results (kinescope image decreases in size). Therefore, after a change of focus current is made, remember to readjust the scanning for normal area before judging resolution of the test pattern. You must, of course, readjust the beam-focus (voltage) control for optimum electrical focus each time you vary the current.

Linearity and Shading

The light output of a photoconductive device such as the vidicon is directly proportional to velocity of scan. So long as the scanning waveform is linear, no inherent shading takes place in the vidicon. If the waveform is nonlinear, the velocity, or rate of scan, is different for different areas of the target, and shading occurs.

The best way to check a film camera is to remove the lens and swing the camera to one side so that a flat-light source (such as a 40-watt bulb placed about 4 to 6 feet away) can be used. With a new vidicon (or an old one that has not been abused), it is possible to adjust the linearity controls for absolutely flat shading. In most cases, the image linearity can be made well within 2 percent by this method. If there is shading when the camera is returned to its normal position in the film chain, you know the shading is the result of optics, not the camera. The above procedure assumes the target voltage has been set for optimum operation as described earlier. It is also assumed that you are already familiar with the conventional method of checking camera sweep linearity: first getting the monitor linear by use of the grating generator, then adjusting the camera sweeps (and linearity controls) to obtain a linear presentation of the test-pattern image on the same monitor.

A far better method of exactly measuring geometric distortion (non-linearity of camera sweeps) is the use of the ball chart (RETMA [now EIA] linearity chart). This aid can be obtained in slide form for a film camera, or in chart form for a studio camera. The pattern is designed so that the black outlines of the circles define 2-percent linearity, and the inner portions of the circles define 1-percent linearity.

> NOTE: The use of the linearity chart is illustrated and described in Chapter 10 of Harold E. Ennes, *Television Broadcasting: Equipment Systems, and Operating Fundamentals* (Indianapolis: Howard W. Sams & Co., Inc., 1971).

11-2. GENERAL PREVENTIVE MAINTENANCE

Preventive maintenance in general consists of:

1. Lubrication per manufacturer's specifications
2. Cleaning of all equipment and optics at periodic intervals
3. Performance checks at periodic intervals

The extent of the periodic interval is determined by manufacturer's recommendations, or as dictated by experience with any particular installation.

Cleaning

In general, cleaning procedures can be scheduled at daily, monthly, or longer intervals, as follows.

Daily—Wipe and clean all exposed surfaces of the camera. Use only lens tissue on turret lenses or zoom-lens front surfaces. When necessary, use a cotton ball dipped in isopropyl alcohol, and use a light circular motion over the optical surface. Wipe the surface dry with a wad of lint-free cheese-cloth or diaper cloth. If warranted, clean the faceplates of pickup tubes in the same manner. *WARNING:* Turn the power off when doing this.

Monthly—Inspect all exposed color optics and front-surface mirrors in the optical path. If necessary, clean dichroic surfaces, but use only pure grain alcohol or a 1-to-6 ratio of benzene and grain alcohol. Use a cotton ball and diaper cloth as described above. On front-surface mirrors, use isopropyl alcohol diluted with a small amount of distilled water (to prevent rapid evaporation), and wipe dry with diaper cloth *before* the solution on the mirror surface evaporates.

NOTE: It is very important, as a final step in lens and optics cleaning, to use a camel's-hair "static" brush to remove dust that inevitably falls on the surfaces because of static deposits.

Every Three Months—This interval depends entirely on the cleanliness of the operating area. In the case of portable and mobile use, these operations should be performed monthly (or more often).

Clean all air filters in the blower path of the camera, and in the forced-air paths of the cooling systems of power supplies, etc. Use only the solvent for cleaning that is recommended in the instruction book for the particular installation.

With a low-pressure air blower, blow out the camera head, control panels, and rack equipment to eliminate dirt, dust, and general contamination. Wipe clean with a soft paint brush and cloth.

IMPORTANT NOTE: At this time, it is imperative to make a thorough *visual inspection* of component parts. Observe resistors for discoloration or signs of heating. Inspect capacitors for signs of bulging or leakage. Inspect terminal boards for cracks and loose connections. Be sure all fuse mounts and fuse caps are tight. The time spent here is invaluable in preventing future breakdowns, which is the basic reason for performing preventive maintenance.

Lubrication

Lubrication of modern camera chains normally is scheduled about every six months. For operation under extreme conditions of dust, heat, and humidity, lubrication procedures must be carried out more often. *Caution:* Some of the older camera chains employ fan and blower motors that require lubrication every 200 hours. Always check the instruction book for a particular system, and work the lubrication schedule into a check-off sheet at the recommended intervals.

Items that require lubrication include: turret shaft and detent rollers; turret gears and bearings; camera-head counterbalance linkage; focus, iris, and zoom mechanisms (which usually involve multiple gears, bearings, and shifts); and blower and fan motors. It is quite important to check the manufacturer's specifications for the type of oil or grease to use on the items involved.

It is important not to overlubricate. Always wipe off excess oil or grease, and use any special applicators that are supplied with a camera chain by the manufacturer.

Plugs and Receptacles

There are two main types of plugs and receptacles used to interconnect the various components. The first type of plug is used with a coaxial line and consists of a metal shell with a single, center pin that is insulated from the shell. When the plug is inserted into the receptacle, this pin is gripped firmly by a spring connector. There is a knurled metal ring around the plug; this ring is screwed onto the corresponding threads on the receptacle. The insulation in these plugs is heavy in order to withstand considerable voltage.

The second type of plug is used for connecting multiconductor cables. The plug usually consists of a number of pins that are insulated from the shell. The pins are inserted into a corresponding number of female connectors in the receptacle, although in some cases the plug has the female connectors in it and the male connectors are in the receptacle. This type of plug usually has two small pins or buttons that are mounted on a spring inside the shell and protrude through the shell. When the shell is properly oriented and placed in the receptacle, one of these pins springs up through a hole in the receptacle, firmly locking the plug and receptacle together. When it becomes necessary to remove the plug, the other pin is simply depressed, and the plug can be removed.

Connections between all plugs and their cables are made inside the plug shell. The cable conductor may be soldered to the pin, or there may be a screw to hold the wire to the pin. Remove the shell if it is necessary to get at these connections for repair or inspection. If there is a clamp holding the cable to the shell, loosen the clamp screws. Usually there are several screws holding the shell; these are removed and the shell is pulled off. In some cases, it is found that the shell and plug body are both threaded; then the shell may simply be unscrewed.

Inspect the following:
1. The part of the cable that was inside the shell for dirt and cracked or burned insulation
2. The conductor or conductors and their connection to the pins for broken wires; bad insulation; and dirty, corroded, broken, or loose connections
3. The male or female connectors in the plug for looseness in the insulation, damage, and for dirt or corrosion
4. The plug body for damage to the insulation and for dirt or corrosion
5. The shell for damage such as dents or cracks and for dirt or corrosion
6. The receptacle for damaged or corroded connectors, cracked insulation, and improper electrical connection between the connectors and the leads

Tighten the following items:
1. Any looseness of the connectors in the insulation. If tightening is not possible, replace the plug.
2. Any loose electrical connections. Resolder if necessary.

Clean these items:
1. The cable, using a cloth and cleaning fluid
2. The connectors and connections, using a cloth and cleaning fluid. Use crocus cloth to remove corrosion

3. The plug body and shell, using a cloth and cleaning fluid. Use crocus cloth to remove corrosion.
4. The receptacle, with a cloth and cleaning fluid if necessary. Use crocus cloth to remove corrosion.

Adjust the connectors for proper contact if they are of the spring type. *Lubricate* the plug and receptacle with a thin coat of petroleum jelly if they are difficult to connect or remove. The type of plug with the threaded ring may especially require this.

11-3. TROUBLESHOOTING

Conditions requiring emergency maintenance can be classified broadly into three categories:

1. Erratic video level
2. Erratic black level
3. Erratic horizontal and/or vertical deflection

A dead camera chain or a combination of all three of the above conditions generally points to the common unit for the entire chain—the power supply. Maintenance of power supplies is covered in Chapter 3.

Erratic Video Level (Black Level Constant)

Review Fig. 6-25. The first step is to check the camera chain with test pulses, when these are provided. If the pulse inserted at the preamp input is erratic relative to the reference pulse (inserted at the camera end of the cable), then the trouble is obviously in the preamp, video amplifier, or processing amplifier in the camera head. If both the inserted and reference pulses are erratic, the problem is in the control console or control-room rack equipment. Test points normally are provided at each module output for quick scope observation. If the pulses are stable, but video with the pickup tube looking at a scene is erratic, the problem lies in the pickup-tube circuitry.

In a multiple-channel color camera, trouble of this nature is usually existent in only one of the channels at a time. If the trouble is exhibited in all channels, then it occurs at a point after combination of all channels—in the encoder. In this case, the B, R, G, and M monitoring points will show no level changes on an individual channel.

Review Fig. 6-28 for typical circuitry involved in remote control of video gain. Since the condition discussed here is erratic video level with constant black level, the problem will normally exist prior to black-level clamping.

NOTE: Bear in mind here that if the trouble is following the camera-head output, both video and black level would be erratic (in the typical arrangement illustrated in Fig. 6-25).

Review Chapter 7 for automatic and manual control circuitry.

Erratic Black Level (Peak-to-Peak Video Level Constant)

Now we will consider the case in which the overall amplitude is erratic, but the peak-to-peak video remains constant with the changing pedestal. Again, the first step is checking the test pulses. This problem sometimes results when a clamp-pulse width or timing adjustment is on the edge of the proper setting. The maintenance engineer should familiarize himself thoroughly with all such adjustments for the equipment in the installation with which he is concerned.

Check first for the existence of clamp pulses. Then, using video of varying APL's, check for the effectiveness of clamping at the clamped point. Sometimes the black-level control pulses are fed into a video stage prior to the clamp, and are timed (by adjustment or design) to occur within the clamp interval. Check these points on a dual-trace scope.

Obviously, the wide variety of circuitry employed in different camera chains prevents establishment of a set pattern of testing. The maintenance engineer must study and analyze his particular instruction-book descriptions. Any point that is not clear is sufficient reason to contact a factory representative for clarification.

Deflection Problems

It will be recalled that if either the horizontal or vertical deflection fails entirely, the pickup tube(s) will be biased off, and no video will be obtained. When this condition exists, the test pulse is passed, but no video is obtained from the pickup tube. In this case, the first place to check is the protection stage (review Fig. 7-22); observe the presence or absence of horizontal- or vertical-deflection pulses, and check back from there. In the event of erratic deflection, location of the trouble in the horizontal or vertical circuits is self-evident from the monitor display.

See Fig. 11-3 for a review of a typical four-channel color-camera deflection arrangement. When different types of tubes are employed (for example, an I.O. for luminance and vidicons for chrominance), different deflection amplitudes must be provided. Fig. 11-3 shows a typical arrangement for both horizontal- and vertical-deflection circuits. Master size and linearity adjustments affect all four channels. These generally are set for the monochrome tube. Deflection for the color-channel tubes then is obtained through a waveshape and attenuation network.

We can see from this configuration that the green size control becomes the master for all chroma tubes, and individual controls are provided for red and blue. Linearity controls for red and blue are variable resistors in series with the deflection coils. This arrangement is for the purpose of facilitating registration.

If all channels show erratic deflection, the problem obviously is located prior to the multiple yoke take-off: in the deflection output stage or ahead of this stage. If the erratic deflection is in one channel only, troubleshooting

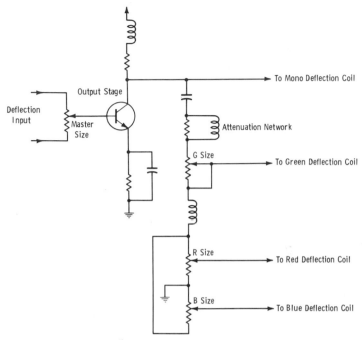

Fig. 11-3. Basic deflection arrangement for four-channel camera.

is isolated to the corresponding network. Review Section 7-4 of Chapter 7 for typical deflection circuitry.

EXERCISES

Q11-1. How can you tell whether shading is introduced by the pickup tube, or by the amplifiers and/or shading controls?

Q11-2. What is the most common cause (other than "optics") of shading in the vidicon film camera?

Q11-3. How many adjustments, in addition to the shading controls, affect shading in the image-orthicon camera?

Q11-4. Why should nearly all adjustments that affect shading in the image orthicon camera be made with the lens capped?

Q11-5. What type of tube checker should you use for preventive maintenance, and what should you check for?

Q11-6. If a slightly excessive beam current is used on a vidicon and a split image results, what are the two most likely sources of trouble?

Q11-7. You have noticed that when setting up I.O. cameras outdoors with long camera-cable runs, you seem to be continually "running out of" image-focus (photocathode focus) range, particularly after the cables have been heated by direct sunlight. Can you do anything about this?

Q11-8. What are the problems most likely to be associated with the image-orthicon tube itself, rather than amplifiers?

Answers to Exercises

CHAPTER 1

A1-1. (A) 30 W/ft². (B) 90 W/ft².

A1-2. (A) 15 W/ft². (B) 50 W/ft².

A1-3. (A) 720 foot-candles. (B) 640 foot-candles. In solving this problem, use the information of Table 1-2. Note that the multiplying factor for illumination is 4 (15 feet is one-half of 30 feet, and light decreases as the square of the distance). Thus, for (A) the chart of Fig. 1-2B shows 180 foot-candles, and 180 × 4 is 720 foot-candles. Also, since the distance is halved, the dimensions are halved, and the off-axis dimension to use with Fig. 1-2B becomes 10 × 0.5, or 5 feet. Since the chart in Fig. 1-2B shows 160 foot-candles, then the answer is 160 × 4, or 640 foot-candles for (B).

A1-4. (A) 45 foot-candles. (B) 20 foot-candles.

A1-5. $I = W/E = 5000/115 = 44$ amperes (approx).

CHAPTER 2

A2-1. Remember the color triangle; green and blue form cyan. From Table 2-1, for cyan:

$$I = -0.60$$
$$Q = -0.21$$

The vector sum is:

$$\sqrt{(-0.6)^2 + (-0.21)^2} = \sqrt{0.36 + 0.04} = \sqrt{0.4} = 0.63$$

This is the amplitude of fully saturated cyan (unity green and blue inputs, zero red).

Now see Fig. A-1. If you take $-I$ as the "adjacent side," then:

$$\cot \theta = \frac{\text{adjacent side}}{\text{opposite side}} = \frac{I}{Q} = \frac{-0.6}{-0.21} = 2.86$$

$$\theta = 19.4° \text{ from } -I \text{ axis}$$

Since the —I axis is at 303° and cyan is lagging this by 19.4°:

303 — 19.4 = 283.6, or simply 284°

A2-2. If these I and Q signals were applied to the grids of a color kinescope that was perfectly linear (not only to cutoff, but *beyond cutoff* into the negative light region), they would swing equal amounts above and below cutoff, and the net luminance would be zero. But the kinescope is not linear, and it cuts off at black (blanking or picture black, depending on monitor adjustments). Therefore the ac axis is no longer zero, and the bars become visible, since these pulses are occurring during active line scan.

A2-3. Any pulse into the red input of the encoder will cause only the red gun of the color picture tube to be excited. (The same is true for the other inputs and their respective guns.) If this red input occurs at a time when the green and blue inputs to the encoder are zero, only the red gun will be activated; therefore the red is fully saturated regardless of amplitude. If there is no mixture with white (meaning a measurable amount of the other two primaries), then, by definition, the color is fully saturated. This normally occurs only with a signal from a color-bar generator. Color-bar pulse amplitudes can be reduced to 75 percent of unity value at the encoder input, to prevent "overshoots" when transmitter color specifications are being checked. Primary signals and their complements are still at maximum saturation.

A2-4. The value indicated on detail C-C of Fig. 2-10B is the *minimum allowable* front porch. Now study Note 4 of Fig. 2-10A. Put down the stated values of x, y, and z from Fig. 2-10B:

x = 0.02H = Minimum front porch
y = 0.145H = Minimum without front porch
z = 0.180H = Maximum with front porch
x + y = 0.02H + 0.145H = 0.165H = Minimum with front porch

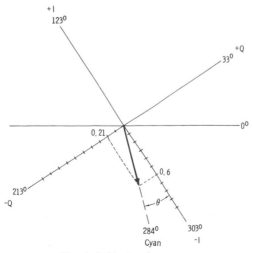

Fig. A-1. Vector for cyan.

The maximum-minimum range is $0.180H - 0.165H = 0.015H$. If $0.02H$ is minimum, then $0.02H + 0.015H = 0.035H$ maximum front porch, *not* accounting for rise and fall times of blanking. But allowance for these times must be made. The maximum specified rise and fall time is $0.004H$. So $2 \times 0.004H = 0.008H$ can be taken up by this time. Then $0.035H - 0.008H = 0.27H$ is *maximum* front porch width. To convert to microseconds:

Minimum $= 0.02H = (0.02)(63.5) = 1.27\ \mu s$
Maximum $= 0.027H = (0.027)(63.5) = 1.71\ \mu s$
Standard $= 0.025H = (0.025)(63.5) = 1.59\ \mu s$

The Tektronix Type 524 oscilloscope provides a $0.025H$ marker for the purpose of setting the front-porch width.

A2-5. The color-sync burst normally is maintained at the same peak-to-peak amplitude as the sync pulse. If the sync is adjusted to 0.3 volt (above blanking), then the burst amplitude is 0.3 volt peak-to-peak. The FCC specification is 0.9 of sync amplitude to 1.1 of sync amplitude.

A2-6. No. It is eliminated during the 9H interval of equalizing and vertical-sync pulses.

A2-7. 8 to 10 complete cycles.

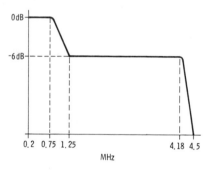

Fig. A-2. Ideal detector-diode curve.

A2-8. Note from the statement concerning this subject in the FCC rules that the color-transmitter response is tightly controlled at the color subcarrier frequency of 3.58 MHz. Note also that the applicable specifications are tied to the "ideal detector-diode curve" shown in Fig. A-2. The first 0.75 MHz of the modulation frequency range is double-sideband; hence the diode response is 100-percent in this region. Due to vestigial-sideband transmission, the diode response at 1.25 MHz will be down 6 dB *if the transmitter has 100-percent response to this frequency*. In essence, then, the actual attenuation-vs-frequency response of the color transmitter must be within plus or minus 2 dB for frequencies up to 4.18 MHz (using response at 200 kHz as a reference) to meet FCC requirements.

A2-9. No. The camera head normally employs a gamma of less than unity (black stretch).

A2-10. Yes. If not, luminance distortion (which also affects colors) will occur.

A2-11. Phase sensitivity.

A2-12. Multipath reflections with attendant phase shift at various frequencies in the radiated signal. Also, of course, the receiver circuits can affect color reproduction.

A2-13. 1. Yellow would go greenish.
2. Cyan would go bluish.
3. Green would go bluish (cyan).
4. Magenta would go reddish.
5. Red would go yellowish.
6. Blue would go toward purple (magenta).
7. White would remain white.

A2-14. Errors in the input-signal amplitudes at the encoder (all inputs must be equal for white). Improper I or Q white balance adjustments. Carrier unbalance. I or Q coefficient errors in the receiver.

A2-15. Low overall chroma gain.

A2-16. Good human "flesh tones."

A2-17. A red with reduced luminance value. This could be caused by lack of proper gamma correction (black stretch) in the camera chain, or by video unbalance in the encoder.

A2-18. Differential gain, in which the chroma gain decreases with an increase in brightness.

CHAPTER 3

A3-1. The panning head, or cradle.

A3-2. (1) Center-of-gravity adjustment. Adjusted for camera balance. (2) Pan-drag adjustment. Adjusted for proper friction in panning. (3) Tilt-drag adjustment. Adjusted for proper friction in tilting of camera head. (4) Pan brake. Locks mount in azimuth. (5) Tilt brake. Locks mount vertically.

A3-3. Two or more prompters on which the copy moves together, line for line, regardless of speed in forward or reverse direction.

A3-4. The purpose is to cause the rectifier output to change, in coordination with the output voltage, in such a direction that minimum voltage drop across the series regulator(s) occurs. This reduces power dissipation in all series regulator elements.

A3-5. Remote-sensing leads connect from the supply to the load end of the camera cable, and therefore permit sensing the voltage at the load rather than at the supply terminals. Thus, the load voltage can be held constant regardless of the length of the camera cable.

A3-6. So that any RFI components occur during blanking and do not appear as beat patterns during active scanning.

CHAPTER 4

A4-1. To magnify electromagnetically the image charge on the target area.

A4-2. By means of the magnetic field produced by the external focusing coil, and by varying the photocathode voltage.

A4-3. By the magnetic field of the external focusing coil, and by the electro-static field of grid 4.

A4-4   Yes, the grid-3 voltage adjustment facilitates maximum possible collection by dynode 2 of the secondary electrons from dynode 1. It is adjusted for maximum signal output consistent with best freedom from shading.

A4-5.   Grid 5 serves to adjust the shape of the decelerating field between grid 4 and the target. It is adjusted to obtain best uniformity of electron landing over the target area, that is, for best corner resolution and minimum shading.

A4-6.   No, not for the same signal-to-noise ratio. More aperture compensation can be used with the vidicon before noise becomes troublesome than is the case for the image orthicon.

A4-7.   Sweep linearity adjustment.

A4-8.   Dark current increases with an increase in target voltage.

A4-9.   Dark current is practically the same at any target voltage.

A4-10.   None. (*Plumbicon* is the registered trademark of N.V. Philips of Holland for their lead-oxide vidicons.)

CHAPTER 5

A5-1.   Because the input coupling from the pickup tube to the preamp is at high impedance, with shunt capacitance (both output capacitance of the pickup tube and input capacitance of the preamp). Also, peaking circuitry to flatten the amplifier response is frequency-selective, and RC or LC coupling causes a phase shift between low and higher frequencies. Hence, correction circuitry is needed to make all signal delays for low and high frequencies the same.

A5-2.   Only when it feeds the coaxial line in the camera cable. When another amplifier after the preamp is located in the camera head, the output impedance usually is 93 ohms.

A5-3.   No. Peaking circuitry only corrects the frequency response of the amplifier itself. Aperture correction takes place in a following video-processing amplifier.

A5-4.   The dynode-gain control for the image orthicon, or the target-voltage control for the vidicon.

A5-5.   An attenuation of 40 dB is a voltage ratio of 1/100. This amount of attenuation is most convenient in feeding pickup-tube video preamps, since it allows normal output levels from the test equipment.

A5-6.   Insufficient beam current to discharge the highest high lights.

A5-7.   Excessive beam current causes loss of resolution and, under some conditions, a "splitting" of the image. In some cases, an effect similar to "target flutter" in an I.O. occurs in the corner of the picture. This also can be caused by improper beam alignment.

CHAPTER 6

A6-1.   Phase shift is the additional phase angle between the input and output signal voltages over and above the normal 180° phase reversal of the stage. This may be expressed by the expression:

$$\text{Phase shift} = \theta - 180°$$

In this equation θ is the total displacement, expressed in degrees, between the input and output voltages. The low-frequency phase shift can be calculated directly from circuit constants as follows:

$$\phi = + \tan^{-1} \frac{X_C}{R_G}$$

where,

ϕ is the phase shift,
X_C is the reactance of the coupling capacitor,
R_G is the grid resistor of the following tube.

This formula simply states that the phase shift is an angle whose tangent is the ratio of the capacitive reactance to the grid resistance. The plus sign indicates a leading phase shift; this is true because capacitive reactance causes the current to lead the applied voltage, and a leading voltage is developed across the grid resistor. It can be seen that the larger the capacitance (less reactance), the smaller will be the resultant phase shift.

A6-2. The coupling capacitance.

A6-3. Low-frequency degeneration in the cathode bias circuit; insufficient time constant (RC) in the coupling circuit; degenerative feedback through the power supply; changed components in negative-feedback circuits; faulty clamping in clamp-type amplifiers.

A6-4. Increased stray capacitance to ground (detrimental to high-frequency response) and a tendency to "motorboat" at a low frequency.

A6-5. If the level of the 60-Hz square-wave signal is excessive, the resultant clipping or compression will remove the tilt, and a poor response may appear to be satisfactory.

A6-6. Since you know that the fundamental frequency for a T pulse (one picture element) for a 4-MHz system is 4 MHz, the duration of two picture elements corresponds to one-half the T frequency, or 2 MHz. The 2T pulse contains one-half the harmonic spectrum of the T pulse.

A6-7. The repetition rate is the line-scanning frequency, 15.75 kHz.

A6-8. Yes, plus an extra equalizer for the viewfinder feed.

A6-9. Not necessarily. It is used only in the luminance channel in systems fixed for operation with one luminance channel and three chroma channels. In some film systems, however, voltage may be removed from the luminance tube, and the system may be operated with three channels (luminance derived from the chroma channels by matrixing). In the latter case, aperture correction is available for all channels.

A6-10. Emitter voltage $= +1.8$ V (approx)
Emitter current $= 1.8/500 = 0.0036$ A $= 3.6$ mA (approx)
Collector-load voltage drop $= (3.6)(3.3) = 11.8$ V (approx)
Collector voltage $= +20 - 11.8 = +8.2$ V (approx)
Voltage gain $= 3300/500 = 6$ (approx)

A6-11. Output peak-to-peak signal $= (5)(0.2) = 1$ volt (or slightly less, depending on the input impedance of the following stage).

A6-12. Q1 base voltage $= +0.7$ V (approx)
Q2 emitter current $= 10.7/10,000 = 0.001$ A $= 1$ mA (approx)
Voltage across $R_f = (0.001)(3000) = 3$ V (approx)

Q2 emitter voltage $= +0.7 + 3 = +3.7$ V (approx)
Therefore, Q2 base voltage (Q1 collector) $= +3.7 + 0.7 = +4.4$ V (approx)

A6-13. The maximum current swing would be ± 1 mA in 75 ohms. Therefore, the maximum peak-to-peak output voltage is (0.002) $(75) = 0.15$ V.

A6-14. Nonuniformity of target or photocathode sensitivity across the useful scanned area; nonuniformity of light incident on the chip chart; defective or dirty chip chart; defects or improper adjustments in optics, shading, multiplier focus, or beam alignment.

CHAPTER 7

A7-1. So that the cable delay between the control unit and the camera will fall within the duration of horizontal blanking. (Camera blanking is formed from the drive pulses.)

A7-2. No. If the RC time constant is long, the output-pulse duration is the same as the input-pulse duration.

A7-3. The output-pulse width is the product of the capacitance of the coupling capacitor and resistance of the base resistor. The output pulse is not narrower than the input pulse if the RC time constant is long compared to the input-pulse duration.

A7-4. Undelayed.

A7-5. Undelayed.

A7-6. No, not within the usual operating range of the target voltage.

A7-7. Yes.

A7-8. Yes. In most late-model cameras, the same circuitry is used in either mode; the difference is in the source of the error voltage.

A7-9. Slight differences between cameras in spectral distribution of the dichroic systems. Differences in accuracy in achieving a white balance. Differences in I.O. landing and shading characteristics. Differences in characterstics of amplitude versus gray scale. Also, since two or more cameras cannot have the same viewing angle for any subject, slight color differences in the scene *actually do exist.*

A7-10. Black level, gamma correction, multiplier focus, shading, and dynode gain.

A7-11. Black level and amplifier gain.

CHAPTER 8

A8-1. No more than 20 to 1 for ideal control. The range can be determined by measurement with a spot-brightness meter. Also, the whitest material should give the same amplitude on the CRO as that given by reference white on the chip chart under the same light. The blackest material should match step 9 on the chart; this is 3-percent reflectance. Thus, the ratio of 60 percent (white) to 3 percent (black) is 60/3, or 20 to 1.

A8-2. 15 IEEE units.

A8-3. Optical and electrical focusing. Amount of aperture correction used (determined by signal/noise ratio). Scene lighting and contrast ratio. Luminance-to-chrominance ratio. Camera registration. (You should be able to see 400 to 450 lines of horizontal resolution at the center of a test pattern.)

A8-4. On three-channel systems not employing the cancellation technique, adjust the final registration for 400 lines minimum horizontal resolution at the center of the test pattern. In the cancellation technique, the polarity of all video signals except green is reversed. Thus, when a negative picture from any one channel is combined with the positive picture from the green channel, adjust the channel being compared to green so that *complete* cancellation takes place at least through the large center circle of the registration chart.

A8-5. How it looks in monochrome. Does it have "snappy" contrast?

A8-6. Yes, it is. On the small area covered by the tight shot, the color temperature of the light can be different, especially with old incandescent lamps. Also remember the effect on skin tones of backgrounds; a tight shot (little background) can eliminate this effect.

There is a way you can rebalance the camera on such shots (if necessary) during operations, *if* you have a reference white and black in the scene. Have the camera monitor on NAM monitoring, and carefully adjust for NAM balance on black and white (black and white balance controls). This takes some experience and is not recommended unless the balance is very noticeably bad.

A8-7. Degree of saturation (purity) of background. Area of background relative to face: for a facial closeup, the background has minimum effect; if the background is comparatively large (skin area small relative to background), the effect is maximum.

A8-8. ±100 K, or ±10 volts.

A8-9. Yes. In fact, more dimmers sometimes are necessary for color than for monochrome. This is because areas in which skin tones exist (where performers will be facing the camera) should not be dimmed. Lighting in other areas can be manipulated by dimmers for special effects and mood scenes.

A8-10. (A) ½ to 1 times base light.
(B) A maximum of 1½ times base light.

A8-11. In the blue region.

A8-12. It helps by increasing the response toward the red end of the spectrum.

A8-13. In parallel.

A8-14. In series.

A8-15. Skew is a condition in which the raster in a channel (or channels) of a color camera does not have the same shape as the raster in the reference channel. It results from slight differences in deflection yokes. It is corrected by introducing a small amount of vertical sawtooth (of the proper amplitude and polarity) into the horizontal sawtooth.

A8-16. 60-percent reflectance.

A8-17. They adjust the ratio of resistance to inductance to achieve similarity (linearity) of the horizontal-deflection circuits.

CHAPTER 9

A9-1. No. It requires the higher-frequency trigger input only for "locking" to the subharmonic.

A9-2. The subcarrier frequency is too high; the sync generator is on crystal control (instead of subcarrier-frequency control); or the sync generator is on free-run operation.

A9-3. No. The subcarrier-frequency oscillator is a temperature-controlled crystal circuit. It is improbable (after normal operating temperature is reached) that an error greater than ±50 Hz will exist. (Tolerance is ± 10 Hz.) But even assuming a ± 100-Hz error, since the total division is 113.75, the final error would be $100/113.75 = 0.87$ Hz. So once the counters are properly centered, the frequency adjustment can be made at any time, as the count is quite broad.

A9-4. The frequency counted down from the 3.579545-MHz oscillator is 31.46852 kHz. The countdown usually is done by three divider stages $(5 \times 7 \times 13 = 455)$ and a $\times4$ multiplier, to make the total division 455/4, or 113.75, times. Double check this by multiplying: 31.46852 kHz $\times$ 113.75 = 3.579545 MHz (to the nearest hertz).

A9-5. No. The phase of the color-subcarrier burst is adjusted in the encoder. The flag pulse influences only the *timing* (position) of the burst.

A9-6. The best way is to use a time base of 0.1 μs/cm and use 0.1-μs markers. Trigger the scope with horizontal-drive pulses. Or, you can use a time base of 1 μs/cm if scope-triggering instability occurs on the extremely short time base. Also, you can use 10 μs/cm with $\times$10 sweep magnification. There should be five of the markers between the trailing edge of horizontal sync and the first (full) cycle of burst.

A9-7. If bursts occur during the entire 9H interval, obviously the 9H-eliminate pulse has been lost completely. However, if some of the bursts are eliminated, but not for nine complete lines, the burst-eliminate pulse has too much slope, and the eliminate keyer (V3) does not remain off for the entire interval. The most likely cause is insufficient clipping at V2. This could result from a low output level from V1, leakage in coupling capacitor C3 or bypass capacitor C4 (either of which would reduce the bias), or leakage in coupling capacitor C5. The first check should be to observe the pulse at the V1B plate to see if the multivibrator is capable of obtaining a 9H pulse width. If it is, a badly sloping pulse from the clipper is likely, as noted above.

A9-8. 18 peaks.

A9-9. Not at all. But the 9H-eliminate pulse would be lost, and bursts would continue throughout the vertical interval.

A9-10. If all inputs are tied together, starting with the first (white) interval, a continuous "white" pulse is sent for the line duration, since at any time at least one of the multivibrators will be sending a pulse. So with Y off and I and Q on, if the encoder is properly balanced, only a thin line representing the subcarrier zero axis should be observed at the output. (The subcarrier should be zero for any "monochrome" condition.)

A9-11. The switches for the Y, I, and Q channels control only the respective video channels. Thus, with all channels off, there still is a subcarrier output if the subcarrier is not properly cancelled out for a no-picture (black) condition. Remember this means both the I and Q subcarrier-balance controls are involved.

A9-12. The white-balance control adjusts the gain of the matrix signal inverter so that all three signals are amplified identically. Note that this implies identical amplitudes at the input, as well as maintenance of proper values of the precision resistors used in the matrix network.

A9-13. (A) For the I-only chroma signal, the two maximum-amplitude signals are cyan and red. The I video component of red and cyan is 0.6 of unit luminance. So when the subcarrier is modulated, the sidebands reach twice this value, or 1.2 of unit luminance. Since 100 IEEE units = 0.714 volt, $1.2 \times 0.714 = 0.85$ volt peak to peak. This is 120 IEEE units.

(B) For Q only, the maximum-amplitude peaks are green and magenta. The Q video component of green and magenta is 0.525 of unit luminance; with modulation, the sidebands reach 1.05 of unit luminance. Since 100 IEEE units = 0.714 volt, $1.05 \times 0.714 = 0.75$ volt peak to peak. This is 105 IEEE units.

Whenever you are in doubt concerning problems with I and Q gain ratio, you can set up the encoder gains by this method to determine whether either one is in error. When you then turn all channels on and adjust the composite gain for 0-100 IEEE units on the white pulse, the chroma signals should be correct in absolute amplitudes as well as gain ratios.

A9-14. If sync were inserted before aperture compensation, it would be over-compensated and would contain spikes on the leading and trailing edges. If sync were inserted after the Y delay, it would need to be delayed externally to match the encoder delay.

A9-15. No. But the effect of luminance can be observed if the Y channel in the encoder is turned off with both I and Q on. Any change in amplitude of the individual vectors indicates *differential gain*. A change in phase of the vectors indicates *differential phase*.

CHAPTER 10

A10-1. Since the demodulation of color-difference signals is narrow-band, only the chroma frequencies below 500 kHz are used. Hence, identical delays occur in the color channels, and only the Y signal need be delayed to achieve time coincidence.

A10-2.

Primary	Complementary
Red	Cyan
Green	Magenta
Blue	Yellow

A10-3. The chrominance signal for the primary color has the same amplitude as the signal for the complementary color, but differs in phase by 180°.

A10-4. To synchronize the receiver or monitor local 3.58-MHz oscillator with the subcarrier-frequency oscillator at the sending point.

A10-5. To amplify and pass only those frequencies between approximately 2.3 and 4.2 MHz, where the color-subcarrier sidebands lie.

A10-6. To gate off any input to the chrominance section during a monochrome transmission.

A10-7. It detects the amplitude variations of one phase of a multiphase modulated carrier.

A10-8. The basic difference is a 33° phase shift relative to the color burst. The cw reference for I leads the one for R—Y by 33°, and the reference for Q leads the one for B—Y by 33°.

A10-9. We know that when the system is normal, white or gray areas appear during a signal interval of zero subcarrier (I and Q are cancelled out). If either one of the modulators (I or Q) becomes unbalanced during the active line interval, a white or gray area will become colored as a result of the addition of the subcarrier vector during this interval. Also, during intervals of the scan, the subcarrier may be cancelled by the unbalanced carrier vector, and the color may become desaturated or white. The overall result is a white-to-color and color-to-white error that changes with picture content and is quite objectionable.

A10-10. Yes. We know that the unwanted carrier has a constant amplitude that is added vectorially to every color vector. A positive unbalance in the I modulator shifts all hues toward orange; a negative unbalance shifts them toward cyan. A positive unbalance in the Q modulator shifts all hues toward yellow-green; a negative unbalance shifts them toward purple.

CHAPTER 11

A11-1. Any shading component originating in the pickup tube (particularly in the I.O.) will vary as a function of the light amplitude, which becomes video amplitude at the output. The shading-generator signal remains fixed at all light levels or video amplitudes.

A11-2. Nonlinear sweeps. A perfect sawtooth current through the deflection yoke has a constant rate of change and, hence, produces a constant velocity of scan across the target. Any departure from a sawtooth current waveform means the rate of change is not constant, and shading will result.

A11-3. Assuming the tube is operating in the proper mode of focus, the adjustments for grid 3 (multiplier focus) and grid 5 (decelerator grid) are the most critical. (Adjust the grid-3 voltage with the lens capped.) Proper use of the alignment controls also may improve shading characteristics.

A11-4. Good shading characteristics depend primarily on uniform collection of the secondary electrons as the return beam scans a small area (about ¼ inch) of the first dynode. Variations of the secondary-emission ratio over the first dynode are amplified in the remaining dynode sections. The most severe case of shading is represented with the lens capped, since the beam is completely returned to the first dynode. Since the amplitude of the shading component is a function of the return beam, it is greatest in dark areas, where the return beam is maximum.

A11-5. Use a reliable dynamic-transconductance type with provisions for checking high-resistance interelectrode shorts, and provisions for testing voltage-regulator tubes. It is not necessary to keep a record of measured tranconductance. Use the GOOD-BAD scales, which automatically include the manufacturer's tolerance (even on new tubes). Run the shorts check, and test all voltage-regulator tubes for firing voltage and regulation within the specified current range. This type of tube testing should be done about every 90 days. After tube replacement, remember to run a complete performance analysis on the units involved, making any necessary adjustments. Recheck after a four-day run-in time.

A11-6. This is a result of low emission caused by either low filament voltage or a weak vidicon tube.

A11-7. The first thing you can do is to lower the focus current for the duration of the remote telecast in order to get a reasonable picture. For example, if you normally use 75 mA of focus current, reduce this to between 70 and 72 mA, and readjust sizes, alignment, etc. The picture sharpness may suffer slightly but not as much as it would as a result of exceeding the image-focus range. The problem occurs as a result of a drop in insulation resistance in some camera cables when heated. The critical conductors are the two leads that go to one side and the arm of the photocathode-voltage potentiometer, variously termed IMAGE-FOCUS or PC FOCUS on the camera control unit. For instance, these leads are connected to pins 15 and 21 in the RCA TK-11 camera chain. Each of these conductors is one of a group of seven conductors arranged in a circle of six with the seventh in the center. If you have this problem, remove the protective bell cover at each end of the camera cable, and check to make sure that a center conductor is used for one pin in one group and for the other pin in the other group. If it is not, simply interchange the conductor with whatever pin is connected to the center conductor. This will increase the conductor-to-ground resistance.

A11-8. These can be outlined as follows:

Spots

(A) Spots that defocus when the photocathode focus (image focus) is varied can be caused by dirt on the faceplate. Open the lens iris. If the spots grow in size and contrast decreases, clean the tube faceplate and/or other optical components in the system. If no change in size occurs, the photocathode itself is blemished. Adjust for the best point to minimize the effect.

(B) Spots that remain unchanged when the pc focus is varied, but defocus when the beam focus (orth focus) is varied result from defects on the target or field mesh. You might be able to return the tube to the factory for correction; check with the manufacturer.

(C) A large white spot near the center of the raster, if it is observed with the lens capped and does not change with adjustment of the focus control, is an ion spot. You must return the tube to the factory for reprocessing. This sometimes occurs in tubes that have not been operated periodically. Be sure to rotate I.O.'s, including all spare tubes, at least once a month.

Portholing (dark corners in picture)

Usually, portholing is a result of improper adjustment. Open the lens to operate over the knee of the transfer curve. Then adjust the voltage on grid 5 (decelerator grid) for best beam landing (best corner brightness). You might need to align the beam on a different loop (mode) of focus and readjust the grid-5 voltage. You also may need to readjust the grid-3 voltage (multiplier focus) slightly to provide maximum uniform signal output. This adjustment normally should be made with the lens capped.

If the above procedure fails to affect portholing, change the lens and note whether the picture improves. If it does, the lens may be a vignetting lens (highly improbable). Also, the yoke might be magnetized. Demagnetization methods are described in Chapter 4.

Noise in Picture

The first step when the picture is noisy is to cut off the beam and check for amplifier noise with the control-unit gain control at reference operating level. If no noise is apparent, turning the beam up will bring in the noise; this indicates a noisy I.O. If there is some noise present with the beam off, sometimes you can handle this situation temporarily by increasing the orth-gain (dynode voltage) control (when provided on the camera) to override the amplifier noise. This is only a temporary solution for use when time does not permit servicing the amplifier.

If the noise definitely is coming from the I.O., check the target-voltage adjustment, and see if adjustment of the grid-3 voltage (multiplier focus) for maximum signal output will minimize the noise. Be sure you are using a sufficiently high lighting level to allow operation over the knee at normal lens $f/$ stops.

Coarse Mesh Pattern in Picture (In Field-Mesh Tubes)

A coarse mesh pattern in the picture is caused by alignment of the beam on the wrong loop (mode) of focus. If you are unable to obtain the proper voltage mode of operation, try using a different focus-coil current, within 4 or 5 mA of the normal current. Remember that a field-mesh I.O. will "align" properly on only one mode of operation (operable grid-4 volts from near center to minimum).

Soft Picture (Poor Resolution)

The only way you can be sure whether a soft picture is caused by the I.O. or by the associated amplifiers (unless lengthy video sweep procedures are used) is to interchange the tube with one that gives a good picture in another camera. Always be sure that the lens and faceplate are thoroughly clean. Also, the blower motor and filters must be in good condition so that the tube is not too hot. To be sure there are no magnetic-field problems, turn off the blower motor and other adjacent electrical machinery temporarily and note the effect; if the resolution changes with the location of the camera, this is almost certainly a clue. Never forget to double-check the settings of such controls as filament-voltage switches (usually in the camera) and cable-length switches (usually in the camera control or processing unit). Whenever the cable length is changed, these switches should be reset properly. Otherwise, the resolution can suffer, and other problems can be encouraged.

Index